The Ecology of Reproduction
in Wild and Domestic Mammals

The Ecology of Reproduction
in Wild and Domestic Mammals

R. M. F. S. SADLEIR

*Department of Biological Sciences,
Simon Fraser University, Burnaby,
British Columbia, Canada*

METHUEN & CO LTD
11 NEW FETTER LANE LONDON EC4

First published 1969
© *R. M. F. S. Sadleir 1969*
Softcover reprint of the hardcover 1st edition 1969
Richard Clay (The Chaucer Press) Ltd.,
Bungay, Suffolk
ISBN-13: 978-94-011-6529-7   e-ISBN-13: 978-94-011-6527-3
DOI: 10.1007/978-94-011-6527-3

Distributed in the U.S.A.
by Barnes & Noble, Inc.

TO MY
PARENTS

Preface

It is inevitable that as the width of biological knowledge increases fields of research must become more and more specialised. The unfortunate corollary of such specialisation is the increasing inability of biologists to keep themselves abreast of the information and ideas accumulating in those fields even quite closely related to their own. As a result of contacts with a number of reproductive physiologists and ecologists studying breeding phenomena, I came to realise some years ago that each of these sub-disciplines had much to offer to the other. This book is the outcome of that realisation combined with a general inquisitiveness on my part as to just how mammals manage to breed and rear young from the high Arctic to Equatorial jungles. I would also like to know how other vertebrates manage likewise, but time and my own ignorance of the relevant literature precluded the inclusion of other vertebrates in this book.

It will become apparent to the reader (or really to that miniscule proportion of readers who actually read prefaces) that the majority of detailed information on the effect of the environment on reproduction comes from domestic species. Agricultural biologists have achieved considerable insight into a number of aspects of reproductive ecology and it is to be hoped that biologists studying wild mammal species will be able to carry out investigations at similar levels with the use of recent techniques. There can be no doubt that much of what has been discovered in domestic mammals is directly relevant to feral species.

I suppose that every writer of a review book hopes that his contribution is worthwhile and will lead other researchers into new and profitable fields. I feel strongly, though perhaps naively, that it is very important to understand the various ways in which mammals respond reproductively to environmental situations in the hope that we may see indications of how another mammal, the human species, could respond. The human race is in the process of rapidly throttling itself by its reproductive prowess and it is possible that, despite mans belief that he can control his own environment, the environment he 'created' is catching up to the extent that it may soon be drastically affecting his ability to reproduce. If this does happen, and if man does respond in a basically mammalian manner, it is to be hoped that the responses of other mammals may indicate to us the types of reproductive phenomena we may expect to see develop in the human species.

R.M.F.S.S.

Acknowledgements

The major environmental factors affecting the conception and gestation of this book have been the academic stimuli which I have received from my various teachers and colleagues. I would like to thank particularly the following persons for their intellectual contributions to the development of my approach to the ecology of reproduction, while holding them in no way responsible for its outcome (i.e. this book): Bert Main, Harry Waring, Hugh Tyndale-Biscoe, Glen Storr, Dennis Chitty, John Eisenberg, Charlie Krebs, Adam Watson, Idwal Rowlands, Peter Jewell, and Gilbert Manley. I also thank Peter Foreman for his work in preparing the figures. The courtesy and service of the librarian and his staff of the Zoological Society of London assisted me considerably in the preparation of the original manuscript. I am grateful also to my wife, Carrie, for coping with my accentuated idiosyncrasies while the book was being written.

I wish also to thank the following publishers and organizations for permission to reproduce figures originally published by them:

	Figure numbers
Academic Press, New York	54, 55
Aktiebolaget Svenska Bokforlaget, Stockholm	15
American Society of Animal Science	1, 8, 18, 53
American Society of Mammalogists	4, 47
Arctic Institute of North America, Montreal	50
Blackwells Scientific Publications Ltd., Oxford	33, 37, 38
Cambridge Philosophical Society	7
Cambridge University Press	14, 17, 21
J. A. Churchill Ltd., London	23, 24
Commonwealth Scientific and Industrial Research Organization, Australia	2, 19, 20, 26, 30, 34, 42, 48
Duke University Press, Durham, North Carolina	40
Logos Press, London	52
Macmillan and Co. Ltd., London	31, 32
Museum of Comparative Zoology, Harvard	28, 29
Museum National d'Histoire Naturelle, Paris	13
New York Academy of Sciences	56
New Zealand Department of Agriculture	14, 44
New Zealand Department of Scientific and Industrial Research	27, 43
Polska Akademia Nauk, Bialowieza, Poland	46

	Figure numbers
C. C. Thomas, Publishers, Springfield, Illinois	39
United States Department of Agriculture	25
Wildlife Society, Columbia, Missouri	35
Wistar Institute of Anatomy and Biology, Philadelphia	22
World Review of Animal Production, Rome	36
Zoological Society of London	9, 41

R.M.F.S.S.

Contents

PART ONE

Puberty

CHAPTER I

Ecological Factors in the Attainment
of Puberty: Introduction

A mammal can be considered as having reached puberty or sexual maturity when it is capable of producing gametes, the fusion of which will result in the production of viable young. This process is complicated in many species by the presence of recurring intervals (the non-breeding or anoestrous seasons) when gamete production is interrupted. The initial onset of puberty is a period in the life of the individual which is difficult to define except in terms of the discipline under consideration. From the point of view of genetics, an individual has not reached puberty until fusion of the gametes produced has actually occurred, whereas from the standpoint of reproductive physiology, the initial production of viable gametes would suffice to mark the onset of reproduction. From the standpoint of population dynamics, the female may be considered as having reached puberty when she first gives birth or the male when he first causes a conception which will subsequently result in birth, whereas the ethologist may consider an animal as being pubertal when it first exhibits patterns of behaviour associated with sexual maturity. These approaches conflict in certain mammals which, though physiologically able to produce young, are prevented from doing so by social or ecological factors (see page 34). Thus the state of sexual maturity is an easily acceptable concept to biologists but the definition of the initial onset of this state is complex and largely depends on the discipline concerned.

An understanding of the factors involved in the attainment of puberty is of great importance in any study of the ecology and reproduction of a mammalian species. The timing of puberty relative to the life span of the species concerned governs the total number of young produced by the female and is thus relevant to investigations of population dynamics. In species with definite breeding seasons a study and comparison of the ecological factors operating at the onset of breeding of the population and the onset of reproductive maturity of the individual will give insight into the general relationships between the animal and its environment. In continually breeding species, puberty may be related to environmental factors which express themselves in the efficiency of reproduction rather than its onset and cessation. In either case the comparison of the effects of environmental variables on animals which have been repeatedly breeding and on animals which are just attaining reproductive status for the first time will be of importance in the understanding of exactly how the variables act on the individual.

Puberty is usually measured in terms of the age of individuals of the species concerned. It can be extremely variable, depending on many factors, such as the season of birth, the state of nutrition, sociological factors, etc. Maturity, here the production of gametes, is shown externally by the development of the secondary sex characters. The age or state of development of the ovary does not control this onset. Foa (1900) demonstrated that it was the somatic environment which controlled the development of the ovary. He transplanted ovaries from juvenile rats into ovarectomized adult females and found that they exhibited follicular development and ovulation, while the reverse transfer resulted in a cessation of ovarian activity. The work of Price and Ortiz (1944) and Ortiz (1947) has shown that the gonads can react to hormonal injections well before the time of pubertal development. Parkes (1966) has reviewed this and other confirmatory work, and further mentions that ovarian grafts in guinea-pigs have resulted in the development of a feminizing influence about six weeks after grafting into young animals but only two to three weeks after grafting into old animals. Observations such as these have led to the so-called 'law of puberty', i.e. the idea that the onset of puberty is not determined by changes in the gonad itself but by changes in the maturation of the soma. In effect, this has been demonstrated to mean changes in the anterior hypophysis of the pituitary gland.

There is, however, evidence that the prepubertal gonad has a low level of activity as has the immature pituitary which does secrete gonadotrophic hormones. Donovan and Harris (1966) have reviewed the evidence for a reciprocal relationship between the pituitary and the gonad in juvenile animals and also that from parabiotic experiments (in which the circulatory systems of two animals are joined) which have suggested that the immature gonad can inhibit secretion from the pituitary. Therefore the idea that the puberty is entirely dependent on the maturation of pituitary tissue must be questioned. Donovan and Harris describe experiments in which anterior pituitary glands from immature rats were transplanted into the subarachnoid space near the hypothalamus of hypophysectomized adult females. These host females exhibited oestrous cycles, even though the donated anterior pituitaries came from rats which were only from 12 to 29 days old. A result such as this strongly suggests that the *pars distalis* of infantile animals is capable of supporting adult reproductive functions but that some stimulus to initiate such activity is lacking before puberty.

The remainder of this part will show the relationship between certain environmental variables and puberty. In view of the evidence above, it will be assumed that these variables act through the anterior pituitary but that the method by which they act, and the relationship of their direct action on the gonads to their action through the pituitary, is largely unknown.

CHAPTER 2

The Effects of Light
on the Onset of Puberty

The earliest reported experiment on the reproductive effects of altering photo-periods was that of Bissonnette (1935b) on immature male ferrets (*Mustela putorius*). He maintained one group of three animals on a period of $8\frac{1}{2}$ hours of light, starting on 10 November (Northern hemisphere) and a second group of three on normal daylight plus 9 hours of artificial light (added after sundown) starting from the same date. Puberty was delayed in his 'long-day' animals and hastened in two of the 'short-day' ferrets, but delayed in the third. These results suggested that light did have some effect on puberty, but the small sample precluded any major conclusions. A considerable amount of research since Bissonnette's report has confirmed the importance of the light regime on puberty attainment, and much of this is reviewed below.

(a) Rodents

Photoperiod has been found to affect puberty in non-seasonal breeders kept under otherwise constant and reasonably optimal conditions. Laboratory rats (*Rattus norvegicus*) continuously lighted from birth or from 21 days of age, show the first opening of the vagina an average of 6 days earlier than females kept under normal conditions of laboratory lighting (Fiske, 1941). That this opening results from an increased level of circulating follicle-stimulating hormone (FSH) was confirmed by the increased ability of the pituitaries from treated animals to induce ovarian growth in test rats. Lower concentrations of luteinizing hormone (LH) in the treated pituitaries were shown by their reduced ability to cause growth in seminal vesicles of test males. Therefore the increased light regime had resulted in an earlier than normal secretion of gonadotrophic hormones from the pituitaries of treated animals. Truscott (1944) found that light had a general effect on the prepubertal hypophysial–gonadial relationship in that he obtained a higher count of non-vesicular follicles in the ovaries of lighted females. The decline in numbers of vesicular follicles was also reduced in the lighted females, and a marked delay in puberty was found in females with severed optic nerves. Fiske's results were confirmed by Jöchle (1956, 1964) who found that the effect of constant illumination started quicker in the juvenile

B

than in the adult rat, and the earlier the continuous lighting regime was established, the higher the proportion of females undergoing continuous oestrus. The observation of Donovan and van der Werff ten Bosch (1959b) that laboratory rats born in the spring on increasing light reached puberty at an earlier age than those born in autumn also suggests an effect of light ($40 \cdot 3 \pm 0 \cdot 6$ days, cf. $46 \cdot 1 \pm 0 \cdot 7$ days).

The opposite effect of darkness was also investigated by Fiske (1941). Rats kept in continuous darkness from birth reached puberty, on the average, some 16 days later than continually lighted animals. The picture of secretion of gonadotrophic hormones from the pituitary was reversed from that of lighted females, in that the FSH content was much lower and the LH content was higher. It is interesting that the LH content of males similarly treated did not show any significant variation.

The laboratory rat and the wild rat of either species breed continually under the normal annual cycle of light changes. Another rodent, the field vole (*Microtus arvalis*), is normally a seasonal breeder which starts breeding in spring. The effect of light on puberty in this species has been the subject of a series of papers by Lecyk. In a very well-designed experiment beginning on 22 September (Northern hemisphere) Lecyk (1962a) set up five groups of young immature voles. The first and second groups were put under gradually increasing light (one natural, one artificial) which approximated to the natural change occurring from 21 March to 24 June. The third group was placed under uniform 16-hour daylength, the fourth was put on a uniform 8-hour day until 5 November, when the daylength was increased in steps to 13 hours by 6 December and the fifth group was kept as a control under natural lighting conditions. By early December the first four experimental regimes had all induced puberty. Sexual behaviour was first seen on the following dates: group 1, 23 October; group 2, 2 November; group 3, 25 October; group 4, 6 December. Over 60 per cent of these females were pregnant, and the males had well-developed testes and auxiliary glands. The control group of voles did not exhibit any signs of reproductive development. These results indicate clearly that in voles light increases act as a stimulus to the onset of reproduction. In a second experiment, reported in the same paper, Lecyk attempted to inhibit the maturation of young voles caught 7 December by putting them on a uniform short daylength of 7 hours or by allowing them normal daylight until 22 May (the equinox) and then artificially decreasing the daylength. This experiment failed and the voles reached puberty at about the same time as controls. It would appear that there is an effect of the age of the animals on their response to light-stimulated sexual maturation. These animals would have been born the previous breeding season and would thus have been some months old at the beginning of the experiment. When Lecyk (1963) carried out a similar experiment starting on 10 May, but using known 10 day-old voles, he was able to completely inhibit the attainment of puberty by putting the voles on to gradually decreasing light or uniform 7-hour daylength. This relationship between the onset of sexual maturity in voles and the light regime may explain the phenomenon of delayed maturity in this and other seasonally breeding species (see page 36).

Males of the same species (*M. arvalis*) reached puberty earlier when maintained on 15 hours of light than when kept under 10 hours (Martinet, 1963). Continuation of this work by Thibault *et al.* (1966) showed that the actual quantity of light was important only above 10 hours. These workers kept voles on 5, 10, 15, or 20 hours of light from birth and found a very great difference in the testis weights at different ages. Figure 1 shows that light can also affect the testis weights when altered after 70 days of age. It is a pity that no data are given

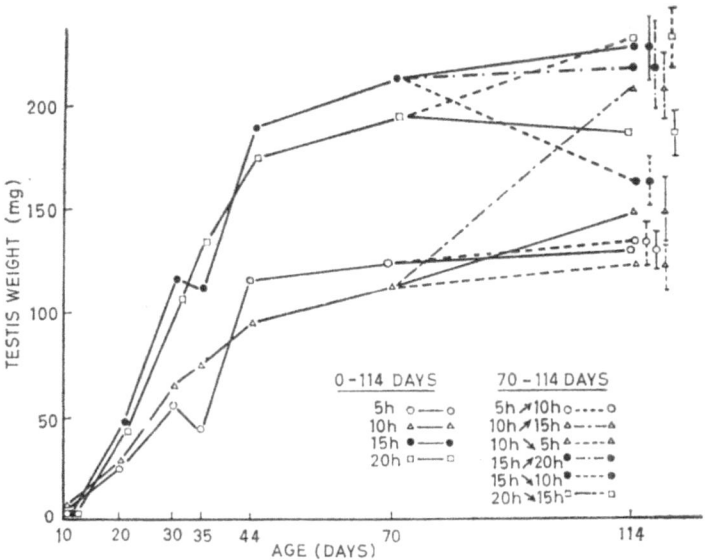

Fig. 1. Influence of different daylight ratios on testis weights in the growing field vole (*Microtus arvalis*). From Thibault *et al.* (1966), *J. Anim. Sci. (Suppl.)* **25**: 119–42.

on the body weights of the voles in this experiment, because growth can also affect the actual testis weights achieved in experiments such as this (see page 41).

Another rodent whose sexual development is affected by light is the cotton rat, *Sigmodon hispidus*. Meyer and Meyer (1944) found that keeping immature individuals of this species under continuous light had little effect on the timing of puberty, but under continuous darkness maturation was significantly delayed.

The effects on puberty of altering the intensity and quality of light have been investigated by Lecyk (1962b) for *Microtus arvalis*. During the non-breeding season young (15–20-day) voles were placed under 16 hours of light of varying intensity for 82 days, with the results as shown below:

Light intensity (lux)	Males Mean testis weight (g)	Females Proportion pregnant (%)	Time after beginning of exp. sexual behaviour first seen (days)
200 (70 f.c.)	193	73	36
10 (3 f.c.)	182	52	43
0·2 (0·6 f.c.)	122	0	63

The interesting part of this result is that although all the males showed complete spermatogenesis, females did not become sexually mature at the lowest intensity. The difference in the onset of sexual behaviour does demonstrate that the light intensity has a graded effect over the range studied. Males at the 0·2 lux intensity had testes much better developed than control unlighted males at the same time of the year. Lecyk also gives data showing no discernible differences in the time to puberty between young voles subjected to light of 390–500 mμ (violet-blue) wavelength or 570–760 mμ (orange-red) wavelengths.

The influence of light on puberty in rodents is thus well demonstrated. An increase in the quantity of light in the time sense results in earlier maturation in at least two species and probably others.

(b) Domestic sheep (Ovis aries)

This species differs from rodents in that its normal breeding season occurs during a period of decreasing light. The field observations and experiments relating the effect of light to the onset of breeding in mature sheep will be discussed in some detail later (see pages 65, 72), and it will suffice to say at this point that sheep are perhaps the most studied animal with regard to photoperiodic reproductive effects.

The evidence of these effects on sheep is not of an experimental nature, but depends on observations on flock animals, relating the season of birth of lambs to the time of their first oestrus. Hammond Jr. (1944) studied in detail the onset of maturity in Suffolk crosses at Cambridge, England. He found that there was neither a fixed age of puberty nor a fixed time of the year for the first heat. Those born early in the breeding season reach puberty at an older age than those born a short time later, but animals born very late in the year do not mature for a long period of time. The threshold of stimulation required falls in the lamb as its age increases, until at about 300 days it reaches the adult level. Hammond suggests that the intensity of stimulation required before the individual can reach puberty increases as the amount of daylight increases. It is maximal in or before the middle of the breeding season and thereafter decreases, but at a greater rate than that at which the threshold for the lamb falls with increasing age. Therefore the minimum age at first heat comes in the middle of the breeding season with lambs about 180 days old. If this age is not reached later in the season, heat will not occur until the subsequent season, when the lamb is about 400 days old. Similar conclusions were reached by Hafez (1952b).

Watson and Gamble (1961) compared the onset of puberty in groups of Australian Merino ewes born at three different times of the year. Figure 2 shows that, independent of the time of birth, most lambs experience their first oestrous period in February and March, on a declining light regime (Southern hemisphere). The actual length of time from birth to first oestrus thus depended on the season of birth. Lambs born September–October were of the order of 5 months old at the first oestrus, while those born January–February or May–June were 9 or more months old. The Merino breed is known

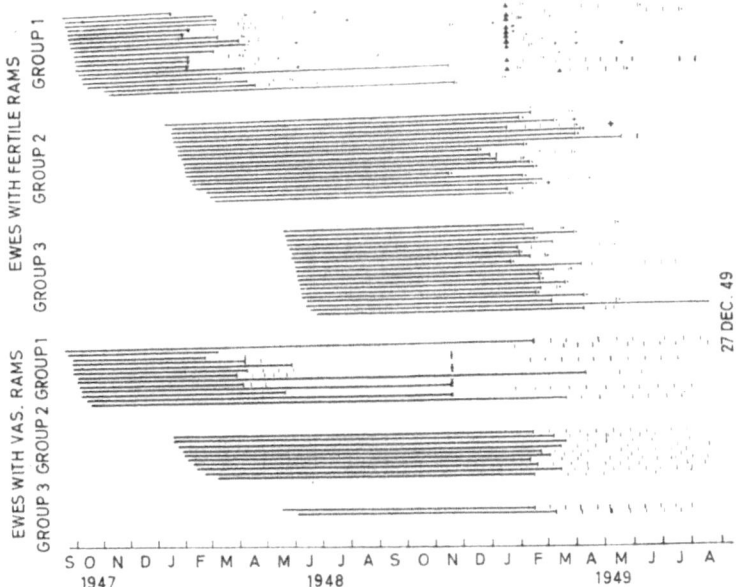

Fig. 2. The occurrence of oestrus and conception in young ewes in relation to season of birth. Each row represents observations on a single ewe. The line in each row represents the interval from birth to the first record of mating. A small vertical stroke indicates a definite mating or oestrus. If this was doubtful the stroke is broken. The '▲' indicates oestrus as demonstrated by a vasectomized ram. From Watson and Gamble (1961) *Aust. J. agric. Res.* 12: 124–38.

as one which seems least affected by changing light regimes as far as its breeding season is concerned, and yet there is here a clear effect of light on the attainment of puberty.

(c) Other Mammals

There appear to be a few other studies on the effect of light on the attainment of puberty. Crowcroft (1964) describes a small experiment in which he put half of a group of six immature shrews (*Sorex araneus*) on a uniform 16 hours of light per day between 31 January and 4 March (Northern hemisphere). At the end of the lighted period the experimental animals were in full reproductive condition while the controls were not.

The effect of light on the attainment of puberty in Welsh stallions has been investigated by Skinner and Bowen (1968). Three stallions which were born in the spring reached puberty between $11\frac{1}{2}$ and $14\frac{1}{2}$ months of age on the basis of the appearance of spermatozoa in their semen. Seasonally, this occurred just as daylight was starting to lengthen. In a fourth stallion which was born at a different time of the year, these workers noted that the reaction time lengthened, the ejaculated volume declined, and citric acid content of the semen declined during the early winter, and yet when daylight started to lengthen in the early spring these parameters improved. Skinner and Bowen noted the correlation in previous work between the androgenic function of the testes and citric acid of the semen and suggest 'that androgen secretion in the stallion is strongly influenced by daylight length, whereas spermatogenesis is not influenced to any noticeable extent' (page 135). They therefore suggest that, because of the photoperiodic influence, puberty in the young horse can only really start after the first winter of life.

Particularly interesting observations with regard to a possible effect of light on sexual maturity were made by Laws (1961) in his study of the reproduction of the southern fin whale (*Balaenoptera physalus*). This species apparently attains puberty at two different times of the year. Approximately 70 per cent of juveniles ovulate for the first time during the normal breeding season of April to November after the northward migration from the Antarctic. The rest ovulate between October and February after the southward migration. Laws suggests that the fin whale is responding to increasing light regimes during its migrations. Most juveniles are subjected to a period of light increase as they migrate north, while the remainder are also subjected to a period of light increase as they migrate south. It would seem that puberty in this species is initiated by a combination of age and increasing daylength, the light threshold becoming lower with increasing age.

Finally, there is a possible relationship between light and the onset of puberty in humans. Zacharias and Wurtman (1964) have shown that blindness is associated with an earlier menarche, especially if the blindness is accompanied by a total loss of light perception. The 50 per cent incidence of menarche is reached in normal girls at a mean age of 149 months in the United States, whereas in blind girls it occurs at a mean age of 140 months. Thus it would seem that, unless some other physiological mechanism is concerned in blindness, the absence of light in humans acts to promote an early puberty. This is in contrast to many other mammals, and should be investigated further before any major importance can be attributed to this observation.

CHAPTER 3

The Effects of Temperature on the Onset of Puberty

(a) Rodents

The reproduction of laboratory mice (*Mus musculus*) kept at low environmental temperatures has been studied in very great detail by Barnett and his co-workers in Glasgow, and this work can be traced through the bibliographies in the papers by Barnett and Coleman (1959) and Barnett (1962). Mice were maintained at −3° C and their reproduction compared with mice kept at 21° C as controls. At the colder temperature the mean age of vaginal opening was 33 days (controls at 26 days), but the first oestrous smear in the cooled mice was seen on average at 61 days, compared with 38 days for the control. The vaginal openings in both groups occurred at the same body weight of 13 g, so that the temperature reduction had a marked effect in reducing the rate of growth of the female mice. A similar effect was also noted in mice with an ancestry of several generations in the cold. Barnett suggested that the slowing down of puberty may be attributed to the prior demands for catabolism but could not account for the reproductive patterns of 21° C females placed under −3° C temperatures, which first altered to the −3° C pattern but then reverted to their original pattern. There can be little doubt, however, that the delay in reproductive maturity in experiments such as this must be related very closely to the reduced rate of growth, and should therefore be considered as a secondary phenomenon. It is also possible that the selection of −3° C as the lower temperature, one considerably lower than normally met with by wild housemice, has unduly emphasized the effect. Knudsen (1962) was unable to find any difference in the age of puberty in groups of mice kept at 18°, 25°, and 32° C.

The effects of reduced temperature on growth, and through this puberty, were noted by Laurie (1946) in his excellent study on the reproduction of wild housemice in several man-made environments. Mice living in cold meat stores at temperatures of from −21° to −10° C were generally much heavier than mice from other environments, such as unheated food stores or urban dwellings. This was due to an increased body size and not just fat deposits. Unfortunately this effect cannot be related to temperature alone, as the populations in the cold stores were feeding on a very rich diet of frozen meat and sugar. At the lower body range of from 9·0 to 9·9 g only 20 per cent of cold-store females were pubertal, compared with over 65 per cent for the other environments. There was an interesting sex difference, however, with regard to male mice as between 11·5 and 12·4 g; 40 per cent of the cold-store mice, 38 per cent of the food-store mice, and 64 per cent of the urban mice had reached puberty.

It would seem that the wild rat (*Rattus norvegicus*) is more resistant than mice to the effects of cold. A population of rats living in a very extreme environment outside Nome, Alaska, showed no difference in the size or age at which puberty is reached when compared with populations under more normal conditions in Baltimore, U.S.A. (Schiller, 1956). This is rather remarkable, as the population reflects its position of being at an extreme of the range by an exceedingly high winter mortality (over 60 per cent of individuals having frostbite damage in January) and by the interruption of breeding over the winter in this normally continually breeding species.

An effect of temperature at the other end of the scale has been suggested by Errington (1963) in his book describing the biology of the muskrat (*Ondatra zibethicus*) in North America. He found considerable differences between years in the proportion of young females which become pregnant in the year of their birth (usually 0 per cent, but maximally 5·3 per cent in a 21-year study). Apparently this changing incidence is related to the temperatures prevailing in the various years, as Errington points out that in the warmer southern states (i.e. Louisiana) a much higher proportion of females breed in the year of their birth. There is an effect of latitude involved here, as the southern populations have a longer breeding season (see page 64), so that relatively more young can reach puberty inside the breeding season of their birth.

The evidence for an effect of temperature on the onset of puberty in rodents would thus seem to vary with species and also with factors related to the depression of growth rates. It does not seem possible at the present time to distinguish between the primary effects of temperature on reduced growth (and its consequent secondary retardation of puberty) from any direct primary effect of temperature changes on the onset of puberty.

(b) Other mammals

Other than rodents, there is again very little published research into the effect of temperature changes on the onset of puberty. The age at menarche in humans has been known for many years to vary somewhat between tropical, temperate, and colder areas. Mills (1939) has considered that these differences are a function of different rates of body growth and has suggested that temperature variation is the direct cause. He quotes the following mean ages at menarche for different parts of the world (see first table on next page).

Mills interprets these data as demonstrating delays in puberty in both very cold and very hot areas of the world. Similarly, Eveleth (1966) has claimed that U.S. schoolgirls living in Rio de Janiero show delayed menarche compared with similar girls in the United States. However, much of the interpretation of this human data must be treated with extreme caution. In the first place statistical analysis is necessary to demonstrate differences of the order indicated by the data presented and this does not seem to have been attempted. Secondly, as well as temperature variations in the different localities, there are marked dietary, racial, economic, and social differences between the areas under consideration which could all effect the age at menarche. Therefore, until a more definitive study is carried out, the relationships between temperature variation and development of puberty in humans must remain as only suggestive.

	Years		Years
Eskimos	15–18	Finland	15·9
Montreal	14·9	Norway	14·5
Minneapolis	13·0	Germany	14·0
Cincinnati	13·0	Paris	13·5
New Orleans	13·6	Milan	15·2
Panama	14·2	Asteria (Spain)	15·9

Of a more scientific nature is the evidence presented by Dale, Ragsdale, and Cheng (1959) for an effect of environmental temperature on the onset of puberty in cattle (*Bos taurus*). The mean age at first oestrus was compared in three breeds maintained either at a constant temperature of 50° F (10·0° C) or at 80° F (26·7° C) or under control conditions outside in normal temperatures for Missouri. The experiment was started with calves of one month, and the mean ages at puberty are shown below:

	Outside controls (days)	50° F (days)	80° F (days)
Brahman	397	307	463
Santa Gertrudis	306	290	290
Shorthorn	280	303	440

The data show clearly the importance of distinguishing between breeds in experiments such as this, for the variation between the breeds in marked. The Brahman cattle matured much more slowly at the higher temperatures and faster at the lower temperatures. Shorthorns also matured slower at high temperatures, but the other combinations showed relatively smaller alterations. This may reflect the different abilities of the breeds to grow under the temperature conditions imposed, since the authors report that Brahmans under 80° F reached sexual maturity at an average of 60 per cent of the predicted mature weight compared with all other experimental groups, which reached sexual maturity at 39–45 per cent of the predicted mature weights. It would seem that all experimental animals have thus grown faster than would have been predicted from the controls.

The effect of temperature changes on puberty of humans and cattle, as on rodents, is thus very closely related to the effect of temperature on growth rates. More experimental work is required to determine if temperature has any direct effect on the attainment of puberty.

CHAPTER 4

The Effects of Nutrition on the Onset of Puberty

The nutrition of mammals is a factor in their ecology which is more difficult to quantify and thus evaluate under natural conditions than the more easily measured and thus experimentally varied physical factors, such as light and temperature. Because of the large number of separate units, such as proteins, fats, carbohydrates, minerals, and vitamins, which constitute an adequate nutritional regime, the study of the effects of alterations in nutrition becomes complex, as many of the individual elements have separate effects on the reproductive processes. Not only do the separate parts of the diet have individual effects but some of the elements involved also do not act entirely independently of each other. Change in the level of intake of a single factor (i.e. calcium) may therefore result in changes in the animals' physiological mechanisms which are really the result of alteration of a proportional relationship (i.e. the calcium: phosphorus ratio). Thus it becomes difficult to interpret and even more difficult to compare much of the published experimental work on dietary effects on reproduction because of the numerous combinations of the factors involved. As well as alteration of the quality of the diet, the amount of total food intake (the plane of nutrition) is of considerable importance in the normal physiological well-being of the animal, which is in turn reflected in its reproductive performance. Qualitative alteration of the diet often involves altering the plane of nutrition as well, so that there are really two variables affected. The plane of nutrition is also important with regard to the energy value measured in terms of calorific intake that is supplied to the animal.

It will soon become apparent to the reader that the evidence relating nutrition to reproductive processes in wild mammals is of a very different nature to that for domestic or experimental animals. There are few studies of the nutrition of wild mammals which go beyond trying to relate the nutritional status of the population by changes in the body weights of the individuals sampled or by rather crude evaluations of the type of nutrition available, though not necessarily consumed. Because of the practical difficulties involved, there is little information on qualitative variation in the diet of wild animals and practically all information is on the plane of nutrition. As field techniques improve, it should become possible to measure the variations in the proportions of the various dietary elements in the nutrition of wild mammals. Evidence from domestic mammals will therefore be presented here for later comparison.

It is important to point out here the general relationships between nutrition and growth. While it is possible to study directly the effect of varying the nutrition of the *adult* animal on its reproduction, any alteration of nutrition *before* puberty alters the rate of growth of the experimental animal. The attainment of puberty is a function of the size of the mammal as well as of its age, so that alteration of the growth rate will alter the attainment of any definite size. Hammond and Marshall (1952) have reviewed work on growing pullets which demonstrates that birds on a high plane of nutrition produced their first egg at an earlier age than birds fed on a low-plane diet; but the low-plane birds, if hatched at the right time of the year, began to lay at lower body weights than the well-nourished birds. It seems probable that below a certain minimal size an individual cannot reach puberty independent of its nutritional status. Joubert (1963), in his review of puberty in farm animals, suggests similarly that they cannot reach puberty until they have reached a certain degree of physical development (usually expressed as live weight) which at that stage of life is typical of their kind.

The curve of growth of animals has been shown by Brody (1945) to be divided into a self-accelerating phase of increasing slope and a self-inhibiting phase of decreasing slope. Between these two phases is the *point of inflection*, which in mammals seems to coincide fairly closely with the time of puberty. Joubert (1963) has plotted these curves for experimental heifers of different breeds and demonstrates that animals showing unrestricted growth exhibit the point of inflection at approximately the same time of the year, even though the animals are of different ages. This coincided with spring, which suggested that the residual or natural tendency towards a breeding season precipitates puberty when conditions are most favourable. A low plane of nutrition delayed the inflection, but it then occurred, without exception, the following spring. The time of first oestrus in Joubert's cattle usually preceded the time of inflection, but under conditions of low nutrition the time of first oestrus was considerably earlier than the time of inflection. The lack of a fixed relationship between the inflection and first oestrus suggests that the age at first oestrus is not delayed by poor nutritional conditions to the same extent as the pubertal inflection. Details of Joubert's experimental data will be given below (page 18). Despite the lack of a fixed relationship between the timing of inflection and the onset of puberty, there can be little doubt that any factor affecting the growth of an animal will also affect, through alteration in the speed of development, the onset of puberty. It will therefore be almost impossible to separate out the effects of varying the nutritional regime on the growth processes (and through this on the onset of puberty) and the direct effects of varying the nutritional regime on the maturation of the reproductive system independently of growth.

(a) Domestic and laboratory mammals

(i) *Rats and other laboratory species.* The whole subject of diet in relation to reproduction in these animals has been excellently reviewed by Russell, F. C. (1948), and the reader is referred to this paper for detailed references for the following. A reduction of the plane of nutrition causes a marked delay in puberty. For example, rats fed on $\frac{1}{3}$ to $\frac{1}{4}$ of their normal calorific intake did not reach sexual maturity inside an interval of 12 weeks after weaning, compared with controls which did. Reduced protein intake, or too *high* concentrations of protein, and restricted intake of some minerals, such as phosphorus or sodium, can all cause a considerable delay to the onset of puberty. Other effects are also mentioned by Moustgaard (1959). One which is apparently sex and species specific is that caused by a deficiency of Vitamin E. Male rats raised on this sort of diet from birth show irreversible degeneration of the seminiferous epithelium at puberty. Female rats show no apparent impairment of puberty, and neither do male mice or rats on similar diets. On the other hand, a deficiency of Vitamin A acid (retinoic acid) was found to inhibit also the sexual development of juvenile rats (Thompson, Howell, and Pitt, 1964). Scott and Scott (1964) reported that a Vitamin A deficient diet inhibited puberty in laboratory cats, 3-kg males having prepubertal testes. Parental gonadotrophins, however, stimulated immediate testes development in these animals.

Moustgaard (1959) has pointed out that for laboratory animals generally, gross underfeeding reduces the gonadotrophic potency of the adult pituitary. Disturbances resulting from a dietary regime of this sort seem to be due to lowered production from the pituitary rather than a lowered level of response of the target gonad. Similar effects can be suggested for lowered food intake in juvenile animals. It is also possible that reduced food intake, while affecting the growth rate by reducing the primary elements of nutrition, may affect the output of pituitary-based growth hormones, and thus have a double affect on the attainment of the size necessary for puberty.

(ii) *Pigs.* The relevant section of the review by Duncan and Lodge (1960) on the effects of diet on reproduction in pigs will now be summarized. Although there is no doubt that the plane of intake of foodstuffs can affect the onset of first oestrus in gilts, there is some disagreement about the interpretation of results from experiments of different workers. In some cases, gilts which were fed to appetite reached sexual maturity earlier than pigs on restricted planes of nutrition, whereas other workers report that in similar experiments there were either no significant differences in age of the two groups or the high-plane diet animals reached puberty at a later age. Apparently even in these experiments the gilts fed to appetite did reach puberty at heavier weights than underfed pigs. This difference is interpreted by Duncan and Lodge as follows. They suggest that the first group of workers were in actual fact not overfeeding their high-plane animals but were underfeeding their low-plane gilts, whereas the second group of researchers were overfeeding their high-plane animals (overfatness causing reduced reproductive efficiency in pigs), but maintaining the low-plane animals on adequate and not nutritionally deficient diets. Alteration in the energy level of the diet also showed somewhat equivocal results, in that different workers reported different effects of altering the calorific intake on puberty. Reduced protein delayed the onset of puberty, but the gilts matured at the same body weight.

It would appear that the evidence relating the onset of puberty to nutritional intake is of a general rather than specific nature for the pig. Although the general relationship of reduced intake of gross nutrients causing a delayed puberty is documented, there appears to be little work on the relationships between the growth rate and puberty. This is rather astonishing, because, as Duncan and Lodge have pointed out, among domestic animals the sow is somewhat unique in that its sole purpose is to produce piglets. An understanding of the dietary requirements to puberty, and thus the economics of pig production, would therefore seem to be of utmost importance.

(iii) *Sheep.* Among domestic mammals the dependence of breeding phenomena on light regimes is best understood in the seasonally breeding sheep (see pages 8, 62). As Hammond (1949) has pointed out, the age and size at puberty in sheep depend on the time of year of their birth, as the output of gonadotrophins is seasonal. Nutritional effects seem to have little role in the attainment of puberty in the sheep, although there is some evidence (Thomson and Aitken, 1959) that the level of winter nutrition may alter the fertility of ewes at their first oestrus.

(iv) *Cattle.* Observations of a number of rearing conditions have shown that cattle are affected by nutrition in the reaching of puberty. Under conditions of normal feeding one group of experimental heifers reached puberty at an average of 11·3 months, whereas heifers on 65 per cent of the normal diet reached it at 17·3 months and on 140 per cent of the diet at 9·4 months (Moustgaard, 1959, quoting Reid *et al.*). Similarly, Hansson (1956) reports the following mean ages to puberty in Swedish Red and White heifers reared on different proportions of a normal feed intake:

Level of nutrition (%)	Mean age at first oestrus (days)
40	399
60	375
80	327
100	312
120	318

Other dietary factors which affect puberty in cattle have been mentioned by Moustgaard (1959). In Scandinavian countries, where cattle in winter are kept on a protein-deficient diet mainly of sugar beets and sugar pulp, heifers emerging from winter showed absolutely no symptoms of heat. Cattle on soils which are deficient in phosphorus show marked delays in the timing of puberty. It would seem that cattle show a graded response to phosphorus dietary content, as far as their re-production is concerned, in that over a range of phosphorus intakes the effects in cows range from complete anoestrus to normal heats. The Ca/P ratio enters into this picture, in that cattle on pastures excessively high in calcium show im-paired fertility and probably delayed puberty due to the concomitant reduction in phosphate utilization.

The effect of nutrition on the development on puberty in male cattle has been investigated by Mann, Rowson, Short, and Skinner (1967) using the excellent experimental technique of placing twins on differing levels of nutritive intake. That member of the pair which was placed on $\frac{1}{3}$ of a normal ration showed marked inhibition of the androgenic function of the testes and some inhibition of sperma-togenesis. The androgen deficiency was clearly reflected in a reduced production of fructose and citric acid by seminal vesicles.

Returning to heifers, Joubert (1954) experimentally varied the winter nutri-tion of cattle in the Transvaal by feeding one group on a supplemented diet (high-plane) during winter and leaving another group on normal winter grazing (low-plane). The low-plane members of the pairs of heifers used during the experiment showed considerable retardation of growth, there being differences, however, in the responses of the four breeds of cattle used. The experiment was started with 6–11-month-old heifers and Joubert tested the development of puberty in an exceedingly interesting manner. The heifers were divided into two groups, one of which was bred when it reached a desirable *weight*, which seems to be about $\frac{2}{3}$ of the mature weight, while the other group was served at a stage when both high- and low-plane animals could calve at approximately the same *age*. This was achieved by mating the low-plane member of the pair as soon as signs of puberty were seen and by mating her counterpart during the earliest possible following heat period. The results of this fascinating experiment are shown below:

	High-plane		Low-plane		Difference	
	Age (mths)	Wt (lb)	Age (mths)	Wt (lb)	Age (mths)	Wt (lb)
Bred on 'weight' basis						
Shorthorn (at 800 lb)	21·8	808·5	31·9	782·5	10·1	26·0
Afrikaner (at 750 lb)	21·3	768·0	31·3	779·5	9·9	−11·5
Friesian (at 800 lb)	22·4	847·0	32·0	823·0	9·6	24·0
Jersey (at 500 lb)	15·0	515·5	23·9	512·5	8·7	3·0
All breeds	20·1		29·7		9·6	10·4
Bred on 'age' basis						
Shorthorn	24·1	910	19·7	550	−4·4	260
Afrikaner	30·7	940	29·3	748	−1·4	192
Friesian	21·8	841	20·2	755	−1·6	86
Jersey	18·4	517	17·0	380	−1·4	137
All breeds		777		608	−2·2	169

The results are given in detail because this experiment attempted to separate the effects of age and body weight on puberty attainment during a nutrition experiment. Joubert reported later (1963) that on average the age at first oestrus was more affected by dietary restrictions than was the weight. As a basis for this idea he presented the following means for the whole experiment:

	Mean weight at first oestrus (lb)	Mean age at first oestrus (days)
High-plane	530 ± 40·0	440 ± 35·1
Low-plane	579 ± 41·6	710 ± 62·1

In all breeds the animals on a high-plane diet reached the desired weight at a significantly earlier age than those on a low-plane diet. In the second the low-plane heifers conceived at slightly earlier ages but much lower weights than the high-plane group. However, it should be remembered that these animals were served at the first sign of oestrus and their high-plane partners were mated only when the mating was carried out in the low-plane animal. In actual fact, the high-plane heifers had undergone on average some 8–17 oestrous cycles before mating.

It would appear from the above survey of the literature that by far the most research has been based on a simple reduction of intake and that little has been done on the effects of changing the quality of the diet. With the exception of Joubert's work on cattle, there seems to have been little attempt to separate the effects of size and age or, in fact, to document the growth relationships caused by nutritional reductions in these experiments. It is also apparent that most experiments are designed on a high/low basis, whereas more information would surely come from experiments in which graded intakes were studied. Generally the picture that emerges is that starved animals begin to breed later, which is only a part of the picture, as will be seen in the section on fertility (i.e. that starved animals breed at lower rates). There is a very real need for more definitive experiments indicating whether or not there are crucial levels in intake of nutrient, below which maturity is impaired, or whether there is a completely graded effect. Questions such as this are of vital importance with regard to the most economic production of animals for foodstuffs.

(b) Rodents

The wild rat (*Rattus norvegicus*) reaches puberty at approximately the same size in different areas all over the world where it is associated with human habitation. Davis (1951b, 1953) has demonstrated in a number of studies of the reproduction of this species, that although puberty is defined in many different ways, it still occurs at approximately the same stage of growth. This probably indicates that rats are able to utilize efficiently the available nutrition. For example, in one study reported by Davis, where farm rats were on a deficient diet, they reached puberty at significantly different weights than did a population sampled from urban areas. Although it was not possible actually to age these rats, Davis suggests that they matured later as a result of the low nutrition available on the farm.

Seasonal differences in nutrition in the wild will alter the timing of puberty in some rodents. Zejda (1962) observed winter breeding in certain years in populations of the bank vole (*Clethrionomys glareolus*) in Czechoslovakia, which was associated with above-average winter feeding conditions due to exceptional fruiting in oaks and other trees. The amount of food available was relatively much less than in the normal spring and summer breeding seasons and consequently the voles grew more slowly during the winter and thus reached sexual maturity at an older age. It is possible that other environmental factors, such as temperature, also delayed puberty in this case. However, no other factors than nutrition could have caused the effect noted by Greenwald (1957) in his study of *Microtus californicus* in California. There was a considerable difference over two years in the level of nutrition available to the voles. The year 1953 was characterized by good even rains, which resulted in excellent feed availability for the whole breeding season, while in 1952 poorer rains meant poorer nutritional conditions. This was reflected in a difference in the weight at puberty in females. In 1953 they reached puberty at 30 g compared with 35 g in 1952. Interestingly enough, there was no difference in the weight at puberty for males. Although there are no data on the *age* of puberty, it would appear that the voles under conditions of good nutrition grow faster and probably reach puberty earlier. In a similar fashion, Koford (1958) has presented indirect evidence that prairie dogs (*Cynomys ludovicianus*) reach puberty earlier where their towns are located near to extra food supplies such as fields of grain. The proportion of yearling females bearing young is highest in those areas where this extra food is available during the winter. Although grey squirrels (*Sciurus carolinensis*) in the United States normally do not breed until over 10 months old, Smith and Barkalow (1967) have reported that an exceedingly heavy mast crop in the autumn of 1965 resulted in 9 of 17 females having bred at 4 months when collected in spring 1966. These observations confirm others in England, where the same species revealed puberty very early in a year of a heavy mast crop.

Hormonal factors in plants, such as plant oestrogens, are known to affect the fertility of domestic mammals (especially sheep), and Bodenheimer and Sulman suggested in 1946 that the fertility of voles could be affected by gonadotrophic factors in the diet. This suggestion has been confirmed by Negus and Pinter (1966), who fed *Microtus montanus* various artificial foodstuffs, such as sprouted wheat, its extract, and spinach extracts. The feeding of sprouted wheat to 3–4-week-old voles caused 9 out of 10 animals to show signs of oestrus in their vaginal smears within 24 hours, while only 1 in 10 of controls did so. Although the ovaries of the voles fed on sprouted wheat were heavier, the rapidity of the response suggests that it may be only demonstrating the effect of an oral oestrogen and not represent the onset of true reproductive development. Until matings are carried out and fertility determined, it is unwise to consider this sort of result as indicating a hastened puberty.

(c) Rabbit (*Oryctolagus cuniculus*)

In their classic study on the reproduction of rabbits in confined pens, Myers and Poole (1962) noted differences in the attainment of puberty of their experimental animals. In two of their enclosures (A and B) females started to breed at a mean age of 7 months and a mean weight of 1,200 g. In a third enclosure females reached puberty at a much older age of 11 months but about the same body weight (1,140 g). Food was more restricted in this third enclosure, with lower nutritional conditions resulting in general starvation. Although density factors entered into this delay in puberty, they are not as important as nutrition in this case, because the actual density was highest in pen B, where puberty was apparently reached normally.

(d) Marsupials

In his study of the breeding season of the quokka (*Setonix brachyurus*), Shield (1964) showed that populations on the mainland of Western Australia breed continually and apparently reach puberty quickly, due to a higher level of natural nutrition being available all the year. On Rottnest Island the breeding season is limited, and young born between February and April do not reach puberty for about 6 months, due to a suggested poor level of nutrition available at this time.

The age at sexual maturity in the red kangaroo (*Megaleia rufa*) was found by Frith and Sharman (1964) to differ between field areas. In their first study area near Toganmain, N.S.W., the 14-inch annual rainfall is evenly distributed through the year, so that pasture conditions remain relatively unchanged and at a high nutrient level. Mt Murchison, N.S.W., has a 9-inch erratic summer rainfall with a variable pasture but many adequate artificial watering points for the kangaroos. Finally, Gilruth Plains in Queensland also have a summer rainfall of 13 inches, but this is extremely unpredictable, so that there are very long drought periods and the average level of nutrition available is low. On the basis of teeth patterns these authors aged their kangaroos. The attainment of puberty in the different areas is shown below:

| Age (mths) | Percentage of females sexually mature | | |
	Toganmain (%)	Mt Murchison (%)	Gilruth Plains (%)
14–17	8	8	0
20–24	64	40	28
28–33	95	89	59
39–46	100	98	95

The relationship between nutrition and puberty in the red kangaroo has been further demonstrated by Newsome (1965) working in central Australia. He compared the mean age of puberty of kangaroos growing during drought periods on

C

poor nutritional regimes with the mean age of puberty of kangaroos growing after rains on good nutrition. His data below showed a highly significant difference:

Mean age at puberty after rains $28 \cdot 6 \pm 1 \cdot 2$ months.
Mean age at puberty during drought $34 \cdot 8 \pm 3 \cdot 4$ months.

There can be little doubt that there is a marked effect of seasonal nutrition on the attainment of puberty in these macropod marsupials. It would be interesting to investigate further the effect of dehydration and to determine the actual nutrient factors in the diet whose absence caused the delay in puberty.

(e) Deer

The number of years elapsing after the season of birth until puberty is reached in a number of species of deer is dependent upon the level of nutrition available to them. Robinette, Gashwiler, Jones, and Crane (1955) suggested that the weather during the winter and the consequent level of nutrition available to mule deer (*Odocoileus hemionus*) fawns governed whether or not they would start reproducing in their first (yearling) summer. For the same species Taber and Dasmann (1957) found in California that yearling females bred at about 17 months of age on most types of range, but on the very poor chaparral, breeding began when the does were over 24 months old.

The degree of winter utilization of browse was found by Pimlott (1959) to be indicative of the level of nutrition available to moose (*Alces alces*) in Newfoundland and to affect their attainment of puberty. In one study area in central Newfoundland only 29 per cent of yearlings were found to be pregnant during the winter and spring, compared with 67 per cent for the rest of the province. Pimlott had previously investigated the utilization of different types of vegetation by moose and found that although there was apparently no gross food shortage in the central area, there was considerably higher utilization of certain deciduous species during the winter. This observation suggests a winter requirement of particular types of nutriment. Pimlott postulates that the lower level of nutrition available to the central-area moose during their first winter of life reduces their growth, and fewer reach sexual maturity in their first breeding season after the season of birth. Nutritional levels are also likely to be the cause of variation in the proportion of yearling moose found pregnant in different areas of Ontario by Simkin (1965). He compared the incidence of pregnancy in yearlings there (17 per cent) with the incidence described by previous workers for Newfoundland (46 per cent) and British Columbia (0 per cent).

It is sometimes difficult to separate the effects of poor nutrition in deer from the effects caused by overpopulation and perhaps social factors caused by high densities. Buechner and Swanson (1955) cite an interesting example for wapiti (*Cervus canadensis*) in Washington State. Normally only about 10 per cent of females become pregnant in their yearling year, but these authors found a population where up to 50 per cent of yearling elk were pregnant. They deduced that this was caused by a severe reduction in numbers in this population the previous year due to overshooting. Although they could see no obvious changes in the vegetation, it seems likely that relatively more nutrition was available. Conception in very young elk has also been reported by Batchelor (1963) in Alaska.

In New Zealand, Daniel (1963) found that red deer (*Cervus elaphus*), which normally breed in Scotland at 3 years of age and calve at 4, were having their first oestrus at 16 months and actually calving at 24 months. This was occurring in indigenous *Nothofagus* and podocarp forests where food was particularly good all the year round. Daniel quotes one record of a fawn kept on exceedingly high protein, lucerne, and clover pasture, showing developing follicles and one mature follicle at only 4 months of age. Valentincic (1960, reference not seen, quoted in Daniel as *Trans. 4th Cong. Intern. Union Game Biol.*, p. 188) found that yearling red-deer hinds from mixed oak and beech forest in the lowlands of Yugoslavia showed a 30 per cent incidence of pregnancy, while no pregnancies were found in yearling hinds from nutritionally poorer alpine pastures of Scots pine and beech. Daniel (*pers. comm.*) has evidence that other species of deer introduced into New Zealand also breed younger as the result of better nutritional regimes. For example, the Javan rusa (*Cervus timorensis*) and the Ceylon sambar (*C. unicolor*) normally breed in Asia and over most of New Zealand when they are two years old, but under exceptionally good nutritional conditions, particularly near forest edges, where they feed on clover and lucerne fields, they can reach puberty at about 6 months of age. Mitchell (1967) reported a female pronghorn (*Antilocapra americana*) pregnant at 8 months in Alberta, but did not relate this observation to food supply.

The attainment of puberty in deer would thus seem to be definitely related to the level of nutrition available to them during their pre-pubertal life. Evidence from a number of different species in different localities would indirectly suggest that puberty is normally delayed in the wild due to the diet available having insufficient nutritional value. It is possible that this is a means of regulating the annual productivity of the species, in that yearlings prevented from breeding by food shortage will not be adding new mouths into the population. However, the detailed studies of herd production in the United States by wildlife biologists have produced a number of examples whereby uncontrolled deer populations have reached such high numbers that starvation has caused a high mortality. It would be very interesting to attempt to correlate the degree of early breeding of juvenile deer and the subsequent changes in population numbers.

CHAPTER 5

The Effects of 'Density' and Social Factors on the Onset of Puberty

The enumeration of individuals in a defined area or volume does not necessarily indicate in any way the environmental factors acting on the selfsame individuals. The most important factor in this context is the presence of other individuals of the same species. The concept of the *density* of a population of mammals results in this sense only from the convenience of quantifying the population in this way, so that a number of populations of the same density may represent a large range of variation in other biological factors operating on the animals. The reader is referred to another interesting approach towards this problem in the paper by Lloyd M. (1967) on mean crowding. In effect, density is usually used, with regard to its action on reproductive processes, to measure what can be vaguely called 'social pressure', i.e. so that the density is used to represent differences in the intensity of social interaction between the individuals constituting the population. Practically, however, density, in terms of the numbers of animals per unit area, can almost never be used accurately to compare one population or sample with another. As more and more evidence accumulates, indicating the considerable variation in behaviour between individuals of all mammal species, it becomes more and more unrealistic to consider any two groups of animals at the same or different densities as being composed of comparable units as far as behaviour is concerned. The natural social organization of the species, whether it is of a solitary or gregarious type (Eisenberg, 1966), will govern to a large extent the nature of its reaction to varying densities. Indeed, the comparability of any two different areas must also be questioned. No two natural environments are the same with regard to the environmental factors operating in them or with regard to their physical structure. Even though experimental populations can be kept under apparently similar conditions, the animals themselves will alter the habitat in the laboratory colony by such means as the variation in individual behaviour at food and water sources affecting food availability to members of the experimental population, or by their own odours, the importance of which is just beginning to be realized (see page 253). The problems, then, in equating any pair or more of density measurements (due to the difficulty of comparing the actual animals and the actual areas) mean that the measurement of density can be used only to give a very approximate measure of the social factors, which vary from location to location. It is possible that the development of new techniques in the study of populations at different densities will produce better methods of actually quantifying the differences in social factors involved. Some field and laboratory studies (Bronson, 1963; Sadleir, 1965b) are indicating possible approaches along these lines.

The lack of any real consistency in results from experimental studies which involve the comparison of different populations on a density basis suggests strongly that density measurements reflect neither the type nor magnitude of the real variation involved. Unfortunately, at the moment, there is no simple method for quantifying 'social pressure', especially in wild populations. It is more than likely that further understanding of behavioural mechanisms operating inside populations will show that there are many individual behavioural characteristics which will need to be quantified before the role of the behaviour of conspecifics on individual reproduction is understood. At the moment, therefore, until better measures of the degree of social interaction and utilization of space are devised, it is unfortunately necessary to use the measurement of density as the comparative yardstick. In this and subsequent sections relating density to reproduction the problems of comparing 'density' effects between separate populations will be re-emphasized in the different examples presented.

I. Natural populations

The house mouse (*Mus musculus*) is a species whose populations show a great range of densities in various habitats. Because of its economic importance, particularly with regard to consumption of stored food, the reproduction of feral mice has been well investigated. Southwick (1958) collected over 4,000 mice from 40 corn ricks of various population densities in England and found variations in the reproductive characters of the different populations. He classified the ricks into four density classes, and the data below show the relationships

	Low density	Medium density	High density	Very high density
No. of populations studied	21	11	7	1
No. weaned mice per cubic metre				
Mean	0·23	1·12	3·50	16·18
Range	0·01–0·48	0·50–1·94	2·69–5·64	—
Average age of rick (mths)	4	6	7·5	8·5
Propn. males fecund (%)				
12·5–17·5 g	92·3	90·6	83·8	84·9
17·5 + g	97·2	100·0	97·6	100·0
Propn. females fecund (%)				
12·5–17·5 g	98·5	93·6	87·9	83·7
17·5 + g	100·0	100·0	100·0	100·0

between the density and the proportion fecund (as judged by ovum and sperm production) in various weight classes. These data are particularly interesting because they do not consist of samples, but are based on the whole population of the rick at the time of threshing. Complicating factors include the varying periods of time that the ricks had been standing, so that it is possible that the stage of population growth in the different aged ricks may have had an effect on fertility. Mice normally reach puberty at 12·5 g, so that it will be seen that there is a tendency for a retarded puberty as the population becomes denser. This was sufficient in the case of the females to render the difference between density classes significantly non-homogeneous.

In a similar investigation, Rowe, Taylor, and Chudley (1964) studied the reproductive characteristics of mice in four ricks either of oats or wheat which were of different densities at threshing. Their data are shown below:

Rick	No. of weaned mice per cub. metre	Weight class (g)	Males		Females	
			Number	% fecund	Number	% fecund
1. Oats	10·7	7·6–12·5	192	3·1	177	4·0
		12·6–17·5	245	55·1	222	36·5
		17·6 +	95	90·5	169	92·9
2. Oats	10·8	7·6–12·5	225	2·2	267	3·4
		12·6–17·5	355	63·6	283	52·3
		17·6 +	127	96·9	214	92·5
3. Wheat	8·0	7·6–12·5	166	7·8	198	22·2
		12·6–17·5	346	80·0	270	82·6
		17·6 +	150	98·7	236	95·3
4. Wheat	3·6	7·6–12·5	68	5·9	61	23·0
		12·6–17·5	151	84·8	115	91·3
		17·6 +	63	96·8	127	100·0

It is important to note that whereas in the previous paper the densities were fairly low, these authors dealt with much denser populations. Again there is some evidence of the attainment of puberty being inversely related to density. The effect is not so marked in males as in females, but males in the middle weight class in the two lower-density ricks showed a much higher incidence of fecundity than in the high-density ricks. A varying attainment of female puberty in the different densities was found in both the middle and lower age classes, a marked increase in the proportion fecund being seen in the smallest class in the two lower-density ricks.

Both the above papers describe other differences in the reproductive efficiency of the mice populations at various densities, and these will be discussed later in the book (see page 241). However, the data have been presented fully here for comparison with laboratory data presented in II below.

Other researchers have suggested that density of wild rodent populations affects the attainment of puberty. Zejda (1961, 1967) found that young bank voles

(*Clethrionomys glareolus*) born in the spring in a very overcrowded population in Southern Moravia did not reach puberty at all in the breeding season of their birth. In his study of various species on small British islands, Jewell (1966) found that *C. glareolus* did not reach sexual maturity in the year of birth in very dense populations on the island of Skomer. It is possible here that the shortened breeding season on the islands allowed insufficient time (relative to the period of increasing light) for young born in the season to mature.

A number of rodent species undergo fairly regular cycles of population density over a 3- or 4-year period. These species offer an interesting comparison as to the relationship between density and the attainment of puberty. There are, however, some problems involved. Chitty (1962) has suggested that not only are the numbers different in subsequent years but different behavioural and/or genetical types of animals could be present at different stages of the cycle, so that it is possible that comparisons based on density alone may give an incorrect picture. There is some disagreement among field workers as to the effect of the different phases of these population cycles on puberty. For example, Hoffman (1958) found no differences in the speed of attainment of puberty at different stages of the cycle of *Microtus montanus* and *M. californicus*. On the other hand, Kalela (1957) describes striking differences in the proportions of juvenile voles (*Clethrionomys rufocanus*) reaching puberty in different years. In his study at Kilpisjarvic in Finnish Lapland, vole numbers were low in the spring of 1954, but had started to increase considerably by autumn. Males and females reached puberty quickly at the beginning of the year, but more slowly after August when the density had increased. The peak in numbers came in 1955. Practically all males remained immature, especially at one locality (Malla) where the numbers were particularly high. Some females did reach puberty early in 1955, but again the proportion which did so was very much less at Malla. Before spring arrived in 1956 the numbers had crashed, and again both sexes reached puberty quickly, especially in the early part of the breeding season. Kalela found that the variation in the attainment of puberty by these voles was not due to differences in meteorological factors and concluded that the fecundity rate of juvenile voles was thus density-dependent. A general retardation of puberty occurred in the year of high numbers, but the evidence suggests that the critical density involved is lower for females than males.

It is probable that this variation in the attainment of puberty with the phase of the population cycle in microtine rodents is a general phenomenon. Similar variations in puberty were described by Krebs (1963, 1964) in his exhaustive study over four years of two species of lemmings at Baker Lake, North-west Territory. Numbers of the brown lemming (*Lemmus trimucronatus*) and the varying lemming (*Dicrostonyx groenlandicus*) were very low in 1959, but reached a peak in 1960, declining in 1961 and 1962 (see figure 50, page 246). Striking differences in the age of puberty, as measured by the median body weight at maturity, were found between different years. Kreb's data are summarized below:

	Males Median weight at maturity (g)			Females Median weight at maturity (g)		
	Winter generation	Y_1 Summer generation	Y_1' Summer generation	Winter generation	Y_1 Summer generation	Y_1' Summer generation
Lemmus						
1959	<30·5	33·8	26·5	<26	<26	20–25
1960	36·7	>61 *	>51 *	51·6	29·1	>41
1961	31–41	>41 *	>41 *	41·2	21·3	24·4
1962	29·3	36–40?	?	32·4	20·3	?
Dicrostonyx						
1959	28·3	?	—	30·5	?	—
1960	43·3	>51 *	—	49·9	>41 *	—
1961	38·8	>31 *	—	49·3	24·0	—
1962	34·1	>36 *	—	40·7	>31?	—

* Actually none of these animals reached sexual maturity during the breeding season. The winter generation refers to animals whose body weights at the beginning of the summer indicated that they had survived over winter. Summer breeding tends to occur synchronously, so that summer young appear in waves. On the basis of body weights the animals were assigned to the first summer litter (Y_1 summer generation) and subsequent summer litters in order.

It will be seen that in every case there was a considerable increase in the median weight at maturity during the peak year of 1960. During the decline year of 1961 a number of different patterns became apparent in that winter generation *Dicrostonyx* males and females and *Lemmus* males had median body weights much the same as 1960, but winter generation *Lemmus* females and summer young *Lemmus* females all declined in median weight. On the other hand, summer young males of both species did not reach sexual maturity at all. Generally speaking, the summers of 1959 and 1962 were most similar in that the most rapid rates of maturation were noted in both species with the exception of the 1962 young females. This rapid maturation took place over 3–5 weeks. Krebs suggested that this change in the pattern of maturation is not caused by changes in density *per se* but by an intrinsic change in the population probably connected with intra-specific strife.

Other cyclic microtine species would seem to show similar inhibition of maturation during years of peak abundance. Chitty (1952) described an inhibition of puberty in years of high density of voles (*Microtus agrestis*) in Wales. Another paper of Zejda (1964) described how in a peak year female *Clethrionomys glareolus* matured very quickly for the earlier part of the breeding season, but from June onwards less than 8 per cent of females reached puberty. By August all reproduction in this peak population was at an end. Similarly, during a peak lemming (*Lemmus lemmus*) year in Finland, Kalela (1961) described how animals grew quickly and reached puberty early if born before July, but after this time puberty was almost completely inhibited. The phenomena for microtines have been reviewed briefly by Christian, Lloyd, and Davis (1965).

Some indications as to the manner by which high densities affect social factors and thus inhibit sexual maturity were reported in the interesting study carried out by Koshkina (1965) on the red vole (*Clethrionomys rutilus*) near Novostroika, a village in the taiga region of the U.S.S.R. Voles were studied from 1958 to 1964, although most of the concentrated research was done after 1962. Numbers were low in 1961 but considerably higher in 1962, due to there being only a small drop in numbers in the 1961–2 winter. Numbers also only dropped slightly during the subsequent winter, so that 1963 was really the peak year. During the winter of 1963–4 numbers decreased rapidly, so that the density of voles was much lower in 1964. It should be pointed out that the degree of fluctuations in density involved was small compared with normal cyclic microtines, as the *average* catch of voles in the taiga at the end of the summer varied only by a factor of three between maximum and minimum years. However, maximum numbers in any year showed considerably greater variation. Another interesting feature was that investigation of food, cover, and meteorological conditions showed them to be extremely favourable throughout the whole cycle, so that they were not considered to be causes of the changes in numbers seen.

Before proceeding to the variation in attainment of puberty it is appropriate to consider the nature of vole territoriality at different stages of the cycle. During the summer of the peak year, 1963, Koskina noted that territorial sizes were much smaller than in low-density years and, more importantly, that the proportion of territorial overlap was greater. On the average only 50 per cent of the range of old females was exclusive during this summer. The degree of wounding indicated that considerable fighting was taking place. In low years there were large areas of ground on the study plots which were not utilized by the voles at all, whereas in 1963 all the available ground was covered with vole territories. These differences in utilization of territories affected the attainment of puberty in the different years. The proportion of animals reaching puberty in the year of birth will be seen from the data below to vary between vegetational types but

Year	Type of forest	Number of red voles per 100 trap-nights (July)		Proportion reproducing in year of birth (%)	
		Males	Females	Males	Females
1952	Taiga proper	30·8	11·2	0	5
	Second-growth mixed forest and aspen	22·0	4·1	7	20
	Coniferous island	11·3	1·8	25	45
1963	Taiga proper	32·5	20·0	0	0
	Second-growth mixed forest and aspen	34·4	19·0	0	0
	Coniferous island	14·4	5·5	19	37
1964	Taiga proper	22·1	3·3	55	63
	Second-growth mixed forest and aspen	15·9	2·0	57	67
	Coniferous island	6·6	1·8	69	72

more markedly with the density (as found in July of each year). During 1963 young voles could not find any available ground to establish a territory, so that in most areas they did not reach sexual maturity at all. In the less dense years higher proportions of young of the year did reach sexual maturity, but that actual proportion seemed to depend largely on the density reached in the different types of habitat, as shown in the table above. Figure 3, which has been modified from Koshkina's paper, shows that there was general relationship between the density of red voles overwintering in May and the proportion participating in reproduction in the year of their birth.

As well as differences in the attainment of puberty found between years in the

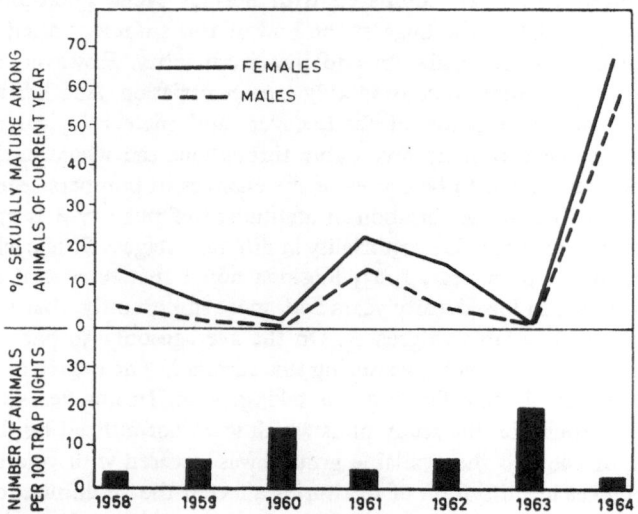

Fig. 3. The relationship between the proportions of red voles (*Clethrionomys rutilus*) reaching puberty in the year of birth and the density of voles in May of the year. Modified from data presented by Koshkina (1965) *Bull. Moscow Soc. Nat. Biol. Sec.* **70**: 5–19.

same type of habitat relative to the density there, Koshkina also presents data which show that during 1964 the proportion reproducing was related to the actual density attained in different habitats at exactly the same period of time. The data below show these differences over two periods at the beginning and end of summer, numbers being low in the first, but then increasing considerably. During the later part of the summer there is a strong relationship between density and attainment of puberty, even though the comparison is made between samples taken at the same period of time. Considered as a whole, Koshkina's work on *Clethrionomys rutilus* presents excellent evidence showing that high density is detrimental to the attainment of puberty in this species. Although the work can be criticized on the grounds that it entirely equates trap success with true density without apparently investigating variations in trappability in the populations, it nevertheless remains as an important contribution, especially in that data have been so clearly presented from different habitats as separate units and also over a considerable number of years.

The nature of cycles of abundance in microtines presents difficulties of interpretation with regard to the effect of density. The variation from year to year in numbers is associated with changes in numbers between months in each individual year, so that inside one year the density can be very variable. Therefore, if a species exhibits reproductive characters which are relatively constant throughout the whole year the relationship between the density and its expression on reproductive efficiency is obviously complex. Moreover, if, as Chitty

Date of capture	Type of habitat	Number of voles per 100 trap-nights	Proportion reproducing in year of birth (%)	
			Male	Female
25 June–13 July	Continuous coniferous and mixed taiga	14·4	68	72
	Second-growth mixed taiga, predominantly coniferous	14·5	66	74
	Aspen and mixed forest, predominantly aspen	9·2	68	73
2–19 August	Continuous coniferous and mixed taiga	35·3	13	25
	Second-growth mixed taiga, predominantly coniferous	29·5	22	39
	Aspen and mixed forest, predominantly aspen	16·0	32	59
	Coniferous islands	7·8	62	68

(1952) has suggested, the genetic types of rodents differ between different phases of the cycle density variations alone cannot be the full explanation for changes in puberty. Finally, alteration in the age at puberty affects the production and recruitment of new young, and thus is a variable affecting the density itself. In species such as the microtine rodents with a very high potential fertility and relatively short period of time between parturitions this means that variation in the age of puberty can very quickly affect the density itself, so that there is a potential counteraction between the two factors.

Finally, there are two more papers which present evidence suggesting that density affects puberty in non-rodent mammals. In his study of the wild rabbit (*Oryctolagus cuniculus*) in North Wales, Lloyd, H. G. (1964) found that the level of nutrition was generally at a higher level than had been reported as available to rabbit populations previously studied in Australia and New Zealand. Welsh rabbit populations reached high densities, and there seemed to be a rough relationship between the minimum weight of does at puberty and density. There were also variations in the level of fecundity of the yearling does. These showed high fecundity in the years of lowest density and low fecundity in dense years. An effect of density was also suggested by Laws (1966) as causing differences in the age at puberty between different populations of the African elephant

(*Loxodonta africana*). In one particularly dense population in the Murchison Falls Park of Uganda he found that the elephants reached puberty at about 18 years of age, whereas in other reported populations the age at puberty was 11–12 years.

II. Artificial populations

The effects of density on reproduction in mammals have been extensively investigated in a number of laboratory studies which have been basically carried out in two ways. In the first, groups of mammals have been put at various densities into very small cages, and in the second relatively small numbers of animals have been put into larger pen-spaces and the population allowed to grow to considerable numbers.

For evidence relating to assembled populations a very considerable amount of research has been done on reproduction by Christian and Davis and their associates in investigations of the interrelationships and effect of density on the adrenal, pituitary, and reproductive systems. Detailed references to this work can be found in the bibliography of reviews by Christian (1963), Christian and Davis (1964), and Christian, Lloyd, and Davis (1965). The grouping together of prepubertal female mice was found in a number of these experiments to result in either a delay or complete inhibition of puberty. A number of other reproductive upsets which resulted from the assembled population cage experiments led Christian (1963) to state that: 'the relative acute and essentially artificial experiments described above, while demonstrating responses to grouping, do not establish that these responses occur in the more chronic situation of a naturally growing population' (page 577). Other workers have reported similar effects in 'caged' experiments. Andervont (1944) found that separated laboratory mice started oestrous cycles earlier than females living 8 to a cage. Petrusewicz (1958) transferred populations from dense cages and found that a reduction in density of a caged population of female mice resulted in the removal of the previously exhibited inhibition of puberty. Ehrlich (1966) reported that puberty was delayed in caged nutria (*Myocastor coypus*) at high densities. This species, which can mature at 4 months, was delayed to 5–6 months in cages and up to 8 or 12 months when kept in very small concrete walled cages. Unfortunately this author does not give any information of the amount of food available, and his statement that the nutria all reached puberty at the same weight, independently of the density, suggests that nutritional factors may have been important.

There have been many reported experiments of mice populations increasing in confined pens with unlimited food. The papers describing these are very varied in their standard, and in some cases the data are either not clearly presented or are hard to interpret. Nevertheless, there appear to be two main reactions in the reproductive efficiency of caged mice as the density increases. The first of these is a general inhibition of reproduction which is reflected in a cessation of oestrous cycles and conception in adults, and in an inhibition of puberty in juveniles. The second is an increase in the mortality of nestlings and juvenile animals. Definite evidence of an inhibition of puberty in increasingly dense populations has been presented by Christian (1956), Christian, Lloyd, and Davis (1965), Terman (1965), and Lidicker (1965). A general decline in birth-rate, which can be assumed also to be reflected in a decline in maturation, has been described by Crowcroft and Rowe (1957), Southwick (1955a) for mice, and van Wijngaarden (1960) and Clarke (1955) for voles. In those studies which included more than one pen the evidence showed very clearly that the type of reproductive alteration observed differed markedly between each population. A general comment from the various researchers was that the factors influencing these alterations in reproduction and recruitment were not related to numbers *per se* but to differences in the behaviour of the animals in the various replicates. The actual degree of inhibition of puberty varied between workers and density, so that it is meaningless to try to compare the amount of inhibition between populations.

III. Comparison of artificial and natural populations.

The very number of mice pen-experiments, which have indicated that puberty is affected by the density of animals in the immediate environment, is impressive enough to suggest a causal relationship. However, the artificial nature of these experiments must be emphasized. As Anderson, P. K. (1961) has pointed out, the density of wild populations very rarely ever approaches the densities involved in laboratory population experiments. The artificial nature of the pen, which permits no immigration or emigration, has been demonstrated by Strecker (1954). His penned experiment allowing movement out of the experimental area showed no detrimental effect on the level or intensity of reproduction. These two factors alone render the application of effects discovered in laboratory populations to wild populations extremely unwise.

Christian and Davis (1964) have attempted to answer this type of criticism by pointing out that even very short exposures of mice to trained fighter mice resulted in increases of adrenal weights greater than when the mice were caged in groups of eight mice per small cage for a week. However, in a later paper Bronson and Eleftheriou (1965), who actually did the work that Christian and Davis mentioned in their review, pointed out that even the 15–20 seconds of fights which occurred in the laboratory situation of these experiments were 'probably too long because of the possibility of escape or avoidance in a field situation' (page 410). As Christian and Davis made no comment on the effect of these short fights on reproductive processes, the original criticism of lack of realism in the artificially high laboratory populations must still be answered.

All that can be said is that under densities in the laboratory, which are much higher than those in the field, a number of experiments which have been carried out almost entirely with mice have shown a marked inhibition of puberty. If the evidence from *Mus musculus* populations, both in the laboratory and in the wild (and even there the association with human activity and food supplies renders the situation hardly comparable with truly wild mammals), was not available, the effect of density on puberty would depend on very few examples indeed, especially as the evidence from cyclic populations can be considered as a special category. There is a very real need for more research in other species of mammals, both in the wild and in laboratory situations, with regard to the effect of density and social factors on maturation and other aspects of reproduction. This evidence will undoubtedly throw light on the problems inherent in the regulation of numbers as densities increase, problems of the greatest importance to the human species at the present time.

IV. The role of social factors in the attainment of puberty

There is considerable evidence that behavioural and social phenomena can affect the attainment of puberty independent of any 'density' factors. For example, Snyder, R. L. (1962) describes an experiment on a natural population of wood-chucks (*Marmota monax*) in which he attempted to alter the sex ratio of the population in one particular area and study the effect of this manipulation on reproduction. This sort of technique is widely applicable to many field studies and has the advantage of having some level of control of the variables without the very artificial effects inherent in a laboratory population study. It has been applied by Sadleir (1965) in a study of adult–juvenile relationships in populations of *Peromyscus maniculatus*. In Snyder's experiment a large number of males and females from one area and females only from another area were removed from the population. This resulted in an increase in population density on the first area compared with a control which did not alter. The population on the area from which only females had been removed also did not alter. There was a marked difference in the intensity of reproduction on the three areas, but the relevant point here is that although the *density* remained much the same in the areas from which females only had been removed, the sex ratio was 57·6 per cent male, a much higher proportion than on the control area which had the normal surplus of females. Also the level of breeding dropped in the area in which females were removed, and many yearling females there showed delayed puberty resulting in a low incidence of pregnancy in that age class. This type of experiment again emphasizes that density *per se* may not be the major variable involved, as in this case sexual maturation proved very different in two populations of the same density but different sex ratio.

Puberty can also be retarded by contact with other members of the same sex and species. Although there is some evidence in the laboratory rat (Drori and Folman, 1964) that cohabitation between groups of males can result in reduction in the size of accessory glands and degeneration in the testes, in most of the examples which follow there is no evidence that the males concerned are physio-logically immature. Despite this, 'behavioural puberty' is retarded, as young males are prevented from mating. For example, in studies of a population of feral sheep (*Ovis aries*) on the island of St Kilda, off Scotland, Grubb (*pers comm.*) has noted repeatedly that young rams are excluded from being able to mate with oestrous ewes due to the attention directed towards these females by older rams (usually 4 or more years old). In their first and second breeding sea-sons the young rams, although physiologically mature, are allowed to mate with the ewes only in the period of time following maximal oestrus. As they get older and achieve a higher degree of social dominance, rams are able to mate more closely to the period of maximal oestrus, with a higher probability of achieving conception. This is thus an example of behavioural factors resulting in a delay in the age of effective puberty. In a recent paper Geist (1968) has considered the delay in social and physical maturation of mountain sheep (*O. dalli* and *O. canadensis*) and notes that the males have a neotenic period after sexual maturation of up to six years. Females do not have a similar period, so that Geist describes them as paedogenic, as they remain at a stage of behavioural development at the level of the young rams.

Very similar relationships between young and old males in social mammals have been reported by other workers. Juvenile elk (*Cervus canadensis*) are driven away from cow–calf groups by the older harem-owning bull elks, and thus suffer what Altmann (1960) has termed 'psychological castration'. Yearling male mule deer (*Odocoileus hemionus*) show interest in does at the beginning of the breeding season, but are driven away from them by the older males. It is possible that they are permitted to mate with the few females which come into heat later in the breeding season when the older males are worn out and are beginning to lose interest in the does (Robinette and Gashwiler, 1950). Although young male impala (*Apyceros melampus*) are physiologically capable of mating at about 13 months of age, they are prevented from doing so for some years after by the adult males (Kerr, 1965).

In the socially breeding seals there is evidence that bulls are physiologically mature well before they are permitted to mate, due to the harem structure of the breeding colony. For example, in a colony of the southern elephant seal, *Mirounga leonina*, Laws (1956) found that bulls were physiologically mature at 4 years, but did not take part in mating on land until their sixth year. The colony that Laws studied was subject to exploitation, so that it may be considered as one of relatively low density, and therefore probably reduced social pressure. Carrick and Ingham (1960) and Carrick, Csordas, and Ingham (1962) have reported on an unexploited colony on Macquarie Island where the bulls were definitely not breeding at 6 years of age, and may even in some cases have been prevented from breeding until 15 years old.

This inhibition of effective puberty due to dominance by older males also occurs in a number of other social mammals, such as carnivores and primates (de Vore, 1965). The reader is referred to Eisenberg's (1966) excellent review of the social organizations of mammals for further information of the phenomena in other groups. However, in many cases there is insufficient information on the actual degree of sexual development of the juvenile males which are inhibited. It is quite probable that further investigation of these problems will show that the younger males are not truly physiologically able to mate and ensure a conception. On the other hand, if they are able to do so, their hormonal state is probably a requisite during the period that they are learning to fit into the complex society of which they are members.

CHAPTER 6

Delayed Puberty

In a number of species of mammal (mainly rodent) the period of time from birth to puberty varies with the actual time that birth takes place during the year. Individuals born in particular seasons have thus a delayed puberty when compared with other individuals born at other times of the year. Evidence has already been presented (page 8) showing how the season of birth of the young lamb governs the time it will take to reach puberty and indicating that this is very probably due to the effect of the changes in daylength controlling the onset of puberty in this most photoperiodic of species. As there is evidence relating light changes to puberty in some rodents, there can be little doubt that the delay in puberty referred to below will in most cases be due to light regimes. However, as well as light changes, other environmental factors effect the attainment of puberty, so the examples of delayed puberty presented cannot be completely explained by this factor alone. Therefore this chapter considers the phenomenon of delayed puberty as a separate entity, while realizing that as knowledge grows about environmental effects on reproduction the various delays will be explained in terms of each individual factor as outlined above. Furthermore, as will be shown later, there are marked seasonal variations in growth rate in the wild species under consideration which affect puberty.

There are numerous examples of rodent species living in temperate climates whose puberty is delayed. In these cases, males and females born early in the breeding season before midsummer grow relatively quickly and reach puberty after a short time. On the other hand, individuals born later in the season do not reach sexual maturity until the first few weeks of the next breeding season, and thus take up to 6 months longer than individuals born only a few weeks before. One of the best-documented field studies demonstrating this was by Breakey (1963) for feral house-mouse populations near San Francisco Bay in California, where females were breeding from April to December. Mice were divided into different monthly age classes on the basis of teethwear; the reproductive condition of aged samples taken throughout the year is shown in figure 4. Females born after October which were nulliparous during the end of the breeding season remained in this condition until the next breeding season. Males behaved similarly.

Delays in puberty have been reported in a number of other rodents. A far from complete list includes *Clethrionomys glareolus* in Sweden (Bergstedt, 1965), on the island of Skomer (Jewell, 1966; Fullagar, Jewell, Lockley, and Rowlands, 1963), and on the mainland of Great Britain (Newson, R., 1963); *Apodemus flavicollis* in Sweden (Bergstedt, 1965) and in Poland (Adamczewska, 1961); *Apodemus sylvaticus* on Skomer island (Fullagar *et al.*, 1963) and on the island of Hirta, St Kilda, Outer Hebrides, Scotland (Jewell, 1966); *Microtus agrestis* (Brambell and Hall 1939); and *Peromyscus maniculatus* and *P. oreas* in the Pacific Northwest of the United States (Sheppe, 1963).

In some other species delayed puberty occurs in temperate localities which is not related to the breeding season. During a study of the Indian gerbille (*Tatera*

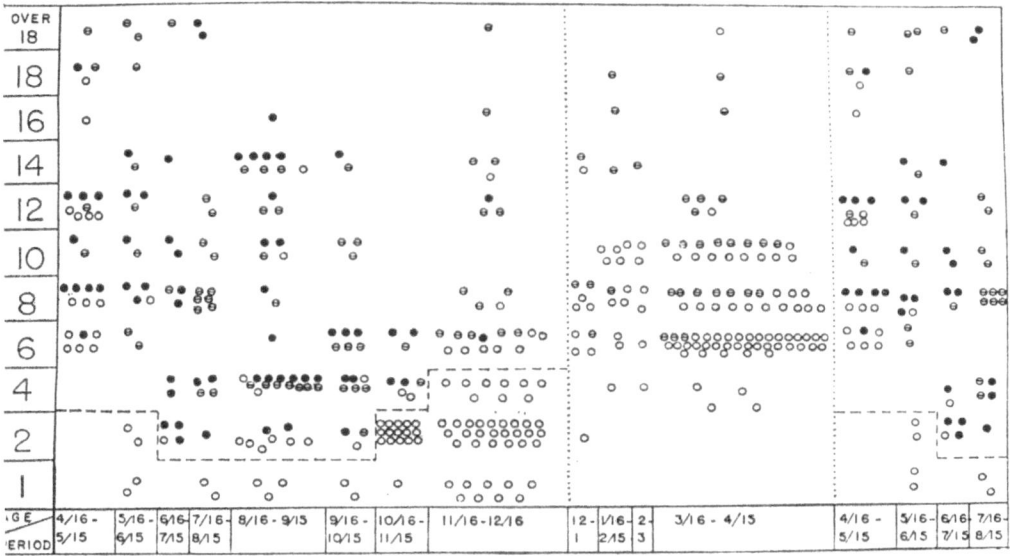

Fig. 4. Reproductive condition of female *Mus musculus* grouped by age and date. Filled circles represent females with foetuses, circles with horizontal lines represent females with uterine scars but no foetus, and open circles represent females with neither scars nor foetuses. Non-breeding season separated by a dotted line, and the dashed line delineates the age for onset of breeding. From Breakey (1963) *J. Mammal.* **44**: 153–68.

indica) in Mysore, Prasad (1956) found that the males were sexually active from July to April and the sexual organs were completely regressed only in May and June. Males born between September and January mature in from 11 to 12 weeks, but males born from February to March reach puberty much later after the non-breeding season in July and August when from 5 to 6 months old. It is possible to suggest here that the environmental factor most likely to be the cause of this delay is the low food supply available from February onwards, when the gerbilles are forced to become omnivorous and even cannibalistic prior to the monsoons. Similarly, Hanney (1965) has suggested that the cause of the delay in puberty that he noted in populations of *Mastomys natalensis* in Malawi was the

D

period of the dry season with its poor availability of nutrition. Comparison of testis length to body length ratio of males collected in February–May showed that all males in the 100–110-mm body-length class had T/B ratios of over 7·9 (i.e. had large testis, indicating sexually mature), whereas of males collected in June and July only 21 per cent of males in the same size class had T/B ratios above this.

There is evidence of a delay in sexual maturity in at least two domestic species. Duncan and Lodge (1960), in their review of the effect of nutrition on reproduction in the pig, mention that this species exhibits pseudo-oestrous cycles before it has really reached puberty. If the onset of puberty is taken when true oestrous

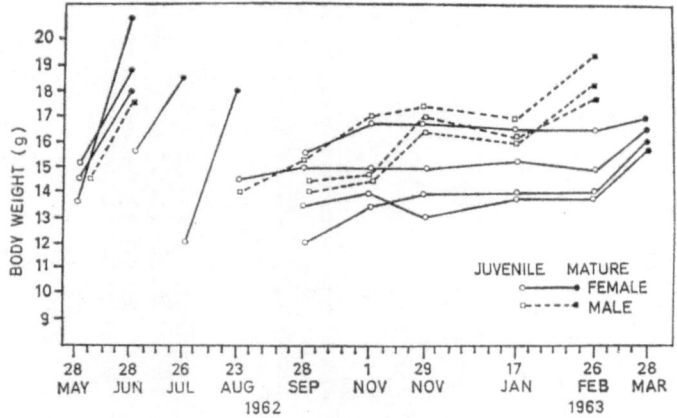

Fig. 5. Body weight changes and puberty attainment in deermice (*Peromyscus maniculatus*). Each spot represents a separate capture record.

cycles set in, then this is found to occur between 3 and 7 months for gilts born in winter, but from 5 to 13 months for gilts born in summer. It is possible that the difference here was partly due to variations in the nutritional regimes at different times of the year. The delay in sexual maturity in sheep has already been referred to.

Before closing this chapter the effect of seasonal variations in growth rates of wild species on the delay in attainment of puberty should be mentioned. Growth rates in wild mammals have been only rarely investigated, due to the difficulties involved in the repeated measurements of wild mammals and in standardization of those measurements. An outstanding example of this sort of study was carried out in admittedly penned populations of wild rabbits (*Oryctolagus cuniculus*) by Dudzinski and Mykytowycz (1960). These workers found seasonal differences in growth rates and differences related to the degree of social dominance of the maternal parent of the individual young rabbit.

In a population study of the deermouse (*Peromyscus maniculatus*) living in coastal fir forest in British Columbia there were great differences found in the growth rates of individual deermice. Other details of this study have already been reported (Sadleir, 1965). The population work involved monthly trappings of marked individuals, and as they were weighed at each capture, their seasonal changes in body weight could be followed. Some individuals were captured monthly over periods of up to 7 months, and their weight changes and attainment of puberty are shown in figure 5. From this it will be seen that up to July deermice grew exceedingly quickly and usually reached puberty by their second capture. However, animals captured after August showed only very small or negligible weight gains until the following spring. During the whole of this time their puberty, as judged by opening of the vagina or descent of the testis, was delayed. As soon as growth started in the spring they immediately came into reproductive condition. The females seen did not exhibit pregnancy, so the weights are not biased by this factor. Although only a few individuals are shown in the figure, similar results were obtained for a much larger number over four different localities. It is interesting that the males started to grow earlier in the spring of 1963 than the females, and thus reached puberty earlier. This example shows that the delay in puberty in deermice is very closely connected with a seasonal change in the growth rate of juveniles. It should be pointed out that nutritional conditions available, as judged by changes in the body weights of adult *Peromyscus maniculatus*, did not seem to be the cause of this inhibition in growth. Whatever the environmental factors responsible, they demonstrated a secondary effect, in that there was also a delay in puberty.

CHAPTER 7

Miscellaneous Factors in the
Attainment of Puberty

There are a number of factors in the description and attainment of puberty which cannot be classified in any of the preceding chapters. For example, it has not been relevant to mention the variability in some species of the size of individuals at puberty. Apart from the complex statistical problems which are involved in the

	Mean figure at which 50% of sample possess character	
	Body weight (g)	Body length (mm)
Males (Scrotal testes)		
Baltimore (farm)	136	—
Baltimore (farm)	157	163
Baltimore (residential)	119	171
Baltimore (residential)	109	155
Baltimore (warehouses)	105	141
England (urban)	190	—
Germany (urban)	93	153
Females (Perforated vaginal orifice)		
Baltimore (farm)	88	—
Baltimore (farm)	72	134
Baltimore (residential)	105	154
Baltimore (residential)	123	150
Baltimore (warehouses)	102	143
Germany (urban)	103	159
Females (Ovulation)		
England (urban)	145	—
England (urban)	146	—
England (urban)	113	—
England (urban)	114	—
Baltimore (warehouses)	153	144
Females (Placental scars)		
Baltimore (residential)	242	224
Baltimore (residential)	305	212
Baltimore (warehouses)	318	216
San Antonio (urban)	178	—

determination of this variability (fully covered in the excellent paper by Leslie, Perry, and Watson, 1946), the reproductive criteria used for the actual assessment of puberty differ so much between worker and worker that it is difficult to attempt any realistic comparison between them. The wild rat (*Rattus norvegicus*) is perhaps the most investigated species in this respect. In his exhaustive review of rat populations, Davis (1953) presents an interesting table (given above) comparing the mean values at sexual maturity for several localities and workers. The original references can be found in Davis's paper. Examination of the figures in the table will show a very considerable variation in the weight and length at which 50 per cent of females reach puberty. Admitedly some will be due simply to statistical variation in the sample, but much will also be due to worker and habitat effects. As a further example of the latter (when worker and statistical methodology variation is minimized), the median body weights at which *Rattus* reached puberty are shown below from the paper by Leslie *et al.* already quoted:

	Median body weight at puberty (g)
Rattus norvegicus	
Corn ricks (30 April–27 May 1940)	113·1
Corn ricks (March–June 1942)	144·3
Liverpool sewers	146·2
Various other sources	145·5
Rattus rattus	
London docks (ships)	88·86
London docks (shore buildings)	91·45

The first set of figures show definite seasonal effects. *R. norvegicus* from corn ricks in early spring reached puberty at significantly lower weights than rats from the other seasons and localities. There is no indication from these data as to the age of puberty. Indeed, we do not know whether or not this species reacts to environmental factors by growing at different rates. The variation in body weights may represent either different growth rates or similar growth rates but different ages at puberty. Despite this lack of information, the data demonstrate very clearly the main point, however, which is the range of recorded body weights at which this well-studied animal can reach puberty. A similar degree of variability can be considered as probably occurring in a number of other mammal species.

Another approach to assessing the variability of the attainment of puberty is to consider the body weight at puberty relative to the normal adult weight. There are, of course, tremendous difficulties involved in assessing the mean adult weight in most wild species, but nevertheless, the proportion of this weight which is achieved when the individual starts reproduction is important in terms

of the energy requirements during the individual's lifespan. It is also relevant
to any discussion of the total energy picture of the species as a whole. There
appears to be little information on this subject, and only one paper on wild
mammals has been found. Twigg (1965) compared the percentage of the total
body weight at which males and females of various rodent species reached puberty. The data on *Holochilus* are from his own study and those of the other
rodents from other workers who can be traced through his paper. In this single

	Males (%)	Females (%)
Holochilus sciureus (British Guiana)	12·9	22·8
Various *Rattus* sp. (Malaysia)	28, 34, 40, 32, and 37	43, 51·5, 76·5, 42·7, and 48
Mastomys couche (Sierra Leone)	43	48
Clethrionomys glareolus (United Kingdom)	47	—

group of mammals there is thus a wide variation in the proportion of the total
body size reached at puberty. This is a potentially profitable field of research
which might be investigated more deeply, especially with regard to the efficiency
of nutritional utilization of each wild species. In the investigation of new and
different species for domestication and other use by man, the relative proportion
of the total energy required to reach puberty and then to reach total body size
is of very great importance economically.

The phenomenon of delayed implantation, whereby the blastocyst in early
pregnancy does not implant immediately after conception but delays over a
considerable period of time, will be mentioned later (see page 205). There is one
interesting reference, however, to the relationship of delayed implantation to the
attainment of puberty. *Mustela erminea* can have two different lengths of gestation. Young are born in the early summer, the exact dates depending on latitude,
but usually about April and May. If the female comes into heat early in the
summer she can conceive and immediately implant the blastocyst and then have
a short pregnancy of 8 or 9 weeks. Mating slightly later in the year results in
conception but delay in implantation, so that females do not give birth until the
following spring. Under these conditions gestations of up to 10 months have
been recorded (Asdell, 1964). This arrangement is of interest with regard to
the length of time taken for juveniles of this species to reach puberty. Hamilton
(1958) reported that the female had been previously said to be able to come into
oestrus and spontaneously to ovulate well before it was fully grown in May or
June of the year of birth. He suggested that normally these females would not
conceive at this time, but would ovulate again the next spring and have a short
pregnancy with no delay. However, blastocysts were found in a very young
female which was killed in July, indicating that females in their first year can
become pregnant. It would thus seem that in this species the possession of the

special reproductive characteristic of delayed implantation has meant that puberty can occur at two entirely different periods of time after birth. In another species, *Mustela frenata*, Hamilton mentions that it is normal for the young to have their first mating when 3-4 months old and to produce their first litter in the following spring.

PART TWO

The Breeding Season

CHAPTER 8

Ecological Factors and the Breeding Season: Introduction

Mammals reproduce successfully over a wide range of latitudes and climatic conditions. In the course of mammalian reproduction, the period from late pregnancy through birth to lactation is a period when the mother and infant are both susceptible to detrimental effects from external environmental factors. It therefore seems logical to assume that mammals will tend to produce their young at a period of the year when environmental conditions are optimal for survival of both mother and young. Baker (1938a) has called the conditions which exist at such an optimum period the *ultimate* cause of the timing of the breeding season. Many writers have pointed out that the mechanism for selecting this timing of the birth period is fairly obvious, in that offspring which are not born at the optimal period would naturally have a lesser chance of survival. Species living in environments which alter so that at certain times survival is better than others will therefore tend to concentrate their birth periods relative to this time, and selection will work to fix such a breeding time in the individual genotypes. However, although most of the mammalian species studied live in temperate regions, where the progression of the seasons is such that the period of time for optimal reproduction is fairly well defined, it must not be forgotten that many more species of mammals live in tropical or semi-arid climates where the optimal season does not hold the same position in the annual cycle.

It is interesting that despite the many field studies of mammalian reproduction, there have been remarkably few reports demonstrating selection against individuals born outside the breeding season. A study of wild sheep (*Ovis aries*) on the island of St Kilda, Scotland, found no relationship between the survival of lambs and whether or not they were born early or late in the breeding season (Jewell, 1966). On the other hand, a study of European hares (*Lepus europaeus*) in Canada (thesis by Reynolds, 1952, not seen, quoted in Flux 1965b) showed that they started breeding in the beginning of winter, and yet all the early litters perished due to cold. This may, however, represent an effect of the introduction of a species into a strange environment. There is apparently a strongly seasonal pattern of human conceptions in certain areas of Mexico due to a local belief (born out by postnatal survival statistics) that infants are more likely to die during the rainy season from respiratory and intestinal infections (Cowgill,

1966d). According to this worker, efforts are made to 'avoid having children born in this period' (page 7). Finally, in their article on the evolution of primate societies, Crook and Gartlan (1966) were unable to suggest why certain of the large apes and baboons should have such markedly seasonal breeding. They state that 'it is as yet unclear whether the timing of the breeding seasons confers advantages during pregnancy, lactation or in terms of food supply for the young animals' (page 1201, see also below). In evolutionary terms, the most plausible reason for this lack of clear demonstration of selective factors operating on young born at the periphery of the optimal breeding season is that studies have been mostly made on species which have already genetically fixed their breeding season. Therefore little selective advantage can be demonstrated because there are so few animals still-born outside the main optimal period. Perhaps the degree of survival of young at the beginning and end of the breeding season may be used as an indication of species stability in terms of the present environment.

It is possible to classify the various types of climatic and environmental conditions met with by mammalian species at the time in the reproductive cycle when they are most susceptible to environmental effects:

(1) *Fixed optimal season.* In temperate regions this falls usually in spring or summer, but the main point is that it is fixed in time relative to the annual cycle.

(2) *Unpredictable optimal season.* In certain deserts and semi-arid environments the optimal season (which is often short) can occur at almost any time of the year.

(3) *Continual optimal season.* In tropical and equatorial regions the seasons vary so little that there is no *optimal* season (*sensu strictu*), so that young may be born at any time of the year and have an approximately equal chance of survival. However, detailed analysis of this type of environment often reveals a more subtle seasonality in factors such as food availability and thus alters the situation to something approaching the first category. Despite this, it is difficult in certain groups of mammals, such as some primates, to associate any particular season with an optimal reproductive period.

A problem inherent in the concept of the optimum period arises concerning at what stage of the reproductive cycle the individual is most susceptible to environmental influences. These difficulties have been clearly stated by Lancaster and Lee (1965) in their paper on primates, and the quotation below may be applied directly to most mammal populations:

'The precise evolutionary significance of a birth season in a primate population is not as immediately obvious as might be expected. Although it is predictable that a species will time its births to take advantage of optimum periods in the environment, exactly when this period will come in relation to the reproductive cycle is not so obvious. In monkeys and apes the length of gestation varies from five to eight months depending on the species. Lactation also varies from about six months . . . to well over a year. If there is only one part of the year that is favourable, such as spring and early summer in the temperate climates, then each species must make an evolutionary choice, so to speak, as to which parts of the reproductive cycle – conception, gestation, lactation or weaning – must be protected and which can come in less favourable times of the year. The rhesus and langur monkeys of northern India give birth during the time of year when temperatures are hottest and wells and tanks are often dry. However, gestation and the later months of lactation come during the monsoon season when food and water are abundant. In contrast the east African baboons give birth at the beginning of the small rains, and gestation and the late part of lactation occur during the six months dry season. Whether any pattern of relationship will be found to hold true for other species of primates is still not clear. It may be that a wide variety of patterns have evolved depending on the lengths of gestation and lactation and the particular ecological complex in which each species or even subspecies lives' (pages 503, 504).

Whatever these patterns will turn out to be, it is nevertheless a measure of the present stage of knowledge about the effects on environmental stimuli on reproduction that in almost no species of mammal are there any detailed data on the relative effects of environmental stresses on survival (i.e. the selective pressures) at the various stages of the reproductive cycle.

The timing of the breeding season, here considered as the period of the year between the first and last conception dates in the population of the species under discussion, is dependent directly on the length of gestation of the species. With the exception of those species which can alter the time of implantation (see page 205), the length of gestation varies but little within a taxonomic group and is most closely correlated with the size of the adult animal. There are some exceptions to this statement however. Certain of the Hystricomorph rodents (Rowlands and Heape, 1966) have much longer gestations than would be judged from their size when compared with other rodents. Some domestic animals can, within a narrow range, alter their length of gestation (Hafez, 1964). The horse in particular has a mean gestation of 335 days, but with a standard deviation of 10 days. Cattle, pigs, and horses all seem to have slightly longer gestations over the winter period than if the main part of pregnancy falls during the summer. Despite these exceptions, the rule would seem to be that members of any species, and of any small taxonomic group, tend to have the same gestation lengths independently of the environmental conditions under which the species are found, so that it would seem reasonable to assume that the length of gestation is a relatively fixed species character which cannot be varied unless under exceptional conditions (for example, in the red tree mouse, *Phenacomys longicaudus*, Hamilton, 1962).

A third factor which will affect selection for the date of conception is the variation from the time of parturition to weaning, and whether or not the species concerned can conceive during lactation (for discussion see Perry and Rowlands,

1962). Some species can have an immediate post-parturition oestrus also. Either mechanism effectively reduces the period of time from one conception to the next, but for simplification this type of relationship will not be considered in the rest of this discussion.

Turning to the possible variations in the types of optimal season and the fixed nature of the period of gestation and weaning, an attempt will now be made to categorize the types of breeding pattern found in mammals. These are shown in figure 6.

TYPE 1. *Length of time from conception to weaning longer than single optimal season.* Birth and weaning occur during the fixed optimal season, but the time

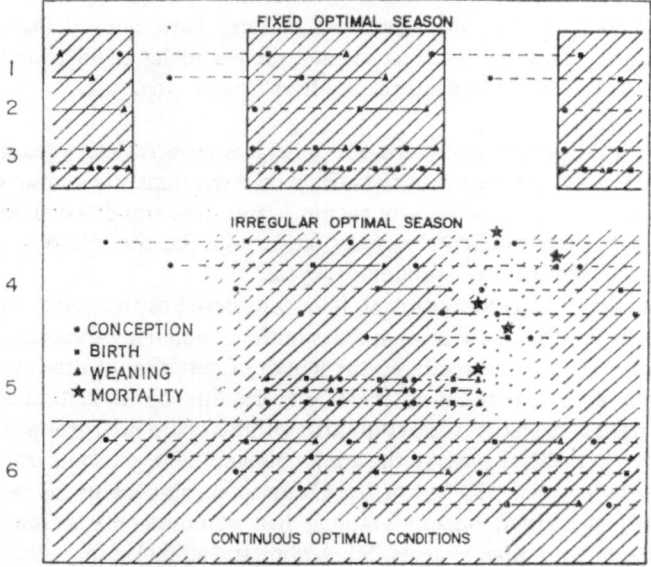

Fig. 6. Breeding patterns found in mammals.

of conception is pushed back so that it occurs either in a poor season or in the previous season. (Examples: *Odocoileus virginianus* conceives in mid-December in New York State, Cheatum and Morton, 1946; *Ovis aries* usually conceives in middle of winter in the temperate zone, Hafez, 1952a, Hammond, Jr., 1944; *Cervus elaphus* conceives in autumn, Lowe, 1966.)

TYPE 2. *Length of time from conception to weaning is short enough to occur inside one fixed optimal season.* The division of this type from the former is rather arbitrary as there is a continuous range of patterns between them, but it is suggested, because conceptions in the former tend to occur on a falling or low plane of nutrition when other environmental factors are becoming or remaining detrimental, whereas in the latter, conception occurs when the environment is rapidly improving. (Examples: *Mephitis nigra*, Pearson and Enders, 1944;

Vulpes vulpes, Rowlands and Parkes, 1935. Many temperate zone bats, Wimsatt, 1960.)

TYPE 3. *Length of time from conception to weaning is short enough for two or more conceptions to occur during the fixed optimal season.* This results in a succession of litters which may or may not be synchronous in time between individual females. (Examples: Almost every rodent, see *Apodemus sylvaticus*, Jewell, 1966; *Peromyscus maniculatus*, Sadleir, 1965b; *Microtus californicus*, Krebs, 1966; also insectivores, *Sorex cinereus*, Buckner, 1966; and lagomorphs, *Lepus europaeus*, Raczynski, 1964.)

TYPE 4. *Length of time from conception to weaning longer than a single irregular optimal season.* This means that the optimal season cannot be predicted or utilised in time so that breeding is continuous, with a considerable degree of post-natal mortality. (Examples: *Macropus robustus*, Sadleir, 1965a; *Megaleia rufa*, Frith and Sharman, 1964.)

TYPE 5. *Length of time from conception to weaning is shorter than a single irregular optimal season.* Effectively the same arrangement as in (3), except that lack of predictability of the end of the optimal period probably means some post-natal mortality due to timing of breeding. (Examples: none available, possibly *Sminthopsis crassicaudata*, Spencer, 1896.)

TYPE 6. *Length of time from conception to weaning relatively unimportant as the optimal season continues throughout the year.* The incidence of breeding may vary but conceptions are found in every month. (Examples: *Hippopotamus amphibius*, Laws and Clough, 1966; *Chaerephon hindei*, Marshall and Corbet, 1959; *Artibeus literatus*, Tamsitt and Valdivieso, 1963.)

The above classification is by no means comprehensive, because there are two physiological phenomena (delayed implantation, delayed fertilization) which alter the length of gestation and thus affect the picture. There are also phenomena, such as hibernation and periods of quiescence without true hibernation, which effectively act to separate the animal from the detrimental section of its environment in the non-optimal season, and are thus utilized to the advantage of the breeding mechanisms. For example, the polar bear (*Thalarctos maritimus*) gives birth in late November in the high Arctic, but as the young are maintained and nourished by the mother in her den, juvenile survival is little affected by the poor conditions outside (Hediger, 1952; Morris, 1965). At the other end of the temperature range, the camel (*Camelus dromedarius*) has a 12-month pregnancy, so that conception and birth occur under the optimal conditions present from January to March. However, lactation takes over 4 months, so that the female camel breeds only every second year (Bodenheimer, 1954).

During the breeding season a female mammal has functional, and usually cyclic, ovaries, whereas during the non-breeding season the female gonads are apparently quiescent. The majority of studies of mammalian reproduction have naturally described in detail the events during the breeding season, but few studies have been made of the actual nature of the reproductive system during the period of non-breeding or, more importantly, of the periods of transition from one state to the other. The expression of the first oestrus or of the cessation of reproduction after the last lactation period are rather singular events in the temporal sense and suggest sudden physiological changes. The few studies which have been made, however, indicate that the appearance of all-or-none activity is rather deceptive.

Attempts to induce oestrus and pregnancy hormonally in the anoestrous ewe (*Ovis aries*) by Robinson (1950) showed that a number of ewes had already spontaneously ovulated during the non-breeding season, although the rams running with them had not detected any oestrous behaviour. Continuation of the experimental treatments, and a maintained programme of slaughtering ewes for examination of ovaries indicated that there was a progression of non-oestrous ovulations which decreased in frequency until the middle of the period of anoestrus when the ewes showed the least number of spontaneous ovulations (although no oestrus), and when the response to hormonal induction occurred

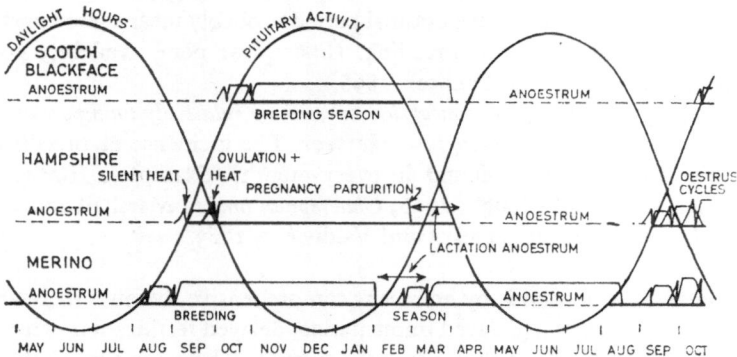

Fig. 7. Graphical representation of the nature of the reproductive rhythm in sheep. The dotted lines represent the period of anoestrum and the heavy lines the breeding season for three breeds. A similarly shaped curve is shown for each, and it is suggested by Robinson that the relative threshold of pituitary activity at which the breeding season is initiated varies between breeds. From Robinson (1951) *Biol. Rev.* **26**: 121–57.

in the lowest numbers of experimental ewes. In this and a subsequent paper (1951) Robinson suggested that the length and timing of the breeding season is a result of the presence of a threshold at which the output from the pituitary is sufficient to result in a complete ovulation and behavioural oestrus. This idea is shown graphically in figure 7. He postulates that there is a continuous curve of pituitary activity which is inversely related to the length of daylight, and that the condition of anoestrus is actually only a relative phenomenon.

Two other observations on sheep support this suggestion. The first is the phenomenon of 'silent heats', wherein, just prior to the first behavioural oestrus of the season, there is an ovulation in the ovary but no oestrus is detected (see also Averill, 1955). The second observation was made by Thibault, Courit,

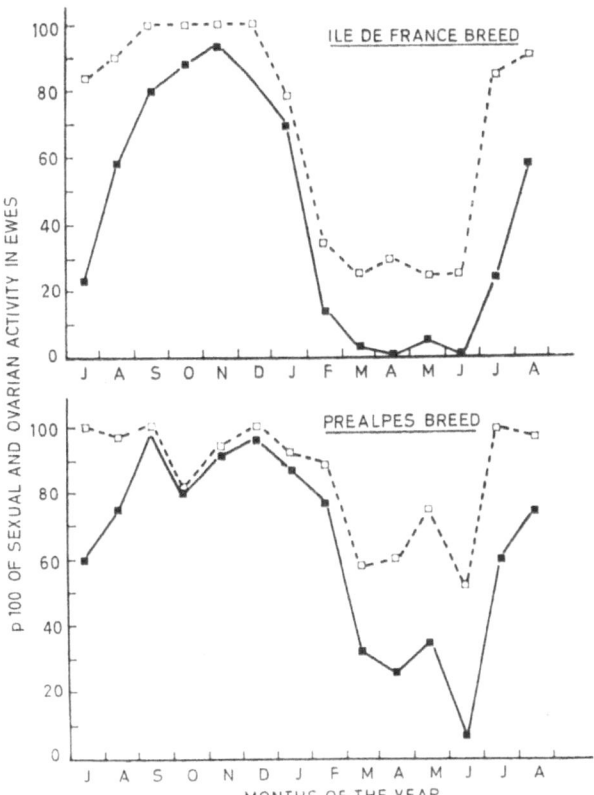

Fig. 8. Seasonal variations in sexual (solid line) and ovarian (dotted line) activities of ewes as detected by teaser rams and endoscopy. From Thibault *et al.* (1966) *J. Anim. Sci.* (*Suppl.*) **25**: 119–42.

Martinet, Mauleon, Du Mesnil du Buisson, Ortavant, Pelletier, and Signoret (1966) in studies of the seasonal variation of ovarian activity. Using endoscopes, they found over 20 per cent of Ile-de-France ewes were undergoing ovarian activity in the middle of the anoestrous period when no behavioural oestruses were being detected by rams (see figure 8). In the Prealpes breed 50 per cent of ewes showed signs of ovarian activity.

On the basis of the arrangement shown in figure 7, Robinson then presents three alternative hypotheses as to why various breeds of sheep have different lengths of breeding seasons:

(1) Breeds with short seasons (i.e. Scottish Blackface) may have a more gradual response of the pituitary to daylight than breeds with long seasons (i.e. Merino, Karakul) resulting in a more pointed curve, so that the level of gonadotrophic hormone produced by the pituitary is only above a postulated ovulation threshold for a short time.

E

(2) The curve shape may be similar for all breeds, but the threshold for stimulation of breeds with short seasons may be greater than for breeds with long seasons.

(3) The shape of the curve for all breeds may be the same, but the axis for breeds with long seasons may be higher than for short-season animals, i.e. at all times of the year their pituitaries operate at a higher level of output of gonadotrophins, so that at comparable levels of ovarian stimulation the long season breeds will exceed the threshold for a longer time than for the short-season breeds.

Further evidence of the relative nature of the anoestrous period is suggested by the work of Lamond (1962, 1964) on Merino ewes. Using more advanced methods to induce oestrus, he found no significant seasonal differences in the number of ovulations occurring during or after short periods of progesterone treatment, but definite differences in the proportion of ewes which did not ovulate at all, and in the time of ovulation of those which did. The proportion of ewes ovulating in January was different when HCG was given 24 instead of 48 hours after progesterone, and this led Lamond to suggest that 'one aspect of the seasonal differences in pituitary ovarian function is the sensitivity of the neural-endocrine system controlling ovarian activity' (page 119, 1962). Later work on the occurrence of oestrus, after its suppression by prolonged progesterone treatments, showed declines in the degree of response to treatment well before there was any decline in the incidence of natural oestrus. Similarly, there proved to be seasonal differences in the effect of the psychic stimuli associated with the introduction of rams on groups of ewes treated with progesterone. On the basis of these, and other observations, indicating that there are seasonal and progressive differences in the pituitary–ovary axis, Lamond concluded that 'a more useful approach to seasonality in sheep breeding than measurement of hormone production may be to examine the sensitivity of the neuro-hormonal complex to hormones, particularly progesterone and oestrogen' (page 119, 1962).

It would thus seem that the onset of oestrus in sheep is the result of either a steady increase in pituitary hormone production or a steady decrease in ovarian sensitivity (although the latter is unlikely from Lamond's results), which finally reaches some threshold level and results in oestrus. There is, unfortunately, very little similar evidence in wild artiodactyls, with the exception of a paper by Simkin (1965), who describes 'silent heats' in moose (*Alces alces*) just prior to the true breeding season. Perry and Rowlands (1962) mention that there is evidence in hedgehogs (*Erinaceus europaeus*), bankvoles (*Clethrionomys glareolus*), and elephants (*Loxodonta africana*) that at the onset of the breeding season the ovaries already contain several sets of corpora lutea, even though no mating had occurred, suggesting the phenomena of 'silent heats' in these species. In cattle (Rajakoski, 1961), horses (Osborne, 1966), and the buffalo-cow, *Bubalis bubalis* (Shalash and Salama, 1961) there is a seasonal variation in ovarian function, but as both these species breed throughout the year, this evidence is only suggestive.

In lagomorphs there is strong evidence that considerable ovarian activity occurs well prior to the breeding season. Conaway and Wight (1962b), in their study of the cottontail, *Sylvilagus floridanus*, in Missouri, described a 'pre-oestrus' an-ovulatory condition in some females. They suggested that:

'There is an overall external stimulus which brings all female members of a population into what may be termed a pre-estrus condition at about the same time. Variable population and environmental factors may then determine whether or not each female will actually achieve estrus and breed. In any popu-lation, all, part, or none of the females may achieve estrus during the first potential breeding period. If any female does not achieve estrus during this brief critical period, regression of the pre-estrus condition follows and an interval of from 14 to 16 days is required before a second pre-estrus occurs (page 285).

In this species there is a high degree of synchrony in the early conceptions, and this apparently results from a pre-breeding-season synchrony in the ovarian activity of the cottontail females. The observation of Brambell (1948) that wild European rabbits (*Oryctolagus cuniculus*) will copulate but not ovulate from November to January, 3 months before true breeding begins in February, sug-gests a similar phenomenon exhibited in a different manner.

In a discussion which follows the results of a detailed investigation into breed-ing of penned wild European rabbits, Myers and Poole (1962) review what is known about the seasonality of ovarian and pituitary function in this species. In isolated domestic rabbits they describe seasonal and regular fluctuations in: (*a*) the incidence of oestrus and proportion of rabbits entering anoestrus after littering; (*b*) oestrogen-facilitated spontaneous ovulations; (*c*) gonadotrophic hormone content of pituitaries; and (*d*) the least amount of pituitary powder causing ovulation within 16–18 hours of injection. Under conditions of dom-estication, despite the rabbit's well-known ability to breed continuously, there are definite seasonal changes in the function of the reproductive system. In wild rabbits (as will be seen later, pages 108, 132) the rabbit is almost entirely dependent on a high level of nutrition before breeding can occur at all, and yet the rapidity of the response to good nutritional conditions indicates a 'sub-oestrous' degree of ovarian activity. Thus in the rabbit, like the sheep, there are strong indications that although breeding is not being exhibited, there is still some amount of ovarian activity.

It is interesting to speculate on whether domestication has altered the breeding season of once wild mammals by affecting the level of interrelations between the pituitary and the gonads. It is a pity that so little is known of the pituitary-ovarian relationships in the non-breeding season of truly wild species of mammals. If the suggestions of Robinson and others apply to other mammals, the idea of complete inactivity during the anoestrous seasons will have to be modified. Their suggestions would certainly bear investigation in species such as the cyclic microtine rodents, who exhibit a very variable onset and cessation of breeding related to the phase of the population cycle (see page 244).

The chapters which follow will consider the effect of light, climate (including temperature and nutrition), and social factors on the onset or cessation of breeding seasons. The ideas outlined above, of a steady alteration in what could be called 'breeding potential intensity', may assist in the elucidation of the effects of the various environmental variables on breeding phenomena.

CHAPTER 9

The Effects of Light
on the Breeding Season

A. THE PATTERN OF SEASONAL VARIATION OF LIGHT AT DIFFERENT LATITUDES

I. Origin of data on light changes

Nautical almanacs carry tables of sunrise and sunset and also information as to the length of twilight. In preparing the graphs which are discussed below, data have been extracted from the most complete of these tables known to the author, which were published as *Tables of Sunrise, Sunset, and Twilight* as a supplement to the *American Ephemeris* of 1946 produced by the United States Naval Observatory (1945). A major advantage of these particular tables is the inclusion of daylight data for very high latitudes.

II. The total hours of daylight at different latitudes and seasons

There are two possible ways of graphing this sort of information. Figure 9, which is reproduced from a paper on birds by Baker (1938b), uses latitude as the ordinate north and south from a central line on the diagram which represents the equator and has an abscissa of the months of the year. The lines on the graph join points of equal daylight length, so that the variation of daylength at any exact latitude can be read by placing a ruler parallel with the abscissa. This diagram requires interpolation between the hourly lines, which is difficult at the higher latitudes. Figure 10 uses the hours of daylight as the ordinate and has a double abscissa, one for the Northern and one for the Southern hemisphere. The curves on the central diagram in this case are of the different latitudes at 10° intervals. This diagram shows a little more clearly the effect of increasing latitude in increasing the magnitude of annual variation in the change in daylight length. In this case interpolation must be made between the 10° units of latitude, but the hours of daylight at any stage of the year can be very easily read.

III. The change in the length of effective daylight

It is very difficult to define what is effective daylight to mammals. As will be seen from the section below on the experimental manipulation of daylight, there is only sketchy information as to the minimal light intensities required to stimulate the reproductive processes. Effective daylength is considered (purely arbitrarily)

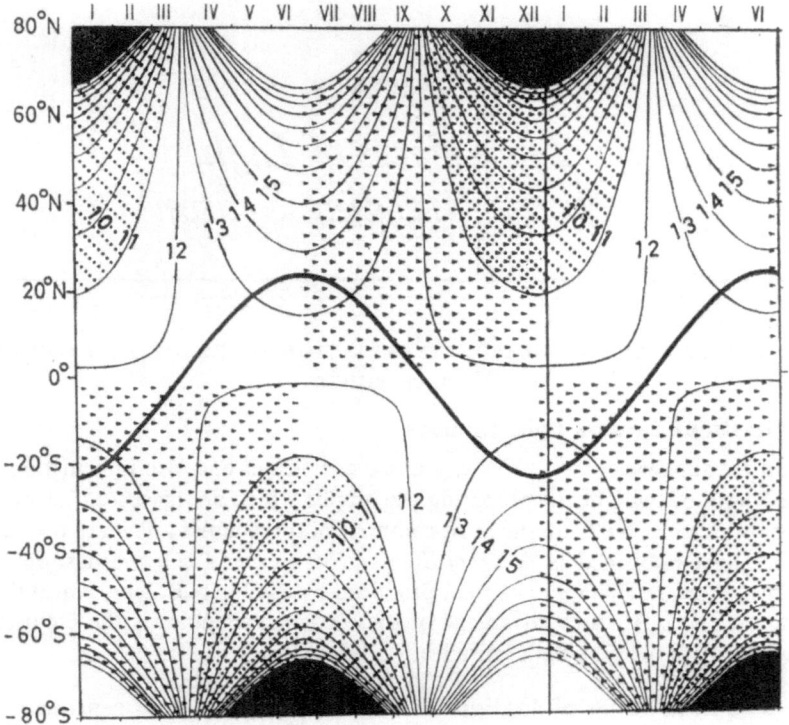

Fig. 9. Diagram showing the seasonal changes in the position of the overhead sun (thick black line) and in length of day (numbered curves). The numbers are the hours of daylight. Triangles represent decreasing daylength and blacked in areas show when the sun is completely below the horizon for 24 hours. From Baker (1938a) *Proc. zool. Soc. Lond.* **108**: 557–582.

in the section which follows, as the period of light from sunrise to sunset (daylight *sensu strictu*) plus twice the length of civil twilight, i.e. twilight before sunrise and twilight after sunset. Yeates (1954) has considered the problem of definition of daylength relative to the animal. He points out that different species will almost certainly have differing sensitivities to what is the minimum amount of light which can affect their reproductive processes. Civil twilight is only one of three recognized types of twilight, and is defined in the *American Ephemeris* as the period of time from sunset to when the centre of the sun's disc is 6° below the horizon. Figure 11 gives curves of civil twilight for the Northern hemisphere at different times of the year at selected latitudes. Up to 40° North it varies little throughout the year, but at higher latitudes it has a very considerable fluctuation. It should be remembered that the variation shown on this figure is daily doubled as far as the animal is concerned.

   The change in the length of effective daylight is shown in two ways in figure 12. The degree of change is considered in monthly units, so that the length of

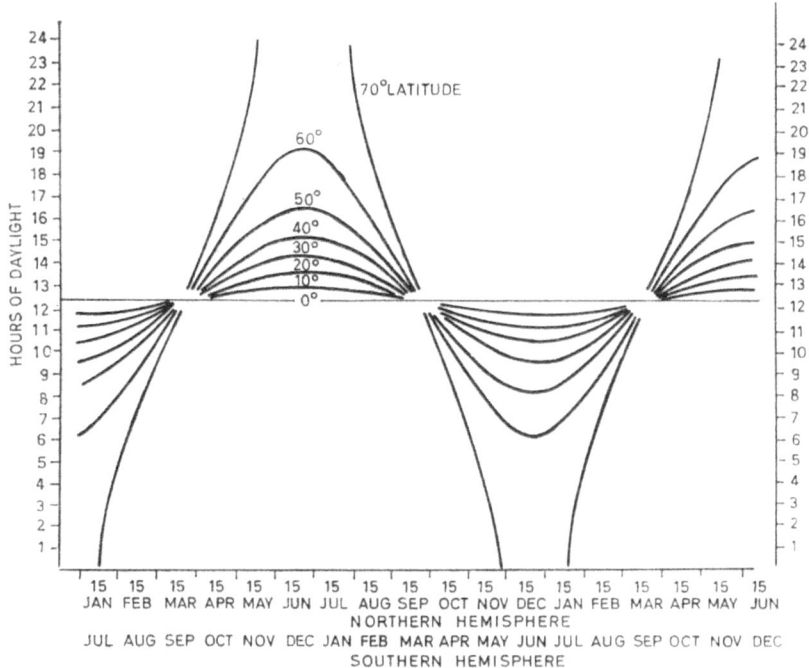

Fig. 10. Hours of daylength at different latitudes throughout the year. Based on data n the *American Ephemeris*, 1946. U.S. Naval Observatory (1945).

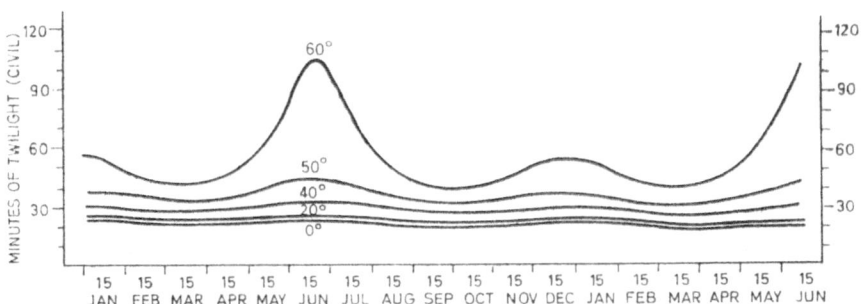

Fig. 11. Seasonal changes in the length of civil twilight at various latitudes in the Northern hemisphere. Based on data in the *American Ephemeris*, 1946. U.S. Naval Observatory (1945).

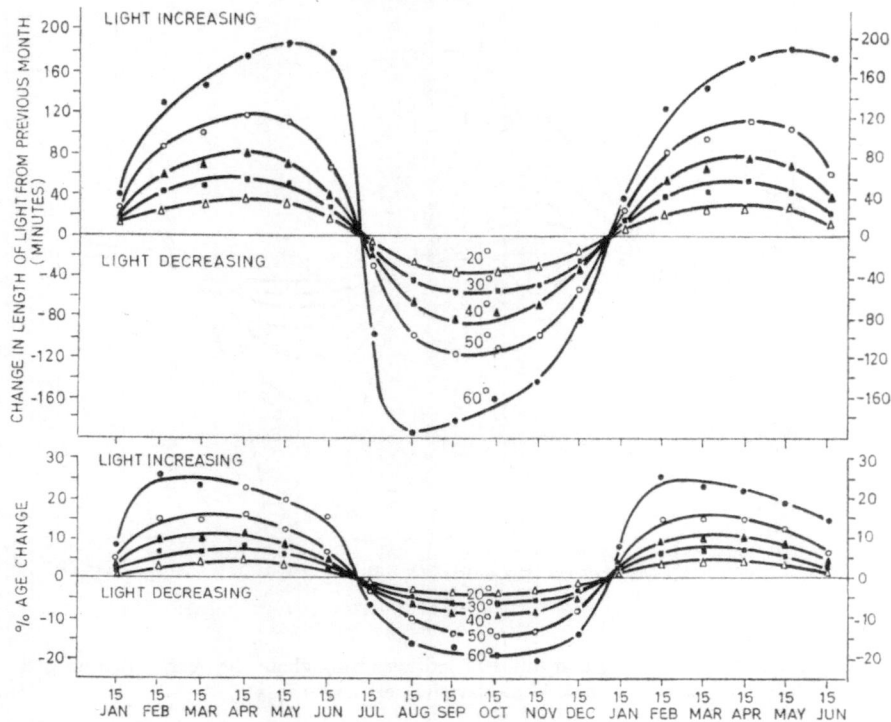

Fig. 12. Relative seasonal changes in length of effective daylight in absolute and relative terms. Based on calculations from data in the *American Ephemeris*, 1946. U.S. Naval Observatory (1945).

effective daylight on the 15th of each month is compared with the length of effective daylight for the 15th of the previous month. The absolute difference is given in the top part of figure 12 in minutes, and in the bottom part of the figure it is considered relatively as the percentage change from the previous month. To determine the daily change in the length of daylight, it is necessary to divide the ordinate scale by a factor of 30. The lines on the diagram are in 10° intervals.

IV. Discussion

It is very difficult, at the present stage of knowledge about the manner in which experimental alterations in the light pattern affect the breeding of mammals, to discuss with any certainty what are the exact features of the natural changes on daylight length to which the animals are reacting. One must, therefore, attempt to analyse the variation in light in a manner similar to that attempted above, and then speculate as to the most likely features detectable by the mammal. At any latitude other than those above the Arctic or below the Antarctic Circles the daily change in the length of light is very small indeed; for example, at 50° the maximum change in the length of day is approximately 4 minutes at the March and September equinoxes. However, the phenomenon of the change-over from increasing to decreasing daylight is a very marked one, again especially at higher latitudes. Examination of figure 12 will show that the annual light cycle can really best be divided into the following periods during which the light changes are relatively constant. For purposes of simplicity, only the Northern hemisphere will be considered.

(a) *February to May*. During this time the rate of change of light is relatively constant. Although there is an alteration in the ratios of light to dark over the equinox, as far as the animal is concerned this is a period of continual and relatively even light increase.

(b) *June and July*. Here the rate of light-increase first decreases and then alters to a period of decrease in light. Depending on the latitude (which controls the amplitude of the change-over), this can be a very considerable alteration in the daylight environment. The curves in figure 12 are slightly inaccurate at this point, as the changes are only considered in monthly intervals. However, they do reflect the very rapid alteration, first in the rate of light increase to zero and then the corresponding rapid increase in the rate of light decrease.

(c) *August to November*. Again the rate of change in light is relatively constant, with a steady decrease in the total daylight.

(d) *December and January*. The opposite to June and July, the change-over being this time from decreasing to increasing light.

It will thus be seen that there are really two major times of the year when it would be relatively easy for a mammal to determine a constant date by evaluation of the light regime. It is possible that mammals may have a physiological mechanism which reacts to:

(1) The cumulative build-up (or decrease) of the total amount of light that has impinged on the individual. This would presumably mean that this build-up would need to be measured from some other landmark in the changing light regime, most likely the solstices.

(2) The daily light reaching some crucial triggering level as it was increasing or decreasing. Either of these mechanisms would mean that the animal would be able to use some date in the periods (*a*) and (*c*) above and would not be so dependent on the light change-over at the equinoxes. (But see comments by Yeates, 1949, page 74.)

(3) As the daylength alters, this alters the lighting regime at some crucial period at a specific time in the mammals' circadian rhythm, i.e. mammals also measure photoperiodic time as suggested by Bünning (Menaker and Eskin, 1967).

B. OBSERVATIONS ON NON-EXPERIMENTAL ANIMALS SUGGESTING THE EFFECTS OF LIGHT

I. The relatively constant onset of breeding at one latitude

In a number of species of mammals, especially those living at higher latitudes, the beginning of the breeding season at any one locality always occurs at fairly fixed dates in the year. Although there are slight fluctuations, it is difficult to envisage any other environmental factor other than light as being related to this fixed phenomenon. In their study of the field vole, *Microtus agrestis*, Baker and Ranson (1933) found a 'remarkable' correlation between the hours of sunlight and reproduction. Over two years voles were found breeding only in those months where there was more than about 100 hours of sunshine. Below this level, breeding was not observed. Similarly, Prasad (1956) found that the Indian gerbil, *Tatera indica*, in Mysore, began breeding only when the length of darkness reached 11 hours and 49 minutes, and was lengthening. Breeding stopped when the length of darkness was reduced below this figure.

Species of mammals living at moderately high latitudes show a very regular onset of breeding. The study by Mizuhara of the Japanese macaque (*Macaca fuscata*), reported by Lancaster and Lee (1965), found that the first dates of births in a population at Takasakiyama were as follows:

1956	7 May	1959	5 May
1957	10 May	1960	3 May
1958	11 May	1961	20 April

A study of female moose (*Alces alces*) in British Columbia reported that the majority of conceptions over 5 years occurred in the same 10-day period of the year (Edwards and Ritcey, 1958). Even in species which migrate over wide ranges of latitudes the mean date of mating is amazingly constant. Laws (1959) calculated the conception dates from measured foetuses of the Southern hemisphere fin whale (*Balaenoptera physalus*) and found that from 1925 to 1958 the mean date of mating varied only by plus or minus $2\frac{1}{2}$ weeks. Farther from the poles, and closer to the equator, the onset of breeding becomes more variable. The introduced rhesus monkeys (*Macaca mulatta*) living on Cayo Santiago, Puerto Rico, have been studied for 5 years by Koford (1966) using methods which allowed detection of births to within 1 day. The date of initial birth ranged from 29 December to 13 February (46 days), but the median initial birth date ranged from 2 March to 5 April (34 days). This worker suggested that much of this variation was due to weather and to the effect of lactation on the onset of oestrus. Perhaps the most constant onset of breeding is shown in European hares (*Lepus europaeus*), as apparently in no matter what part of the world they are studied, they always start breeding very soon after the shortest day of the year. This is shown in figure 13 from a paper by Flux (1965b).

There have been a number of detailed studies in sheep (*Ovis aries*) on the fixed nature of the onset of breeding. Hammond Jr. (1944) studied the breeding season of Suffolk ewes at Cambridge, England. Counting the days of the year in a continuous series from 1 January as day 1 (1 September = 244, 1 October = 274), Hammond presented the following data:

Season	Onset of oestrous season		End of oestrous season	
	Mean	Range	Mean	Range
1935–6	299	294–307	81	78–85
1936–7	288	262–310	82	65–101
1937–8	282	263–305	72	49–95
1938–9	281	272–301	—	—
1932–9	286	262–310	77	49–101

Although there is some individual variation involved, the mean onset of behavioural oestrus is relatively constant from year to year. Averill (1964a, 1965), studying Romney Marsh ewes in New Zealand, investigated this in more detail,

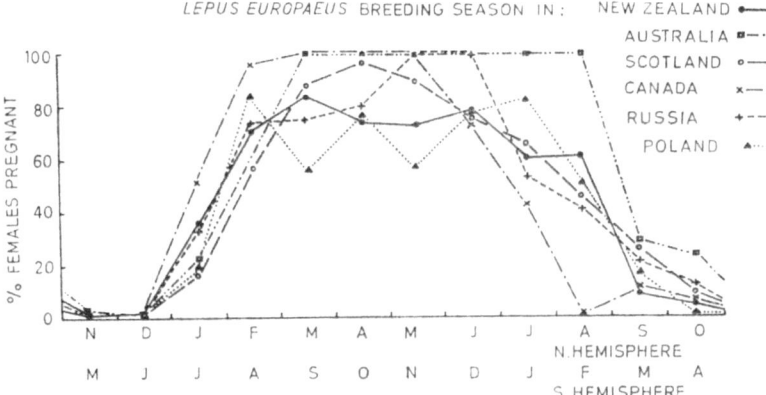

Fig. 13. Breeding season of the hare (*Lepus europaeus*) in six different countries. From Flux (1965b) *Mammalia* **29**: 557–62.

giving the 50 per cent level of ewes which had actually ovulated (as determined by ovarian investigation). The figures are for ewes in Otago:

1957 21 March ± 2·3 days
1958 17 March ± 1·4 days
1959 15 March ± 2·1 days
1960 16 March ± 3·1 days

Averill suggests that this annual consistency is due to light being seasonally constant. He postulates that immediate climatic and nutritional factors will affect the onset of behavioural oestrus, but the onset of ovulation will remain relatively fixed. It is possible that temperature was the cause of a larger variation in the onset of oestrus as described by Godley, Wilson, and Hunt (1966) for Rambouillet ewes in South Carolina. The average date of first oestrus was:

1959 1 July $\pm$ 5·9 days

1960 27 August $\pm$ 7·9 days

1961 29 July $\pm$ 10·5 days

Smith (1966) described a similar degree of variation in Southdown ewes at latitude 27° S in Queensland. Over 5 years the mean date ranged from 11 February to 17 March. Clun ewes at Aberystwyth in Wales showed a very significant correlation between the day of previous lambing and the day of tupping (Lees, 1966), so that the date of first service seems highly dependent in this breed on the previous breeding season.

The effect of the ram on the onset of behavioural oestrus (Schinkel, 1954, and see page 255) is also relevant here.

II. Variation in the breeding season with latitude

Ornithologists have recognized for many years that the breeding seasons of species of birds with wide ranges varied considerably with latitude. The main egg-laying season of most birds is from April to June, between 30° N and 50° N. At latitudes of less than 30°, the egg-laying season becomes much wider, spreading from March to September between 20° N and 30° N, and then extending over the whole year at lower latitudes (Baker, 1938b). A similar narrowing of the season occurs with increasing degrees of south latitude. Another trend is that at extremely high latitudes the onset of the egg-laying season tends to occur later in the year, so that as well as the season shortening, the mean egg-laying date occurs later.

Similar trends have been reported by field mammalogists, but the picture is not as clear. In that very well-studied mammal the white-tailed deer (*Odocoileus virginianus*) of North America the breeding season varies considerably with latitude, but is also affected somewhat by the longitude and local climatic variations. Severinghaus and Cheatum (1956) mention that this species breeds practically all the year round in Florida, but at the same latitude farther east in Texas and Arizona breeding is concentrated in December and January. In the northern United States breeding commences earlier in the year, starting in the first weeks of December in New York State (Cheatum and Morton, 1946) or in mid-November in the central United States (Severinghaus and Cheatum, 1956). Farther north in Manitoba breeding can commence as early as late October (Ransom, 1966). Although this general trend exists, smaller latitudinal differences often do not reveal constant changes in the onset of breeding. Ransom found that white-tailed deer at locations of 49° 5′ N and 51° 35′ N bred at the same time, whereas another population at 49° 44′ N bred significantly earlier in the year.

This could have been due to genetic differences between individuals making up the populations.

The variation in breeding season of another North American deer (*Odocoileus hemionus*) has been looked at in detail by Einarson (1956). He gives the following mating periods in different states:

New Mexico	November– January
Arizona	15 December–15 January
Nevada	15 November–15 December
Colorado	15 November–15 January
California	10 December–27 January
Utah	15 November–15 December
Oregon	9 November–30 November
Alberta	24 October –14 November

Other than in California, where the milder west-coast climate undoubtedly has a major effect, it will be seen that the breeding tends to occur earlier as the latitude increases. It also tends to shorten.

Another ungulate, the domestic sheep, also shows a very close relationship between the breeding season and latitude in some breeds. Fortunately, due to the spread of agricultural practice, this species now occupies a very wide latitudinal range on both sides of the Equator, so that potential latitudinal effects can be easily investigated. The variation in breeding season has been reported (Hafez, 1952a). The breeding season of wild Caprinae is relatively restricted, but again species from northern latitudes, such as the argali (*Ovis ammon*), are reported to have more restricted seasons then southern-living species such as the mouflon (*O. musimon*). Hafez found that the incidence of oestrus in *O. aries* at different latitudes was inversely related to the length of daylight. The peak of sexual activity always coincides with the period of shortest days. At high latitudes the breeding season was very closely related to daylight length, but this was less pronounced at lower latitudes (figure 14). The breeding season gets shorter near the Poles. It has been suggested by Amoroso and Marshall (1960) that the sexual season in sheep is later in the northernmost and southernmost regions because animals living there might be expected to require a more extended interval from the time of the initial stimulus earlier in the year, before the cumulative effects of light were sufficient to stimulate the pituitary to full gonadotrophic activity. Since Hafez's review, Anderson, J. (1964) has published some interesting data on the breeding season of sheep on the Equator. He documents the monthly incidence of 480 lambings in Kenya as follows:

Jan.	*Feb.*	*Mar.*	*Apr.*	*May*	*June*	*July*	*Aug.*	*Sept.*	*Oct.*	*Nov.*	*Dec.*
8·5%	7·1%	8·1%	21·0%	9·6%	6·7%	3·3%	5·8%	6·5%	15·8%	3·3%	4·2%

Although there are two peaks in lambings, the continuous nature of breeding at this low latitude is well demonstrated. Anderson emphasizes the low fertility (44 per cent sterile) of ewes living at this latitude.

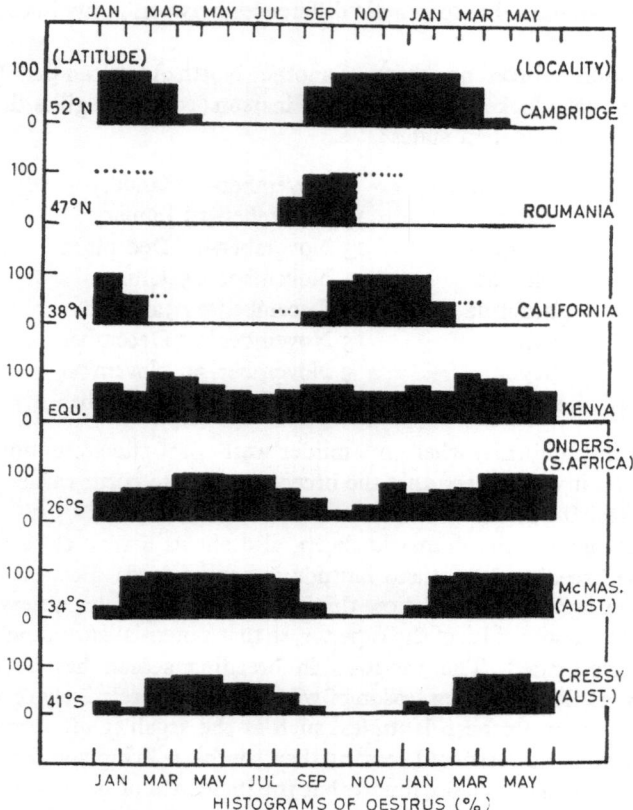

Fig. 14. The duration of the breeding season of sheep at different latitudes. From Hafez (1952a) *J. agric. Sci., Camb.* **42**: 189–99.

In the sheep even slight differences in latitude apparently produce consistent differences in the onset of breeding. Averill (1964a, 1964b, 1965) found that the median date of onset of sexual activity (as determined from examinations of the ovary of Romney ewes in New Zealand) varied with only 3° difference in latitude. Ewes in Otago (lat. 46° S) had a mean onset date of 16 March ± 3·1 days, whereas ewes in Canterbury (lat. 43° S) had a mean onset date of 5 March ± 4·3 days. It should be pointed out that there are very considerable differences between breeds of sheep in their reaction to latitudinal effects. Hafez (1952a) noted that only breeds originating in temperate climates show restricted breeding seasons. Breeds such as Merino (Moule, 1950a) have a breeding season in Australia which is largely unaffected by latitude.

The variation of the breeding season with latitude in various lagomorphs has been reviewed in some detail by Watson (1957): what follows is based on his paper, and the original references may be traced from it. The European hare

(*Lepus europaeus*) breeds in southern Russia from January to autumn, but is much more restricted in northern Russia (around Moscow), where breeding takes place only in late spring and early summer. Later data on the variation in breeding with latitude in this species can be found in a paper by Raczynski (1964). The snowshoe hare in North America (*L. americanus*) starts breeding about the end of March in Minnesota and Wisconsin, but about a month later in northern Alberta. Watson gives a review of the complex of *Sylvilagus* species in America (see also Sowls, 1957). *S. floridianus* breeds through the spring and summer in the central states of Michigan, Missouri, and in the eastern states of Connecticut, Pennsylvania, and Ohio. To the north and west, in Utah and Nevada, the breeding season is much more restricted, occurring from April to July. The picture here is confused, because the latitudinal effects are partially concealed by the diversity in climatic conditions found between east and west of the American continent in the same way as previously described for white-tailed deer. Watson (1957) and Wodzicki and Darwin (1962) found tendencies in New Zealand for the onset of and cessation of rabbit breeding to be affected by latitude, but Flux (1965) found no such effect in the timing of breeding of European hares.

Mammals living on or near the Equator may or may not breed continuously. The group which has been most investigated here are the bats. Some breed continuously throughout the year, such as *Chaerephon hindei* in Uganda (Marshall and Corbet, 1959), and *Artibeus lituratus* in Colombia (Tamsitt and Valdivieso, 1965a). A list of continuously breeding tropical bats is given by Tamsitt and Valdivieso (1965b). Others are characterized by remarkably discrete and short breeding seasons, although the three species which follow are not found exactly on the Equator. *Miniopterus australis* at Hog Harbour, New Hebrides (lat. 15° 15′ S) had young born only in the second half of December (Baker and Bird, 1936). The larger fruit bat, *Pteropus geddiei*, at the same locality copulates only from February to March (Baker and Baker, 1936), and another species in the same genus (*P. giganteus*) has a very sharp breeding season in Ceylon at lat. 7° N, with all conceptions taking place from December to early January (Marshall, A. J., 1947). This same author has reviewed data on the effect of length of daylight in relation to copulation dates in a number of species of *Pteropidae*. In both hemispheres by far the majority of species copulate during periods of increasing light, and these copulation dates tend to get closer to 22 June in the Southern hemisphere as the latitude increases. There are a few exceptions to the effect of increasing light, but all except one of these occur within 10° of the Equator, where the light variation is minimal.

The effect of latitude on the breeding season of smaller mammals shows the same general trends as for ungulates and lagomorphs. For example, the forest dormouse (*Dryomys nitedula*) gives birth in Israel from March to December, whereas in more northern latitudes it breeds only from May until August (Nevo and Amir, 1964). Similarly, investigations by Hoyte (1955) on some small mammals in Arctic Norway (*Sorex araneus*, *Clethrionomys rutilus*, and *Microtus oeconomus*) showed that they all ceased breeding in August, whereas the same species in Southern Norway did not stop breeding until September. *Sorex araneus* in England will often not stop breeding until October. Hoyte's investigations were not started until the breeding season had already begun, but his observations are of particular interest, as they were carried out at a very high latitude (Rosta, lat. 68° N) and at the northern limits of the species involved. Delany and Bishop (1960) compared the breeding season of the bank vole (*Clethrionomys glareolus*) in North-west Scotland with the season for the same species reported by Brambell and Rowlands (1936) in North Wales and Kent. The percentage of females pregnant per month in the two studies were as follows:

	Apr.	May	June	July	Aug.	Sept.
Wales and England (%)	43	84	62	76	53	20
Scotland (%)	0	21	88	12	13	3

The more northerly population will be seen to have a shorter and more intense breeding season.

As pointed out for white-tailed deer, small latitudinal differences often show inconsistencies in breeding variations. In one of the pioneer field studies of mammal reproduction, Baker and Ranson (1933b) found inconsistencies in the effect of latitude on the breeding of the field vole (*Microtus agrestis*). They compared breeding at 52° 45′ N, 54° 15′ N, and 57° 30′ N, and found that although the most northern population began breeding latest of the three, it continued longer into the winter. The population at the intermediate latitude began breeding before the most southerly one. In this, and the white-tailed deer study, the effect of small samples should be remembered, since the size of the sample will govern to a large extent the possibilities of capturing the very small numbers of breeding individuals at the beginning and end of the breeding season. For a proper analysis of small latitudinal effects one would need very large repeated monthly samples for a statistical analysis of the variation in individual onset and cessation of breeding. This does not seem to have been done on any population of small mammals.

The inconsistencies reported in the effect of small latitudinal differences on breeding seasons may be due to selection acting on local populations so as to produce ideal birth seasons which are not directly related to light regimes, yet do fit into the general pattern when much larger differences in latitudes are considered. There have been very few studies on the effect of transporting animals from one latitude to another where relatively small differences are involved, although much has been written on the effect of trans-hemisphere movements on breeding. Adams (1960) considered that genetic effects caused differences of a month in the mean breeding dates of white-tailed deer (*Odoceileus virginianus*) kept in pens at a single latitude in Alabama but originating from northern and southern areas of the state. Ransom (1966) suggested that small inconsistencies in the effect of latitude on white-tailed deer in his Manitoba study could also be due to genetic differences, as one of his three populations belonged to one subspecies, whereas the other two were of a different subspecies. In his beautifully detailed study of the reproduction of farm mink (*Mustela vison*), Hansson (1947) showed that two strains, one from Quebec and one from Alaska, had considerable differences in their mating patterns (see figure 15).

Before closing this section it should be pointed out that because a species breeding pattern varies with latitude, it does not follow that the environmental

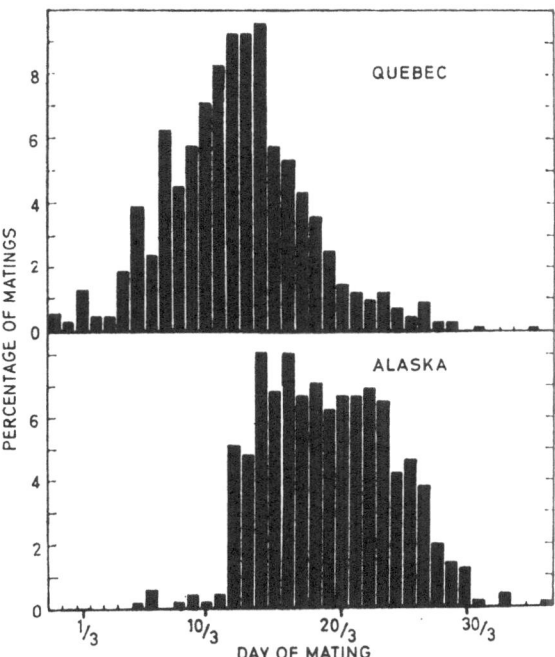

Fig. 15. Seasonal variation of the matings of once-mated Quebec (*N* = 697) and Alaska (*N* = 494) strain female mink (*Mustela vison*). From Hansson (1947) *Acta zool. Stockh.* **28**: 1–136.

variable which has affected breeding is always necessarily the alteration in the light regime. The degree of temperature fluctuation also follows a more or less latitudinal alteration. Temperature effects may account for some of the anomalous data reported above. In the case of species such as *Citellus tridecemlineatus* it is possible that the difference in the length of breeding season from 3 weeks in Wisconsin to $9\frac{1}{2}$ weeks in Texas (McCarley, 1966) is due to temperatures being more severe in the northern state. This would also seem to apply to hibernating species of rodents (see page 122).

III. Alteration of the breeding season by transference across the Equator

The shipping of domestic species and deer from their native country to England, or vice versa, has given considerable information as to the effects of

F

inter-hemisphere transfer on breeding. A series of papers (Marshall, F. H. A., 1937, 1942; Amoroso and Marshall, 1960; Duke of Bedford and Marshall, 1942) has reviewed data on domestic sheep and deer. In 1933, 21 Southdown ewes were shipped to South Africa, where they lambed on arrival in January 1933. They came on heat again and were tupped in May of the same year, and subsequently advanced the tupping season to April in 1934 and finally March in 1935, the complete reverse of their English October breeding season. Similar changes have also been noted in sheep transferred to the Argentine and Australia. There have also been alterations in the rutting time of red deer (*Cervus elaphus*) in movements from England to New Zealand, and vice versa. Two stags and four hinds shipped to New Zealand arrived in November 1907. They began roaring with clean antlers on 2 May 1908, which was about 6 weeks after the New Zealand stags had begun to rut, and 2 weeks after they had ceased. The imported stags had thus shed and grown two sets of antlers in one 12-month period, although the second set did not develop properly. In 1909 the British stags rutted at the end of March, the usual New Zealand time. The hinds arrived pregnant and calved in April 1908. They again calved in February 1909, which meant that they had been covered by New Zealand stags in July 1908, a most unusual time. On the other hand, red deer moved from New Zealand also showed a similar pheno-menon. Females brought in pregnant and arriving in England in March at first maintained their original gonadal rhythm and came on heat in April, which is the New Zealand time. The next year they adapted somewhat and had a sexual season in December and January, and in the third year after transportation they had completely adjusted to the British situation and were on heat at the normal time in October. It is interesting that no pregnancies resulted from the April heat, but this may have been because the hinds were sexually immature. The December–January heat resulted in some calves, but only a few survived. It thus took these New Zealand hinds 2 years to adapt. What is of most interest is that the English stags covered the New Zealand hinds in April and again in Decem-ber–January. As in New Zealand, the stags were able to copulate completely out of the normal season.

A number of other deer species have been imported into New Zealand (Wodzicki, 1950), and Marshall has also reported (1937, 1942) on the changes in their breeding seasons. The alterations are summarized below:

Species	Origin	Northern season of rut	N.Z. rut
Fallow deer (*Dama dama*)	England	October	April
Moose (*Alces alces*)	Canada	Sept.–Oct.	March–Apr.
Wapiti (*Cervus canadensis*)	Canada	Sept.–Oct.	March–Apr.
White-tailed deer (*Odocoileus virginianus*)	U.S.A.	October	Apr.
Chamois (*Rupicapra rupicapra*)	Europe	November	May

The case of the Thar (*Hemitragus jemlahicus*) is especially interesting. This Indian species breeds in the wild in September and October. Transported to

England, the season shifted to early November. A move to New Zealand induced breeding there in April and May, which is the reverse of the English season and not the Indian. This may have been due to the closer similarity of latitudes.

There are four examples of non-artiodactyls reported in the literature which have also altered their breeding season. Cunningham (1905) reported that a pair of Cape hunting dogs (*Lycaon pictus*) which were brought from South Africa to the Dublin zoo reversed their breeding season, as the female had the majority of her pups in January. Ferrets (*Mustela putorius*) transported to South Africa in 1937 bred when they first arrived there in October, but eventually changed over

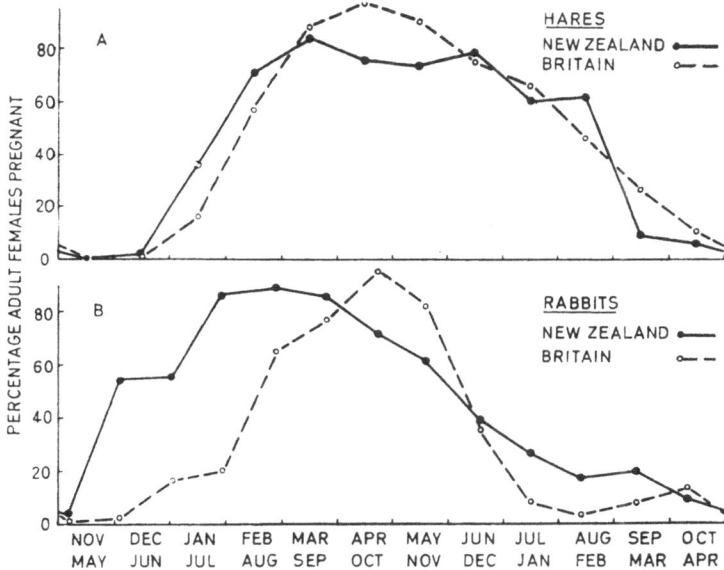

Fig. 16. The breeding season of rabbits (*Oryctolagus cuniculus*) and hares (*Lepus europaeus*) in New Zealand and Britain. From Flux (1964) *N.Z. Jl Agric.* **109**: 483–6.

completely, so as to be adjusted to the southern light regimes (Duke of Bedford and Marshall, 1942).

The wild rabbit (*Oryctolagus cuniculus*), which was introduced from Great Britain to Australia and New Zealand, has been the subject of some of the best studies of reproduction of a wild mammal (Brambell, 1944; Poole, 1960; Watson, 1957). As far as the effect of moving from one hemisphere to another is concerned, the situation is somewhat confused, because in all three locations, under certain conditions, the rabbit can breed over a very large part of the year. However, there does seem to be the same trend to transpose the main peaks of breeding by a 6-month period. Flux (1964, 1965b) has shown that wild hares (*Lepus europaeus*) also breed at the opposite time of the year in New Zealand. Figure 16 gives a comparison from Flux's paper of hare- and rabbit-breeding seasons in Britain and New Zealand.

From the above review it will be seen that if temperate-zone mammals are moved from one hemisphere to another they tend to alter their breeding season so that breeding is displaced by 6 months from the season in the original hemisphere. The timing of this alteration depends largely on the timing of the transportation of the individual animals relative to their last breeding season. As in many cases the shift has been to relatively the same latitude and similar environments in the opposite hemisphere, it is well-nigh impossible to postulate that any other variable than light is involved. In a discussion of the experimental effects of light on the breeding season of sheep, Yeates (1949, 1965) has pointed out that the alteration in the seasons of sheep can only be explained in terms of photoperiod. Amoroso and Marshall (1960) have further noted that understanding of the effect of light will show the most economic time for such transfers of stock, especially where valuable stud animals are concerned.

C. THE EXPERIMENTAL ALTERATION OF LIGHT REGIMES AND THEIR EFFECTS ON BREEDING

There exists a very large literature on the alteration of light regimes and its various effects on different aspects of mammal breeding. It is best to discuss the various experiments under two main headings, namely, those in which the experimental animals were subjected to light changes which approximated to the rate of change which they would experience in a natural environment and those in which light was altered more dramatically by either completely depriving the animals of light or putting them under fixed light–dark regimes.

I. Gradual alteration in the length of light

These experiments attempted to alter the light in a gradual manner, often commencing at the equinoxes, so that the animals were subjected to the reverse type of lighting situations as were occurring in the natural-light regime outside the laboratory. Two common regimes are: (1) gradually increasing the light when the environmental daylight is decreasing, or (2) vice versa.

(a) Sheep (*Ovis aries*)

The pioneer papers of Marshall, F. H. A. (1937) on the effects of moving sheep from one hemisphere to another pointed the way to the experimental manipulation of light to control sheep breeding. Sykes and Cole (1944) in the United States gradually decreased the light received by a group of ewes by placing them daily in a darkened pen from March until May and found that the ewes came into oestrus in May and June and successfully conceived at this time.

This work was followed up by Yeates (1949) using Suffolk sheep at Cambridge, England. There were two main light changes in his experiment which are shown in figure 17. In experiment A both groups of ewes came into oestrus in September, but the experimentals had a much shorter season. The non-pregnant experimental A ewes started coming back into heat and were all in full oestrus by June of the same year, but the controls did not come into heat until the normal period of September–October. The experimental ewes had ceased oestrus by December 1948, whereas the controls were still in full oestrus. Some of the experimental ewes were mated in their first and second breeding seasons. Lambs born to ewes mated in the first season and lambing down in March had normal healthy lambs with a mean weight of 9·1 lb, but lambs born from the second (experimentally induced) breeding season showed poor survival, with a mean birth weight of

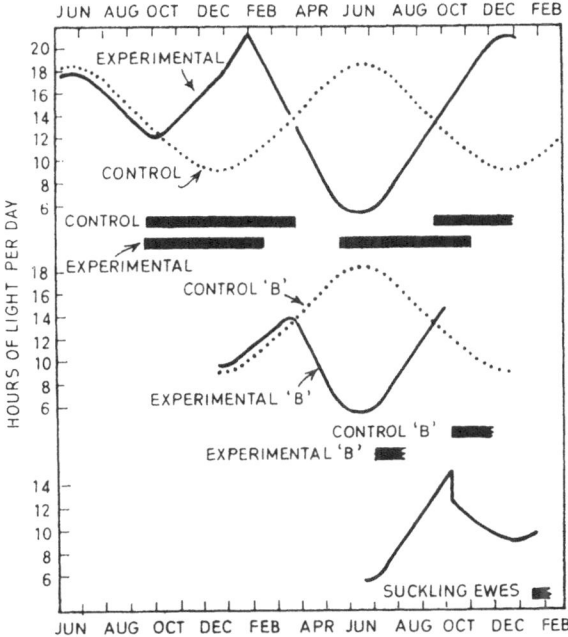

Fig. 17. The relationship of the sexual season to lighting conditions in various groups of ewes. The sexual season is represented by the black band, which, if broken at the end, indicates that data were not available. From Yeates (1949) *J. agric. Sci., Camb.* **39**: 1–42.

only 4·6 lb. Semen from experimental A rams increased in volume in March (while the control rams stayed low) and then decreased in July and August, when the semen volume of the control rams was increasing.

In group B the experimental ewes started coming into heat in June 1948, while the controls started at the normal time of September–October. In his discussion Yeates makes the following important deduction:

'Study of text-fig 5 (figure 17) shows that the experimental A group ewes commenced their summer sexual season on May 21, 1947. But from the spring equinox up to and beyond that date the experimental B group had the same lighting conditions, yet did not breed until July 2. This indicates that whatever stimulus induced the sexual season in the A group, it must have been received prior to the spring equinox. Otherwise the experimental B group would have exhibited oestrus at the same time or nearly at the same time. In addition, up to the time of the spring equinox, experimental B group had received the same light treatment as control B group. But when their treatments diverged after the spring equinox, the experimental B group began breeding July 2 when the control group B showed no breeding response. This proves that the stimulus which induced breeding in experimental B ewes must have been received after the spring equinox. Here then, is an instance of two groups of ewes (Experimental A and B groups), each having commenced sexual seasons as a result of treatment with decreasing light, but each group owing its breeding responses to a stimulus received at a distinctly different level of light. This is considered proof that breeding does not necessarily result from the attainment of a particular threshold of light, or that the stimulus which results in breeding is received at a particular threshold' (page 22, Yeates, 1949).

He goes on to suggest that the ending of the breeding season was analogous except that it occurred after the change from decreasing to increasing daylight. Yeates also pointed out that the start of the sexual season occurred in all experimental and control situations about 13 weeks after the change-over from increasing to decreasing light, irrespective of the quantity of light at the change. Yeates's paper has been quoted at length because his interpretation seems irrefutable and should be born in mind when results from other mammals are discussed later.

Yeates went on to suggest that a possible explanation of the differences between different breeds of sheep with regard to the onset and nature of their breeding seasons may be looked for in terms of selection for the time required to respond to the change-over from increasing to decreasing daylength. For example, in Arctic regions most lambs would be born to individual ewes who did not react either too early or too late to the light stimulus. This would be less important in temperate regions, and may account for the longer seasons of sheep breeds originating there. These will still retain the response to lighting regimes, as Yeates (1956b) showed by conducting a similar experiment to the one above on Merino ewes in Brisbane. It is interesting that he found no significant differences in the birth weights of lambs from ewes tupped at two different times of the year.

Thwaites (1965) carried out a similar experiment at Armidale, N.S.W., using Southdown ewes, but the experimental conditions were maintained for a longer time. Starting at the March equinox of 1962, one of his experimental groups was put on reversed decreasing and increasing light in the opposite phase to his controls for a total of two years, the only difference being that photoperiodic fluctuations were accentuated so that experimental ewes received a maximum of 17 and a minimum of 9 hours light, compared with 11–15 hours for the control animals. His results were similar to those of Yeates, as breeding in experimental ewes was completely reversed. However, during the period which should have coincided with the second experimental anoestrus, some experimental ewes showed signs of oestrus. He interpreted this as an effect of temperature which is known to also play a role in the breeding season of sheep (see page 117).

Another reversed seasonal photoperiodic experiment was carried out on Rambouillet ewes in South Carolina by Godley, Wilson, and Hurst (1966). As part of a large experiment on the interaction of light and temperature they subjected ewes (in April) to October light conditions for a period of 3 years. In each case the experimental ewes came into oestrus and were tupped with a higher rate of conceptions than the controls. Unfortunately insufficient data are given to compare the actual dates of oestrus in each group. A fascinating approach to the

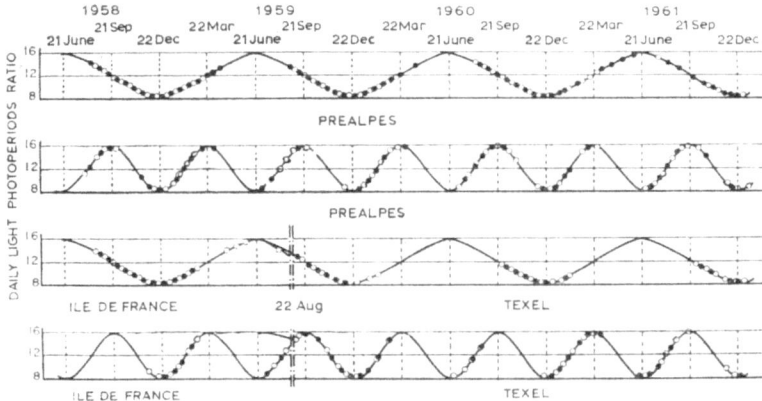

Fig. 18. Occurrence of breeding periods in several breeds of sheep during normal 12-month photoperiodic cycles and experimental 6-month cycles of photoperiodic variation. From Thibault *et al.* (1966) *J. Anim. Sci. (Suppl.)* **25**: 119–42.

problem was carried out by workers in France using Texel, Ile-de-France, and Prealpes breeds (Thibault *et al.*, 1966, reporting work by Mauleon and Rougeot). They placed animals on a compressed light regime, where the normal 12-month cycle was covered in 6 months, and obtained a breeding season every 6 months, as shown in figure 18. Unusually this season began on a period of increasing, not decreasing, light. In tropical areas sheep tend to breed continuously, with little evidence of an anoestrous season. Symington and Oliver (1966) subjected sheep in Rhodesia (lat. 17° S) to a gradually increasing light regime from February till August, while the external photoperiod was first decreasing then increasing. The indicence of oestrus was much higher in the experimental group.

Light alterations also have an effect on the development of the male breeding season. Ortavant (1961) sacrificed Ile-de-France rams which had been kept on gradually decreasing light and then held at fixed photoperiods for over a month. Although the epidydimal spermatozoa reserves and the testicular weights reached a maximum at 8 hours light, the only regime which allowed maximum spermato-genesis was found at a photoperiod of 10 hours following one of 12 hours (see also Ortavant, Maulfon, and Thibault, 1964). Moule (1950b) reported an increase in the libido of Merino rams in Queensland which were subjected to a fairly rapid decrease in light during September compared to controls whose light was still increasing.

(b) Ferret and mink (*Mustela putorius* and *M. vison*)

The ferret was the first mammalian species in which the effect of experimental manipulation of light was demonstrated by Bissonnette (1932). Unfortunately, much of the early work on this species has come under criticism from later workers on the basis of questionable interpretation of results. It should be pointed out that some of the manipulations involved in the work on ferret and mink do not strictly come under this section on the gradual alteration of light; much of the work has been done by adding photoperiods to the length of normal day-light, and not by regular slow changes in the quantity of light.

In a number of papers Bissonnette (1932, 1935a, 1935b, Bissonnette and Bailey, 1936; for a more complete list the reader is referred to the bibliography in Amoroso and Marshall, 1960) described experiments in which ferrets subjected to increased illumination during the winter came into oestrus after 38–64 days. Male ferrets did not show as complete a response until Bissonnette slowly increased the extra amount of light, so that from 2 October to 6 December the amount given after sunset increased from 4 to 8 hours per night. He then kept males (1935a) on an extra 8 hours per night until 30 March. The males then came into complete breeding condition between November and December, but later regressed into the non-breeding state from February onwards. Bissonnette sug-gested that: 'this indicates that some desensitization occurs permitting regression or failure of activity, or that the effective hormones are excreted too rapidly to maintain testis activity, or both together, as has been found for extracts of anterior pituitary and other gonadotrophic agents, or that the lighting passed above the optimum' (page 364, Bissonnette, 1935a). This regression was further investigated (1935b) by placing immature males on normal daylight plus an added 6 hours of artificial light per night from 12 November and then testing their reaction to females after 13 January. These animals still underwent sexual regression before June, a time when normal ferrets were in the height of sexual activity. Despite this, Bissonnette and Bailey (1936) were able successfully to induce pregnancies as a result of copulation by males and females who were both subjected to gradually increasing light after nightfall from November to January.

Bissonnette's (1935a) paper has been criticized in some detail by Yeates (1949), who comments that the return to reproductive quiescence requires confirmation because of the very small number of animals used. He also points out that Bissonnette's implication that the light was gradually increasing in his experi-ment is not correct because, due to the method of adding light on to daylight, the regimes Bissonnette used actually increased the amount of light during

October, from whence it stayed constant for 2 months and then increased again until March. Yeates suggests that the regression which occurred in the single male may have been due to this somewhat peculiar regime. Leaving this problem of regression aside, the pioneer work of Bissonnette showed quite clearly that increasing the light in the early winter brought both male and female ferrets into an early reproductive condition. The ferret has been subsequently used in many physiological and neurophysiological experiments to elucidate the mechanisms whereby light does stimulate the reproductive processes. In these experiments the control animals have repeatedly confirmed Bissonnette's work (for example, see Hill and Parkes, 1933; Thompson, 1951, 1954; Thompson and Zuckerman, 1954; Marshall, F. H. A., 1940; Marshall and Bowden, 1934, 1936).

Hammond, Jr. (1951) also used periods of regularly increasing illumination as a control in ferret experiments on broken daylight regimes. He described the early onset of oestrus under these conditions and comments that mink under the same regime came into induced oestrus ahead of ferrets. Hart (1951) used regularly increasing light in winter as a control. He added half an hour on to the end of the natural daylength beginning 5 December, and increased it by another half hour every 3 days. His animals came into oestrus between 25 January and 9 February the following year. As a result of experiments on changing the sequence of light and dark periods, Hart suggested that a gradually increasing plane of light is not essential in itself in inducing oestrus in the ferret, but is merely the means whereby the change-over from a short-light, long-dark sequence to a short-dark, long-light sequence is effected. This will be discussed further below.

There have been many investigations on the effects of changing light regimes on mink breeding. The stimulus to this work has been the practical improvement of the production of mink on fur farms and has consisted of attempts to induce an earlier and thus longer breeding season and also to increase litter size. Aulerich, Holcomb, Ringer, and Schaible (1963) reduced the lighting of experimental animals between 10 June and 16 July so that they were finally receiving the equivalent light to the December equinox. The light was then kept constant until August and slowly increased. This resulted in the induction of oestrus in females in late summer and autumn, but males were infertile. Extra daylight added to the normal day (in the manner of the ferret experiments) from 21 September to 13 January also hastened the onset of oestrus. Similar experiments were carried out by Holcomb, Schaible, and Ringer (1962), who gave increased light before the normal mating time by adding 8 minutes of extra light per day (to a total of 80) from 1 February to 10 February. This started mating on 15 February instead of the normal early March. The males are apparently ready to mate by 1 February whether lighted or not, so these experiments did not run into the difficulties reported in the later paper. It is interesting that certain breeds of mink were more light-responsive than others. Sapphire females were influenced more by increased light than were dark or pastel females. The workers suggest that this was due to less pigmentation in the eyes of the sapphire breed, but, as Hansson

(1947) has reported that there are differences between the onset of mating in untreated mink of different breeds, it is possible that in selecting for different coat colour, the mink breeders have also been selecting for different sensitivity to light.

(c) Raccoon (*Procyon lotor*)

This is another species which also naturally comes into breeding condition in February. Bissonnette and Csech (1937, 1939) increased the illumination by added extra light from 10 October onwards, and induced three males to become sexually active in December (somewhat later in a second experiment) and mate successfully with two of three females which had also been induced to come into oestrus at this time. These three females had young successfully the previous year. Three other females subjected to the same regime, which were non-breeders in the previous season, also came into heat, but only achieved pseudo-pregnancy.

(d) Goat (*Capra hircus*)

In a similar way to sheep, and unlike the ferret, mink, and raccoon, the goat begins breeding in September and then tapers off its breeding season, which usually finishes in early March. Bissonnette (1941) kept five nannies and a billy on increasing light periods (daylight and extra light at night) from 25 January to 5 April and then diminished the light until 5 July. The experimental animals ceased breeding in February, and then new cycles were induced from May to July. Matings in July, within 10 days of return to normal lighting, resulted in kids being born in December to four of the nannies. The first births in control animals were in February. Results similar to this have been reported by Yoshoika, Awasawa, and Suzuki (1952). Gradual shortening of the daylight hours in spring and summer caused nannies to come into oestrus some 70–80 days after the treatment began. In the opinion of Amoroso and Marshall (1960) these results oppose those of Bissonnette, but this is not apparent to the writer. Both sets of experiments involved shortenings of daylight from the spring equinox onwards, and in both cases this induced oestrus in the nannies 1 or 2 months later. Other observations on goats have been reported very recently by Fraser (1968).

(e) Field vole (*Microtus arvalis*)

The effect of fixed regimes of light on voles was investigated many years ago by Baker and Ranson (1932) and this work will be considered later. However, the effects of gradual changes in light on *Microtus* have only recently been investigated by Lecyk (1962a, 1963) in Poland. In three well-designed experiments he altered the light to investigate the effect of both increases and decreases on breeding. In the first series (1962a) he subjected wild voles to the following treatments starting at the September equinox:

Group (1) Sunlight plus extra increasing light from September to December, which approximated the natural increase from 21 March to 24 June.

Group (2) Same regime, but entirely artificial light.

Group (3) Uniform 16-hour day.

Group (4) Uniform 8-hour day from September to 4 November, and then gradually increasing light to 13 hours on 1 December.

Group (5) Natural decreasing light conditions.

With the exception of group (4) animals, which were killed at intervals, all animals were sacrificed in December. The results are given below:

Treatment	Testis weight (mg)	Auxiliary glands weight (mg)	Percentage of females pregnant	Sexual behaviour first seen
1	188	211	74	23 Oct.
2	197	223	64	2 Nov.
3	174	205	70	25 Oct.
4	154	92	67	6 Dec.
5	60	49	0	Not seen

All experimental treatments thus induced oestrus. The minimum length of light at which this occurred was 13 hours, and sexual activity began about 30 days after the daylight began to increase in length.

In the second series Lecyk (1962a) kept wild voles captured in December under normal daylight until 22 May. He then decreased the light gradually until the middle of July to 7 hours. This experiment started well after the equinox, and was entirely unsuccessful in inhibiting the breeding of the experimental voles. Both experimental and control groups had begun breeding before the experiment began, and the light reduction did not stop full sexual development. However, in later experiments Lecyk (1963) put very young (10-days-old) voles on the following treatments:

Group (1) Daylength shortened from 10 May to 20 June from 12 to 7 hours.

Group (2) Constant 7-hour daylight from 10 May to 20 June.

Group (3) Normal increasing daylight.

When compared with controls, group (1) and (2) males had small and non-functional testes with little spermatogenesis, and females had no mature follicles in the ovaries. Thus reducing light could inhibit the development of sexual maturity, but could not cause the breeding season to stop, once started. These results have been described in detail because very few workers have investigated the causes of the cessation of breeding. With the exception of some of Bissonnette's early work on ferrets and goats and some other work on sheep, Lecyk seems to be the only modern worker to have looked for the effect of gradual reductions in light on already breeding animals. *Microtus arvalis* has been found not to react to this sort of regime. It would be very interesting to attempt to study the effects of decreasing the light at different times in the natural breeding season and also to decrease it at different rates.

(*f*) Rabbit (*Oryctolagus cuniculus*)

The well-known difference between the almost continuous breeding of the domestic rabbit and the seasonal breeding of its wild counterpart has complicated much of the work as to the effect of light regimes on rabbit reproduction. Smelser, Walton, and Whetham (1934) reported that gradually increasing or decreasing the daylength had no effect on laboratory rabbits. Kihlstrom (1958) claimed to have caused significant changes in rabbit-semen characters by giving very small amounts of extra light in October and November. On the basis of a very small sample of animals, he suggested that light had stimulated the accessory glands, but due to the time lag in the results reported, this work needs confirmation before a major effect of light on rabbit reproduction can be definitely considered. The breeding season in the wild rabbit is so variable when compared to the fixed season of the hare (*Lepus europaeus*; Flux, 1965) that this may be considered as further evidence for the lack of an effect of light in the former species.

(*g*) Deermouse (*Peromyscus leucopus*)

Like many rodents, this mouse starts breeding in March and continues until October, being reproductively quiescent during the winter. In a very well-organized experiment, Whitaker (1940) studied its breeding under gradually changing light conditions. Four groups of experimental animals were kept under reversed seasonal conditions in which the light regime was displaced in time by a 3-month period. This meant that at the September equinox two experimental groups were under midsummer light conditions (18 hours per day) and light was about to decrease, whereas the other two experimental groups were under midwinter light conditions (6 hours per day) and were about to have the light increase. Temperature was partly but not completely controlled during the experimental regimes, which were maintained for 2 years. The results were somewhat equivocal. There were alterations in the breeding season, but there was also a drop in the proportion of breeding females when the light was decreasing *and* when it was increasing. Generally speaking, there was more breeding (98 per cent of females) during the days of long daylight than in the shorter days (39 per cent of females), but breeding still persisted while the deermice were under experimental winter daylight. Part of this discrepancy may be due to the fact that Whitaker assessed the fertility of his deermice by palpation of testes or detection of open vaginae. This paper will be discussed in detail later (page 99) because of the possibility that the period of darkness was the significant factor to these naturally nocturnal animals.

(*h*) Prairie dog (*Cynomys ludovicianus*)

The reproduction of the female of this species is apparently completely unaffected by photoperiod (Foreman, 1962). However, it is partially fossorial and hiber-nates through the winter, so that it is difficult to visualize how light changes could be effective in this case.

(i) Cat (*Felis cattus*)

This species breed in laboratories in Massachusetts from January to July, but Dawson (1941) was able to induce oestrus in December by gradually increasing the amount of illumination. Decreased lighting in late August and September, and then increased lighting from October onward, caused cats to come into oestrus in November.

(j) Horse (*Equus caballus*)

Burkhardt (1947) gradually increased light faster than normal from January to March on a group of four New Forest mares in Cambridge, England. The normal oestrus period is November to March, but the lighted mares came into oestrus in February or March, while controls did not exhibit their first oestrus until April.

Nishikawa, Sugie, and Harada (1952) subjected Korean ponies to 5 hours of extra light added at the end of the day from mid-November to the end of February. The ovaries of mares given this treatment began to function some 65–80 days earlier than unlighted controls, and were functioning in late January and February. Similar treatment started in mid-August, when the mares would normally have been entering anoestrus, and carried through until the following March, resulted in the ovaries being still functional in the normal non-breeding season. In both experiments the induced ovarian activity was accompanied by normal fertility. Similar experiments using the same light regimes were reported for the stallion by Nishikawa and Horie (1952). Extra light added from mid-November accelerated the normal increase in semen volume which occurs in April, bringing it forward by about 2 months. Extra light given in August resulted in an increase, and then a gradual decrease in volume, so that this light treatment was not effective in causing a prolongation of the function of the accessory glands.

(k) Discussion

With the exception of the prairie dog, the species described above have all shown a degree of breeding reaction to gradual alteration of the light environment in a manner which attempts to approximate the natural seasonal changes. Taylor (1963) was unable to alter the natural decline in spermatogenesis in grasshopper mice (*Onychomys torridus*) by light (or food, or temperature) manipulations. Amoroso and Marshall (1960) have used the term 'long-day' animals for those species which normally commence their breeding season when the daylength is increasing, i.e. some time after the winter solstice. These species can be induced to breed by artificially increasing the length of daylight at that time of the year when it is naturally decreasing. The 'short-day' species, on the other hand, commence breeding after the summer solstice and are thus stimulated by artificially decreasing the light. A similar classification for domestic mammals was suggested by Ortavant, Mauleon, and Thibault (1966). Only in a very few species is the converse known to be true, i.e. that the decrease of light in a long-day species or the increase in light for a short-day species can cause a cessation of oestrous cycles. Furthermore, despite the very large number of studies on this aspect of light effects on reproduction, there are still a number of unanswered questions. The effects of changing over from increasing to decreasing (or vice versa) light at different times after the last breeding stopped, or after the last change to increasing light, are as yet unknown. Experiments such as this would give insight into the refractability of the light stimulus. As will be noted in the section on temperature, it is now apparent that in some species the effect of temperature changes can almost match that of altering the light regime. In much of the older work on experimental alteration of light this factor was unknown and not allowed for in experimental design, so that some of the early results must now be considered suspect.

II. Maintenance under continual light

The experimental maintenance of animals under conditions of constant light (or darkness) has not been carried out primarily with the intention of altering the breeding season (i.e. in the sense of changing the period of time at which animals mate). This work has really been carried out to find how the continual presence (or absence) of the stimulus which is known to affect seasonally breeding mammals can change reproductive phenomena in the experimental animal. It should be born in mind that some of the work described below has been done on species which do *not* undergo seasonal breeding in the wild state. The effect of continual light on non-seasonal breeders can thus only result in a change in the intensity of the breeding phenomena.

(a) Laboratory rat (*Rattus norvegicus*)

The effects of continual light on the maturity of rats have been discussed on page 5. If mature, but young, female rats are kept under continual light they exhibit extremely long periods of continuous oestrus (Fiske, 1941). Pituitaries of females and males under continual light for 50 or more days were richer in FSH content than similar animals kept in continual dark, and this and other evidence lead Fiske to postulate that the effect of light was to cause an increase in FSH production and release. On the other hand, this effect was reduced if the female rats were kept in constant light for periods of 250–300 days. Continually lit females for the shorter period had heavy pituitaries, ovaries, and uteri, and the state of the uterine epithelium indicated secretion of oestrogen. Male rats kept under continual light had heavier pituitaries, testes, and seminal vesicles, the effect being greatest when the males were about 150 days old. Fiske's results emphasize light as a major stimulus to the pituitary for the production and release of FSH. As female rats on constant light may stay in oestrus for several weeks, Fiske suggests that the effect of light in causing FSH release is stronger for a time than the effect of oestrogen on the pituitary.

Further studies on rats by Jöchle (1964) confirmed that continual illumination caused a cessation of oestrous cycles and resulted in indications of permanent oestrus. Histological examination of ovaries of these rats showed no corpora lutea present, indicating, however, that no ovulations were taking place. There were plentiful enlarged follicles and follicular cysts, the oestrogen from which was causing uterine hyperplasma. Jöchle found some individual variation in this reaction to continual light. The younger a group of rats were when permanent illumination was started, the higher was the percentage of individuals which went into permanent oestrus. On the other hand, the effect of continual light was detected quicker in the smears of older animals. Jöchle suggests that 'from this resistance one can conclude a hereditary pre-formed or an acquired resistance against this action of permanent lighting destroying the estrous cycle' (page 93). Similar results on Sprague-Dawley rats were described by Chu (1965), who found that continuous light for 2–4 weeks resulted in an increase in the incidence of oestrous smears.

(b) Other rodents and lagomorphs

Chu (1965) found differences in the reactions of three other species of rodents to continual light. In-bred mice (*Mus musculus*) of the Swiss strain, kept in light for 2–4 weeks, reacted like laboratory rats, in that their smears showed signs of permanent oestrus. However, randomly bred Swiss mice, an inbred laboratory strain (DBA/2 mice), and Syrian hamsters (*Mesocricetus auratus*) all showed no increase. The guinea-pig (*Cavia porcellus*) showed an intermediate response, in that the vaginal epithelium changed under continuous light but the cytology of the smear did not. Meyer and Meyer (1944) found that the cotton rat (*Sigmodon hispidus*) showed no changes in the length of vaginal oestrus or dioestrus when maintained under continual light. The male rabbit (*Oryctolagus cuniculus*) left in continual light for short and long periods showed no effect on reproductive organ weights or histology (Maqsood and Parsons, 1954).

(c) Ferret (*Mustela putorius*)

Hammond, Jr. (1951) put female ferrets on continual light in January and found they came into oestrus at almost the same time as naturally lit females, but considerably later than a group on 14 hours of light per day. Hart (1951) placed female virgin ferrets on natural daylight plus continual artificial light from 30 October. After 11 weeks this regime was changed to continual artificial light. Whereas control females had a latent period to the mean onset of oestrus of 23·5 weeks, these experimental females had a latent period of 16·6 weeks, which was not as fast, however, as females placed on broken light regimes (see page 93). This experiment is really inconclusive, as Hart pointed out that the light inside the continual light pen varied considerably in intensity through the day.

(d) Sheep (*Ovis aries*)

Sheep kept for 7 weeks in summer on continuous light were found to produce their lambs earlier than controls (Terry and Meites, 1951), but this work cannot be evaluated, due to the limited information given in the abstract. The length of the experimental treatment was also very short for a species with a 16-day oestrous cycle. However, Radford (1961) maintained a group of Merino ewes on continual light (daylight plus extra light at night) from 2–3 months of age for a period of 3 years. Then one sub-group was exposed to continual artificial light for only a year, whereas two other sub-groups had continual light of constant intensity on which daylight was imposed, with and without exposure to wind and rain. All experimental animals were subjected to more or less environmental temperature fluctuations, and in the first part of the experiment the night-lighted ewes were run with control ewes during the day. These details are necessary because, as Radford himself points out, they may account for the somewhat unexpected results shown in figure 19, from Radford's paper. Although sexual maturity was partly suppressed by light treatment (the occurrence of oestrus, as detected by vasectomized rams, was suppressed in the first year, even though ovulation was not), in the second year there was little difference between the continuously lighted and control groups, both showing a definite breeding season. In the second part of this experiment, where one group had continuous artificial light only, there was evidence of increased sexual activity. Inspection of figure 19 will show, however, that before the experimental regime was changed there was already a tendency for the 1958 season to be prolonged in many individuals. It is a pity that Radford did not continue observations on his control group after July 1958, as the same tendency could be seen in them. Similar results were obtained in France, where Thibault *et al.* (1966) reported the effect of placing 8 Ile-de-France ewes under continuous light for 3 years. The breeding season appeared at the 'normal time' over this period, but data were not given and there was no information with regard to temperature changes. Remembering that the sheep is a short-day animal which starts breeding on decreasing light, these results would appear somewhat anomalous, in that seasonal phenomena of reproduction were still expressed; but the lack of temperature control suggests that this environmental factor may have 'timed' the onset and cessation of oestrus (see page 117).

(e) Discussion

The practical difficulties of maintaining a group of animals on continual light of the same intensity for a long period of time, and yet still controlling other variables, such as temperature and social contact, have meant that such experimental work to date can be criticized as not truly maintaining constant conditions. The work on the ferret and sheep outlined above has demonstrated this sort of difficulty. Thus the real effects of a constant and continual light stimulus on any seasonally breeding mammal are still unknown.

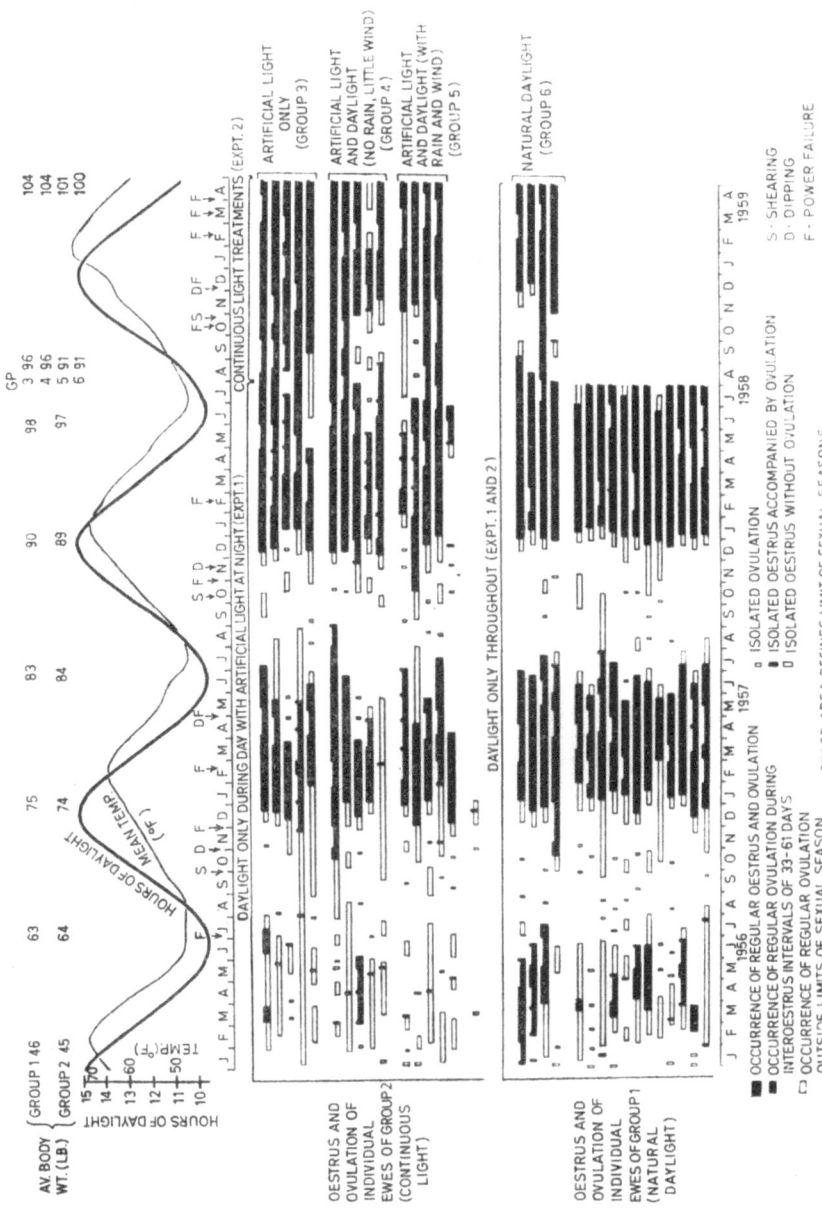

Fig. 19. The effect of experimental alteration of light regimes on oestrus and ovulation in ewes. From Radford (1961) *Aust. J. agric. Res.* **12**: 193–53.

G

III. Maintenance under continual darkness

Keeping mammals under conditions of no light for long periods has not been done with the main intention of altering the breeding season but to determine the effects of complete removal of a stimulus known to affect reproduction.

(a) Laboratory rat (*Rattus norvegicus*)

In experiments already referred to, Fiske (1941) kept rats in complete darkness for long periods. Females exhibited an increase in the number of days on which they had metoestrous smears, and their pituitaries had less FSH than those of females under constant light. However, the pituitaries of constant dark females had more LH than lighted females. Males kept under constant darkness had smaller pituitaries, testes, and seminal vesicles than lighted males, and there was evidence that the proportion of interstitial tissue in the testes was greater in dark males; but as these organs were so small, detailed comparison proved impossible. Fiske concluded that the lack of light favoured a decrease in FSH production and, possibly in the female only, an increase in LH. Chu (1965) found similar effects when Sprague–Dawley rats were kept under constant darkness for 2–4 weeks. Sakai (1963) kept normal male rats in complete darkness for up to 72 days and found that after 48 days the testis weights decreased with an atrophy of the interstitial cells and spermatogenic arrest. The secretory activity of the prostate and seminal vesicles was also reduced.

(b) Other rodents

Inbred Swiss mice (*Mus musculus*) kept under constant darkness behave like rats and show a decrease in oestrous smears, while random-bred Swiss mice, inbred DBA/2 mice, and Syrian hamsters (*Mesocricetus auratus*) show no effects in their oestrous smear of any alteration in the reproductive cycles (Chu, 1965). The guinea-pig (*Cavia porcellus*) also showed no response (see also Dempsey, Meyers, Yonge, and Jennison, 1934). The oestrous cycle of the cotton rat (*Sigmodon hispidus*) was not affected by continuous darkness (Meyer and Meyer, 1966).

Whitaker (1940) kept deermice (*Peromycus leucopus*) in darkness for over two years. They bred continually all this time and showed no evidence of any seasonal anoestrus as did control animals. About 60 per cent incidence of breeding was maintained through the whole period. It would thus seem that there is considerable variation in different groups of rodents as to their response to complete darkness. It would prove very interesting to compare these differences taxonomically within this vast order and also to try to relate the ability to respond to absence of light to the diurnal nature of the species activity in its natural environment.

(c) Ferret (*Mustela putorius*)

The role of darkness in the initiation of breeding in this species has been the subject of some controversy. However, there have been no reported experiments where ferrets have been kept under complete darkness. Hill and Parkes (1934) did put a small group of ferrets under $23\frac{1}{2}$ hours darkness a day beginning at the end of January. The animals came into season at the normal time, and in at least one case mated and produced a normal litter. Hill and Parkes concluded that while additional light can induce oestrus in anoestrous animals, the onset of the breeding season in spring is not dependent upon the increasing length of daylight. Although this conclusion was apparently accepted by Hammond Jr. (1951), it has been pointed out by Yeates (1949) and Hafez (1952c) that basically it is incorrect, as Hill and Parkes overlooked the fact that their experimental animals had been previously subjected to some 5 weeks of increasing light before the experiment began, and thus the onset of oestrus at the normal time cannot be said to be independent of increasing light.

(d) Rabbit (*Oryctolagus cuniculus*)

Kihlstrom (1958) has reported that the maintenance of buck rabbits under almost constant darkness for periods of up to a month, starting in July and in May, had absolutely no effect on semen obtained, indicating no changes in the accessory glands. Similarly, Maqsood and Parsons (1956) found no effect of continuous darkness on the weights of testes or accessory organs or on testis histology.

(e) Discussion

For exactly the same reasons as mentioned in the preceding discussion, there have been few experiments on maintaining species other than the smaller laboratory ones in complete darkness. Despite the considerable difficulties involved, especially from the point of view of nutrition, these experiments are theoretically necessary if an absolute effect of light on the breeding of any mammal is to be postulated. They are also necessary if the controversy about the importance of dark periods, as opposed to light periods, is to be resolved.

IV. Maintenance under fixed ratios of light and dark

The majority of experiments on the effects of light have involved placing mammals under regimes where the light quantity remained constant. Some of these were designed to see what would happen if the species under consideration were maintained for long periods of time at either the natural winter minimum or summer maximum. Others attempted to determine the effects of splitting up the light ration so that the animal has four (instead of two) changes from light to dark in each 24-hour period.

(a) Sheep (*Ovis aries*)

The breeding of sheep in tropical regions has provided the initial stimulus to studies of the effect of equatorial light periods over longish periods of time. The treatment and results of an experiment of this nature carried out by Radford (1961) are shown in figure 20. The development of an anoestrous season while under a constant light regime does not fit in with concepts of a *light*-controlled breeding season in the sheep. However, Hafez (1952c) has postulated that the reinitiation of sexual activity in ewes subjected to supposedly unfavourable light environments was due to a memory of previously experienced favourable light environments. In the section of his paper on the effects of continuous light on reproduction (see page 84) Radford had suggested that once sexual activity is established, it continues to occur unless it is actually suppressed or the reproductive mechanisms become exhausted: 'It has to be concluded, therefore, that it is suppression of activity which is occurring during anoestrus, not stimulation of activity during the sexual season' (page 146). Attention should also be drawn

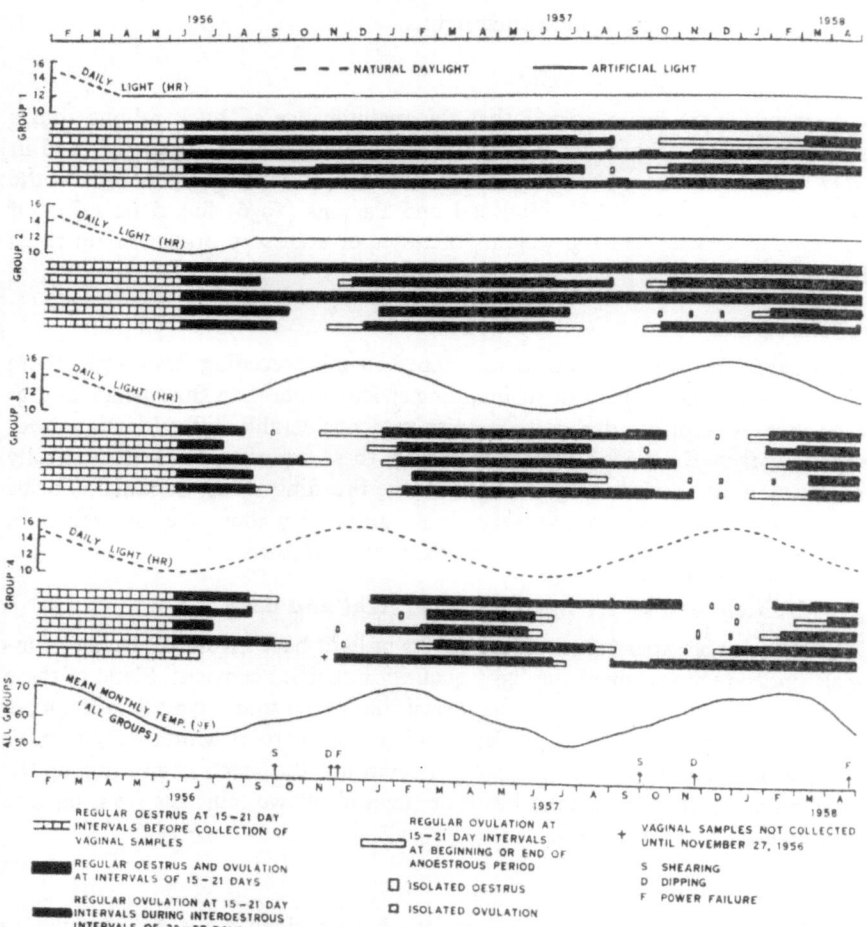

Fig. 20. Light treatment, ambient temperature variation and occurrence of oestrus and ovulation on individual ewes in various groups. From Radford (1961) *Aust. J. agric. Res.* 12: 139–53.

at this point to the curve of mean monthly temperature in figure 20. In the light of later work showing that temperature changes can induce oestrus in sheep (see page 117), it is possible that temperature was the stimulus which caused an apparently normal seasonal breeding rhythm to be maintained.

The effect of equatorial light regimes has been investigated on Southdown ewes by Thwaites (1965), who found that this British breed completely lost its seasonal breeding pattern after one year. The oestrous activity in individual ewes became sporadic and unrelated to any external factor. Although breeding in the experimental group occurred at any month of the year, its intensity at any one month was greatly reduced. These results should be compared with the data of Anderson, J. (1964) on the reproduction of British breeds in Kenya (see page 65).

Experiments by Means, Andrews, and Fontaine (1959) in Indiana were carried out with the following light regimes and results:

Period of experiment	Experimental regime	Result
22 March–12 July (daylight increasing)	10 hrs light : 14 hrs dark	Controls: no oestrus Experimentals: 10/19 showed oestrus
15 March–15 July (repeat experiment)	10 hrs light : 14 hrs dark	Controls: no oestrus Experimentals: 17/20 showed oestrus
27 July–9 Sept. (daylight decreasing)	11 hrs light : 13 hrs dark	Both groups showed same timing of oestrus Controls: 6/8 conceived at first heat c.f. 4/20 of experimentals
15 July–15 Sept. (repeat experiment)	10 hrs light : 14 hrs dark	As above, but even lower conceptions at first oestrus in experimentals

These experiments show that long periods of darkness can induce oestrus during the first half of the year. It is significant that they were carried out under temperature-controlled circumstances where the total variation was from 70° to 80° F.

Comparison of these three papers shows, first, that there seem to be differences between breeds in their susceptibility to fixed light regimes. If the possible effect of temperature can be ignored in Radford's paper it would seem that the Merino may be able to maintain a seasonal cycle under fixed light conditions while the Southdown cannot. Robinson's hypothesis (1950, 1951) is relevant here with regard to the control of breeding (see page 53). His second or third suggestion may explain this difference between the two breeds. A very slight change in light would be sufficient for Merinos to initiate the cessation of breeding in Robinson's second group for the first spring, but the Southdown would need a greater variation in the light stimulus, and so the constant light regime very quickly caused a breakdown in the seasonal pattern of breeding in Thwaites' experiment. Robinson's hypothesis, however, does not explain why Means, Andrews, and Fontaine found that the first heat of ewes subjected to reduced light at the beginning of the breeding season was infertile. This is usually a silent heat, when ovulation but no behavioural oestrus occurs.

Much longer dark periods were used by Clegg, Cole, and Ganong (1965), who maintained Suffolk and Hampshire ewes for 3 years on a 6-hours light, 18-hours dark regime, and compared the incidence of oestrus with a control group. Reproductive activity was prolonged after the first year in experimental ewes and occurred frequently during the California anoestrous season (from April to July). These results do not support the contention that diminution of light acts as a stimulus for oestrus, but suggest that changes in the light: dark ratio can override any 'inherent biological rhythm'. The ewes in this experiment were entirely free from the influence of external light changes, but no mention is made of temperature in the experimental rooms, so that again the results have to be considered suspect, as such changes may have contributed to the maintenance of a breeding season. Fraser and Laing (1966) put ewes for either 2 or 4 weeks on 7 hours light: 7 hours dark after late July. The later group came into oestrus earlier.

It could be predicted that the time in the breeding season at which sheep are put on to fixed light regimes may affect the subsequent changes induced by the altered regime. With the exception of the incidental observation in Radford's paper (see figure 20), there have been no experiments of this nature with regard to putting ewes on to a constant single period of light per day. Hart (1950) placed two groups of Suffolk ewes in May 1948 on first either 4L : 2D : 4L : 14D (group I) or 4L : 8D : 4L : 8D (group II) for every 24 hours, the first light period always commencing at 0900 hours. In 1949 two more groups (group III, just lambed and suckling, and group IV, just lambed but lost lambs) were placed under the same regime as group II, but starting in March. All ewes came into oestrus as a result of these treatments, but at very varying times.

(Days after commencement of treatments)

	First oestrus induced	Last oestrus induced	Mean date of induced oestrus
Group I	8	50	22
Group II	15	57	30
Group III	35	134	91
Group IV	82	146	117

Control ewes in 1949 had a mean 'time to oestrus' of 272 days after the light had been altered for the experimental animals. Groups I and II were significantly faster in the time to mean oestrus than groups III and IV. Hart concluded that: (1) a gradually decreasing plane of light and an increasing plane of darkness is not an essential factor for stimulating the onset of oestrus in sheep; (2) a standard and regularly maintained rhythm of short-light and long-dark will stimulate the onset of oestrus, using these terms in a relative sense only, as their significance is solely a means of supplying the necessary contrast stimulus to the pituitary gland; and (3) a ratio of 1 part of light to 2 parts or more of dark is sufficient to supply the contrast effect. The difference between the short latency of groups I and II, and the long latency of the other two groups, were explained as being due to the first two groups having their experimental treatments starting after a short period of anoestrum (as the breeding season had only just finished), whereas in the later two the treatment started at the end of the long pregnancy anoestrum.

The effect of breaking the light regime into more than one period per 24 hours

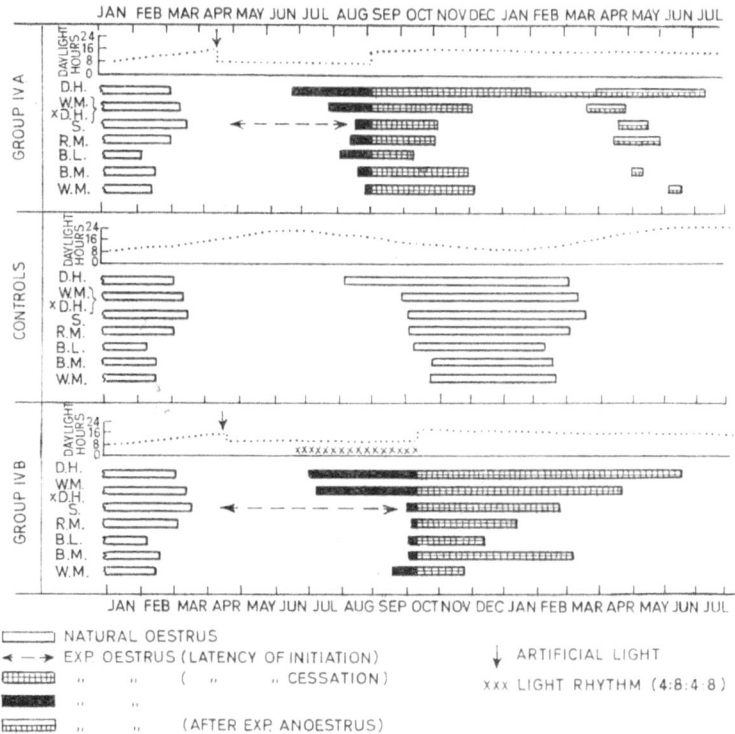

Fig. 21. The modification of the breeding season of different breeds of sheep following artificial light treatments. The breeding seasons are indicated by the length of the bands under the light-treatment curves. From Hafez (1952c) *J. agric. Sci., Camb.* **42**: 232–65.

has also been investigated by Hafez (1952c). The following British breeds were placed under fixed light regimes beginning 15 April: Blackface Mountain, Border Leicester, Dorset Horn, Romney Marsh, Suffolk, Welsh Mountain, and Welsh Mountain by Dorset Horn crossbred. The light regimes and results of Hafez's experiment are shown in figure 21. The onset of the breeding under 8L : 16D was advanced by 57 days over controls, whereas the onset of oestrus of the group first under this regime and then on 4L : 8D : 4L : 8D was speeded up only by an average of 27 days. Long-day treatment (16L : 8D) hastened the cessation of the breeding season by an average period of 15 weeks. Hafez concludes from these experiments that there is no need for a gradual change in light to initiate oestrus experimentally in sheep, but that it can be done by putting sheep on a fixed short-light regime. Hafez considered that his results were compatible with only one of Robinson's (see page 53) three alternatives: namely, that different breeds have a curve of pituitary activity of the same shape but require different levels of activity for the initiation of the breeding season. He also suggests that there may be two thresholds of pituitary activity, one for the initiation of ovulation and one for the onset of breeding.

There can thus be no doubt that light regimes do not necessarily have to decrease slowly to induce oestrus in sheep. Changes to fixed periods of light, when the light period is shorter than the dark, have been shown by a number of workers to stimulate the onset of sexual behaviour. Unfortunately, most of the experiments on sheep use oestrus as the determined reproductive parameter, and there is little information on the ability to conceive at the induced oestrus. There is some evidence (for example, see Means, Andrews, and Fontaine, 1959, and Hart, 1950) that oestrus can be detected by vasectomized rams, and yet a much reduced fertility can occur in experimentally lighted ewes. It would thus be of considerable interest to compare the actual pregnancy rates of ewes with heats induced by gradual, as opposed to sudden, changes in light regimes. It has been pointed out before that some experiments which purport to show that constant light regimes do not interfere with the seasonality of sheep reproduction must now be considered suspect in the light of results obtained by altering temperature (see page 117). Also, some of the research reviewed above should be re-evaluated in the light of recent work on the behavioural importance of social stimulation in inducing oestrus or anoestrus (see page 255). In some of the experiments described, either control and experimental ewes were run together or the use and timing of the introduction of rams to test for oestrus could have affected the results.

(b) Ferret and Mink (*Mustela putorius*) (*M. vison*)

In one of his early experiments, Bissonnette (1935b) put three young male ferrets on $8\frac{1}{2}$L : $15\frac{1}{2}$D from 10 November and tested their reactions towards females after 13 January. In two males the onset of sexual activity was markedly delayed by this treatment but only slightly in the third. Hammond Jr. (1951) placed female mink and ferrets under $3\frac{1}{2}$L : 5D : $3\frac{1}{2}$L : 12D in winter. His experimental animals came into oestrus ahead of controls, which were under increasing light. The mink came into heat ahead of the ferrets, being apparently more sensitive to the broken light treatment. Other female ferrets on short-day lighting, starting in summer, came into oestrus at times related to the period when treatments started, but with a delayed response. From these results, and others showing that ferrets kept on a constant 16L : 8D came into oestrus much sooner than ferrets on 24 hours of light, Hammond suggested that darkness as well as light may be of importance in the regulation of ferret breeding. He postulates that the spring breeding season is a 'short-day' phenomenon, as spermatogenesis can start in December, an idea which seems to be rather at variance with the general concepts of photoperiodic control in this species.

In two series of well-organized experiments Hart (1951) has done much to clarify the effects of light on ferret reproduction. In the first, groups of females were placed under the following treatments:

(1) Normal daylight plus one hour of artificial light at midnight.
(2) As for (1), but duration of midnight lighting gradually increased.
(3) Periods same as for (2), but added all at end of the natural day so that the night was not broken into two periods.
(4) Controls.

Treatments were all started on 5 December, and all experimental animals came into an early oestrus between 25 January and 9 February (50–67 days) whereas controls did not come into heat until 30 March (117 days). It is significant that group (1) came into oestrus on only $11\frac{1}{3}$ hours of light, while groups (2) and (3) came on heat at 18 hours. This indicates very clearly that in the ferret the total quantity of light is not so important as the sequence of long-light and short-dark. This worker also suggests that not enough emphasis has been placed on the significance of the short-dark period. Hart postulated 'that the stimulating factor may be the change-over from long-dark to short-light, or short-dark to long-light in much the same way that nerve muscle is stimulated at the make and break of the current' (page 7). There seems little doubt that in ferrets: (a) the total daily quantity of light is not the controlling factor, but the sequence of light and dark is very important; (b) a gradually increasing plane of light (or decreasing darkness) is not essential in itself, but is merely the means whereby the change-over from a short-light : long-dark sequence to a short-dark : long-light sequence is effected; and (c) these terms are merely relative, being defined from the animals' point of view, by contrast with each other and not in terms of total hours. The reader is referred to Hart's similar conclusions regarding the sheep (page 90) and also to the comments by Yeates (1949) on the total quantity of light not being important in sheep reproduction (page 74).

In a second series of experiments, Hart initially put groups of females on 30 October under the following three regimes:

(1) Natural daylight and constant artificial light.
(2) 8L : 4D : 8L : 4D.
(3) 16L : 8D.

The result from group (1) was a latent period to mean oestrus of 16·6 weeks, and this has been discussed on page 83. Control females had a latent period of 23·5 weeks, and those in (2) and (3) had latent periods of 11·25 and 7·67 weeks respectively. The latter treatments resulted in a very significant acceleration of oestrus. Because the onset of oestrus occurred even faster under (3) than (2) there is no 'tetanus' effect resulting from more than one change-over stimulus in a single 24-hour period. After oestrus was observed in the three experimental groups they were reconstituted under the following two regimes: (4) 4L : 20D and (5) 8L : 16D. The latent period to last oestrus was 3·91 weeks for (4) and 6·3 weeks for (5). In a final reconstruction a group on

$$4L : 2D : 4L : 2D : 4L : 2D : 4L : 2D$$

had a latent period to oestrus of 5·5 weeks. This very rapid change-over had the most stimulating effect on the pituitary. Hart suggests that it may be possible to reduce this even more if the pituitary is really sensitive to the contrast between light and dark and not to the total quantity of light.

Hammond Jr. (1953) has criticized Hart's paper and has found that ferrets will come into oestrus out of season if given only 7 hours illumination divided over a 12-hour period daily or even if given as little as 4 hours when divided over a 14-hour period daily. The further details of this experiment are very difficult to comprehend from the information presented.

(c) Horse (Equus caballus)

Korean mares under 3 to 5 hours of light per day (beginning in mid-November) exhibited a delay in onset of ovarian activity. However, breeding was only delayed and not inhibited in any major fashion (Nishikawa, Sugie, and Harada, 1952). Stallions given short-day treatment starting in early April (when the volume of semen is normally increasing) showed a decline in semen quality. By June and July the semen was like that found in the non-breeding season (Nishikawa and Horie, 1952). Using only a single stallion, Lintvareva (1955) found that the decreased light in a stable affected breeding. If the stallion was kept in the stable except for 1–1½ hours per day, or continually during daylight, the semen quality dropped to low levels.

(d) Goat (Capra hircus)

Nannies in Japan were placed on a fixed daylight length in July equal to October daylength. The majority of females came into heat some 70–80 days later and conceptions occurred (Yoshioka, Awasara, and Suzuki, 1952). Maintenance at long daylengths equal to the daylength at the summer solstice delayed the onset of oestrus, but did not prevent it occurring.

(e) Rodents

The earliest experiments of maintaining a mammal on a fixed light regime were those of Baker and Ranson (1932) on the field vole (Microtus agrestis). In a simple yet elegant experiment, the like of which has still not been carried out on many mammals, they kept two groups of voles, one on 15L : 9D and the other on 9L : 15D. Reproduction was normal in the former but almost completely ceased in the latter group, although both were given the same food and left at approximately the same summer temperature.

Very different results were obtained for Microtus orcadensis (now considered to be a M. arvalis subspecies) by Marshall and Wilkinson (1956). They found no significant differences in the reproduction of two groups of these voles kept on 6L : 18D or 15L : 9D when given abundant food at summer temperatures, and also found that reproduction did not decrease under the same regime but at winter temperatures.

Lecyk (1962a, 1962b), as a part of experiments already mentioned (page 78), kept *M. arvalis* on 16L : 8D from 21 September until 24 December, and had induced oestrus in females and sexual development in males at termination. Normal controls on decreasing light showed no reproductive development. Voles on 7L : 17D from 24 December until 16 May showed no inhibition. This work was carried out using adult voles and showed that a fixed light regime could not prevent the initiation of oestrus. However, this result cannot be explained in a similar manner to the results of Hill and Parkes (1934) (page 87) on ferrets, because the vole experiment was started at the winter solstice, so that the animals had not experienced increasing light. As the temperature was fairly well controlled, increases cannot be invoked as causing the induction of breeding. Thus this result is very interesting, in that it shows that voles can still change from a non-breeding to breeding condition on a fixed light stimulus. It is of even greater significance when it is recalled that Lecyk (1963) was able completely to inhibit the onset of sexual maturation in immature voles by subjecting them to 7L : 17D starting from 10 May until 20 June (page 6). If some biological 'inborn' rhythm is involved this would seem to indicate that it needs to be learned!

The effect of keeping *Peromyscus maniculatus* on long days was investigated by Price (1966). It is difficult from this publication to determine the actual light regime used, but it would seem that long-day treatment prolonged the breeding season so that the deermice were still breeding until September, although no information is given about controls.

In experiments designed to investigate the role of the pineal gland in reproduction, Hoffman and Reiter (1965a, 1965b) placed male golden hamsters (*Mesocricetus auratus*) on 1L : 23D and induced atrophy of the testes. No atrophy was noted in groups of males kept under 16L : 8D. When the pineal gland was removed the dark-induced atrophy did not occur, and this result is explained in terms of a pineal hormone, melatonin, which can suppress gonadal activity.

(f) White-tailed deer (*Odocoileus virginianus*)

Bucks maintained on 16L : 8D from October until April over two winters rutted 2 weeks earlier than controls (French, McEwen Magruder, Rader, Long, and Swift, 1960). Antler growth and shedding were also found to occur earlier in the lighted animals.

(g) Cat (*Felis cattus*)

Laboratory cats in London were found normally to be in anoestrus from October to December (Scott and Lloyd-Jacob, 1959). However, 9 out of 10 cats placed on a 12L : 12D light regime in September came into oestrus in late November and December, compared with control cats which did not show oestrus until January.

(*h*) Discussion

With the exception of the sheep and ferret, it will be seen that few details are known about the effects of fixed light regimes on mammal breeding. As these regimes can induce early oestrus in some mammals, those experiments in which light quantities were gradually altered can now be seen to have affected breeding, because at some stage or other they achieved the conditions found to be effective in the fixed-light experiments. At the same time it should be emphasized that this is not necessarily equivalent to the attainment of a minimum (or maximum) quantity of light. There is much evidence from the papers reported above that the actual stimulus which induced breeding was independent of the total length of the light period. Indeed, it has been suggested that the amount of darkness is the crucial factor for two 'short-day' species (sheep, ferret). Perhaps suppression of anoestrus during the breeding season and not stimulation of oestrus during the non-breeding season needs the most explanation if we are further to understand the effects of light. At the present stage there is no unified picture of the definitive experimental light regime by which breeding seasons can be manipulated. Many of the experimental conclusions reached have been based on the exhibition of behavioural oestrus as an indication of ovarian activity. Evidence discussed above and below indicates that the two events are not necessarily synchronous, so future work on the effect of light regimes on mammal breeding should use more definite parameters as indicators of experimentally induced reproductive activity.

V. Investigations into different intensities and qualities of light

The previous four sections have described changes in reproduction when the length of the light period was altered. These manipulations have also involved changes in the type of light the animals have been subjected to, and this section will consider these experiments, which deliberately investigated these changes. It is appropriate first to mention the meaning of the light measurements used, then to look at some of the regimes imposed. Light strength is measured by reference to the output of a standard candle. One lumen is the flux emitted in unit solid angle by a uniform point source of one candlepower, i.e. it is the quantity of light emitted in unit time into such an angle by such a source. An intensity of illumination of 1 foot-candle corresponds to a flux of 1 lumen per square foot. In metric terms these parameters are 1 centimetre candle or phot and the intensity of illumination is 1 lumen per square centimetre = 1 metre candle = 1 lux (Daish, 1954). A final important fact to remember is the inverse-square law, which describes how the intensity of illumination of a surface is inversely proportional to the square of the distance from its light source.

The effects of irradiation with different wavelengths of light have been investigated by Marshall and Bowden (1934). They put pairs of female ferrets on to extra light in winter, each receiving light of a different wavelength. Ferrets on ultraviolet light (λ 3650) showed vulval swellings in early December, ferrets on green (λ 5200), red (λ 6500), or yellow (λ 5770) light showed vulval swellings in

February while ferrets on infrared (λ 7500), violet (λ 4368), or with no extra light at all showed their first vulval swellings in April or early May. Marshall and Bowden concluded that heat rays and light near the infrared are relatively inactive in stimulating ferrets. In a later paper Marshall and Bowden (1936) found females were most sensitive to ultraviolet irradiation and heat rays, and those of long wavelength had little accelerating effect on breeding. Lecyk (1962b) put two groups of voles (*Microtus arvalis*) on 16 hours per day of either violet-blue light (390–500 mμ) or orange-red (570–760 mμ) from November to January, well outside their normal breeding season. He found no significant differences between the breeding of these two groups (testis weights 190 g, cf. 187 g: proportion of females pregnant, 52·1 per cent, cf. 46·6 per cent).

The response of these two species can probably be considered as demonstrating a general mammal pattern, although it is a great pity that so few have been investigated in this regard, especially in the light of recent work on the penetrability of the brain to light (Van Brunt, Sheperd, Wall, Ganong, and Clegg, 1964). Breeding can be induced by all wavelengths of visible light, but not by infrared or heat-wave sections of the spectrum. The sensitivity of mammals to ultraviolet is of particular significance when daylength and latitudinal factors are considered, because daylength will be seen to operate on the cloudiest days, so that the only important factor is the actual time from sunrise to sunset. There appear to be no published data on the change in level of ultraviolet light with the exact times of sunrise and sunset.

The finding that most of the spectrum is effective in stimulating breeding means that experimental work using filament lamps is valid, although it is astonishing that there has been so little mention of the spectral emission of the lamps used. Hart (1951) did show that the fluorescent lamps he used had almost exactly the same percentages of total light in the different wavelengths as natural daylight. Radford (1961) found no differences in the onset of oestrus in ewes maintained entirely on light with a normal seasonal rhythm from filament bulbs over two years when compared with control ewes receiving natural daylight (see figure 20, page 88).

The intensity of light used in experiments altering its length has varied considerably from worker to worker. The table below gives some of the intensities used:

Type of lighting	Intensity (f.c.)	Experimental animal	Reference
100-W bulb	3–18	Sheep	Yeates, 1949
Natural, dull autumn day	10–25	,,	,,
Artificial	4–9	,,	Means et al., 1959
25-W bulb	7–24	Ferret	Bissonnette, 1935a
300-W bulb	7–18	,,	Bissonnette, 1935b
40-W fluorescent tube	22–24	,,	Hart, 1951
40-W fluorescent tube plus daylight	29–100	,,	,,

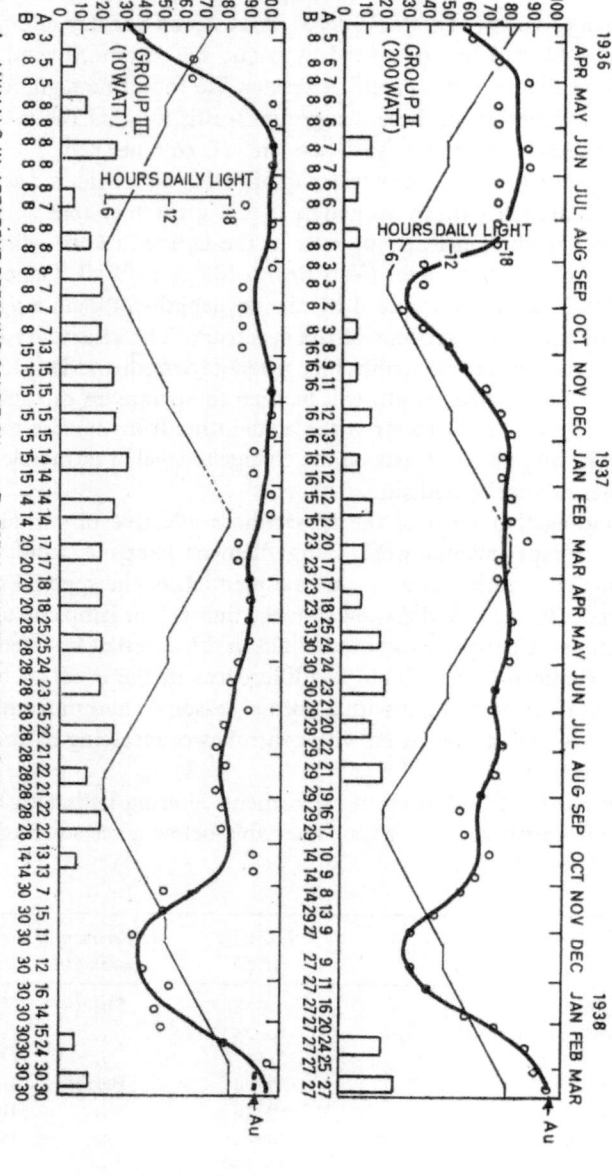

Fig. 22. The effect of variation in light intensity on the breeding season of white-footed mice (*Peromyscus leucopus*). From Whitaker, W. L. (1940) *J. exp. Zool.* **83**: 33–60.

The effects of changing the intensity were first investigated by Marshall, F. H. A. (1940). Pairs of female ferrets were placed at varying distances from a 1,000-W lamp and given extra light after sunset in winter. The distances used were 1, 2, 4, 6, 14, and 22 ft from the bulb, and females 6 ft and less from the light exhibited their first oestrus in December or earlier. Marshall's claim that the speed at which females came into heat was roughly related to light intensity is incorrect because of the inverse-square law, as the time to oestrus was actually related to the distance from the lamp.

During a study of the effects of light on breeding of *Peromyscus leucopus*, Whitaker (1940) put two groups of white-footed mice on to regimes where the daylength was changing in a seasonal fashion but at two levels of light intensity. His results are shown in figure 22. The group lit by a 200-W bulb had a light intensity of from 15 to 32 f.c. in the cages and that under the 10-W bulb had light intensities from less than 1 to 2 f.c. at the cages. The dimly lit mice showed no decrease in fertility while under very short daylength during the first year, but did show a decline after the same period during the second year. Whitaker comments that this species is nocturnal and most active when it is dark. Animals under the stronger lighting were seen less frequently in the lighted parts of their cages than were the dimly lit mice. This sort of comment points to a major deficiency in the work on light in mammals. As Lloyd and Weisz (1966) have pointed out, there has been no comparable work to that on birds where the individual has a choice of being lit or being in darkness. These free-choice experiments would give a great deal of information in mammals about daily rhythms of activity as well as being extremely informative as to how far the species can control its own light ration. The importance of the circadian clock in photoperiodic time measurement has been beautifully demonstrated for a bird species by Menaker and Eskin (1967), and similar types of experiments could easily be carried out on mammal species to determine if the Bünning hypothesis, which they were testing, is also applicable to mammalian photoperiodic control of reproduction.

Kellas (1955) pointed out that in the tropics although the length of daylight does not alter appreciably, the intensity of illumination (both in the visible and ultraviolet regions) does change and has two annual peaks in some regions. Seasonal breeding in tropical mammals could potentially be related to these changes in light intensity.

Because of its extreme variability (as any photographer is aware), it is difficult to evaluate the effect on reproduction of the natural daily variation in light intensity. Hart (1951) gives some figures for measurements of intensity out of doors at Cambridge, England, during March. On bright days the intensity was above the reading power of his meter (i.e. over 250 f.c.) but at dusk (about 1700 hours) he obtained readings of 135 f.c. and even when it was completely overcast at dusk the readings were still 125 f.c. Although daylight varies considerably in its intensity, the minimum intensity on even very dull and overcast days is still well above light intensities which have been found to be experimentally effective in altering the breeding of mammals.

D. THE ROLE OF AN INHERENT ANNUAL RHYTHM IN THE MAINTENANCE OF THE BREEDING SEASON

Since the observations of Marshall, F. H. A. (1937) as to the manner in which the breeding seasons gradually altered in mammals which had been transferred from one equator to another (see page 70), the concept of an annual inherent breeding rhythm has repeatedly arisen in the literature (Baker, 1938a; Baker and Baker, 1936; Bissonnette, 1935b; Bullough, 1951). The subject has been reviewed excellently by Amoroso and Marshall (1960) for both mammals and birds, but since that time some new evidence has been published which indicates that the validity of the hypothesis should be reassessed. This will be done by presenting the lines of evidence for and against the various reasons that such an inherent annual rhythm has been invoked. The reader is referred to Amoroso and Marshall's review for references to the positive lines of evidence.

The first line of evidence for the existence of an annual rhythm came from the observations of Marshall already mentioned. Animals transported across the Equator occasionally retained their original breeding season for a year or so, despite the fact that they were then subjected to a reversed seasonal light cycle (Marshall, F. H. A., 1936; Bedford and Marshall, 1942). It was suggested that there were two antagonistic tendencies at this time; one being the inherent rhythm which tended to keep the animals breeding in time with their original hemisphere light changes, and the second being the new photoperiodic conditions. The contrary suggestion has been made by Yeates (1954), who maintains that the situation can be explained without involving any inherent rhythm. He postulates that the actual rhythm of breeding which was induced by light changes in the original hemisphere simply had a certain refractoriness, and as it took some time to die out, the old season was maintained for a short time.

The second line of evidence for the existence of an inherent rhythm comes from experiments when animals have been maintained on continuous darkness or light or on fixed photoperiods for long periods of time. Under these conditions some workers have reported that seasonal breeding was maintained despite the lack of variable external stimuli, and have thus invoked the inherent annual rhythm to explain these observations. There are two contrary explanations for these observations. In the first place, for obvious logistic reasons, it is difficult to maintain mammals under these regimes for very long periods. However, as the unit under investigation here is the breeding season, the conditions must be maintained for a number of years, and the breeding rhythm demonstrated to remain over this long period, if a truly inherent rhythm is to be postulated. A majority of experimental work has kept animals under constant light conditions only for periods of less than three years, so that for at least half the experimental period the dying-out of a previous light-induced rhythm will affect the expression of seasonal breeding. Secondly, the effects of the variation of environmental stimuli other than light have often been ignored, so that the maintenance of a seasonal breeding can be more easily explained by the uncontrolled nature of these variables. For example, recent work on at least two species (hamster, *Mesocricetus auratus*, Hoffman, Hester, and Towns, 1965; sheep, Dutt and Bush, 1955) has shown that temperature changes can induce breeding in the absence of light variation. These observations invalidate much of the earlier work on light control. Another important variable often ignored by, or more correctly unknown to, the earlier workers was the role of social factors in inducing oestrus. Within certain limits, not only can the presence of other females on heat induce oestrus in non-cycling females when placed with them but in a number of species males can definitely synchronize and possibly induce heat in females. This situation forces re-evaluation of early experiments where experimental and control animals were housed together or where males were used to determine oestrus.

The third line of evidence for the existence of an inherent rhythm was the observation that certain animals which breed early in the spring become sexually quiescent before the summer solstice, i.e. before the days stop lengthening. This has been interpreted as indicating that the cessation of the breeding season is not controlled by the external light stimulus. This line of reasoning assumes that these mammals are entirely stimulated by increasing light. It is quite possible that the stimulus is actually a pattern of decreasing then increasing light. Of particular interest here is the suggestion of Radford (1961) that photoperiodicity may act as a stimulus, not to initiate the breeding season but to end it. He feels that the onset of anoestrus may be due to a temporary exhaustion or suppression of adenohypophyseal activity which is synchronized by the photic environment. Oestrous activity, once it begins, would thus continue to occur unless actually inhibited by light. After the anoestrous season, oestrous activity commences again independently of the daylight environment.

A fourth line of evidence is based almost entirely on observations on ferrets whose optic nerves have been sectioned, or who have been blinded in some other way. Amoroso and Marshall (1960) and Thorpe (1967) have pointed out that these experiments have produced very conflicting results (see also Farner, 1961), as some workers have reported oestrous cycles as occurring only during the normal breeding season, whereas others have reported no cycles at all. Again, conclusions from some of these experiments can be questioned, as there was no consideration in the experimental arrangements of environmental factors other than light.

The fifth and final line of evidence indicating an inherent annual cycle come again from ferrets, whereby it has been suggested that in laboratory colonies with no access to direct sunlight a definite breeding season is still maintained. The suggestion that light is actually constant in these circumstances can be largely discounted, especially as it is known how very low light levels can affect breeding (see page 99).

In actual fact, when the total literature available to this reviewer was listed it was immediately apparent that the presence of an inherent annual rhythm in mammals was almost entirely based on data collected from sheep and ferrets. Before assessing the theory as a whole, recent relevant work on these two species will be presented.

H

The longest investigation on ferret reproduction has been recently reported by Thorpe (1967). Figure 23 from his paper shows that if female ferrets were kept under continuous light for a period of over 5 years there was no demonstration of any inherent rhythm. After the first year oestrus was exhibited at almost any

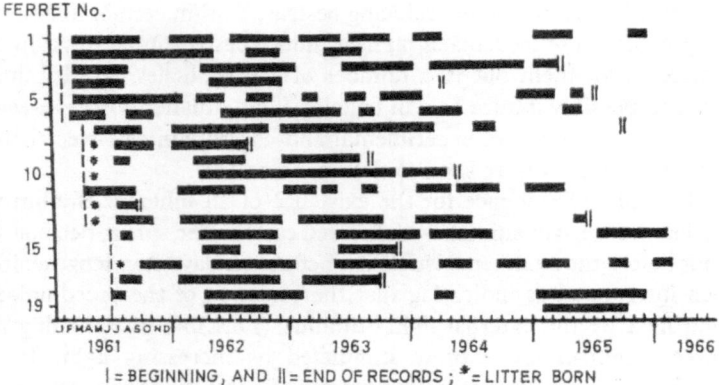

Fig. 23. The times of oestrus (black bars) in mature female ferrets (*Mustela putorius*) kept under continuous illumination. Single and double vertical strokes indicate beginning and end of records and asterisks indicate birth of a litter. From Thorpe (1967).

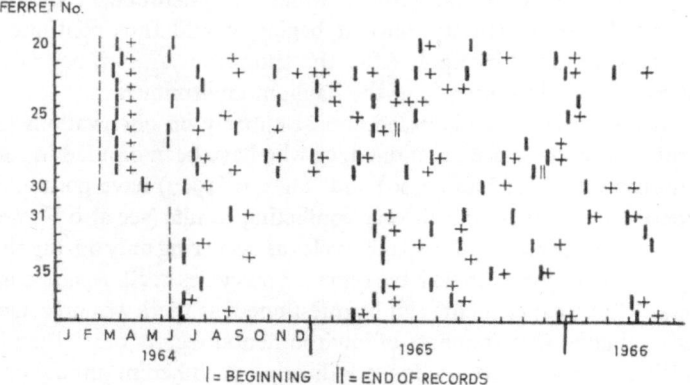

Fig. 24. The times of oestrus (thick vertical stroke) and mating (plus sign) in mature female ferrets (*Mustela putorius*) kept under conditions of continuous illumination. From Thorpe (1967).

time of the year. In another experiment females were again kept under continuous illumination for over 2 years, but were mated during each full oestrus. Figure 24 shows that if the experimental lighting was started after the normal breeding season there was evidence in the subsequent spring of a seasonal breeding, but this dropped away as the experiment was prolonged. Thorpe concludes that 'these observations provide no support for the proposition that there is an endogenous rhythm controlling the time of oestrus in the ferret' (page 65). He points out that there was often a prolonged transition in his experiments before the females exhibited oestrus which was entirely out of phase with the seasons. The length of this transition period depended on the time of the year at which the period of continuous lighting commenced. He further suggested that this may be the reason why previous ferret workers interpreted their data as demonstrating the presence of an inherent rhythm. Exactly the same reasoning can be applied to interpretations of similar short-term experiments on other mammals. Indeed, these results of Thorpe's suggest that at least 2 and possibly even 3 years must be considered a minimal time before the previous light-induced rhythm will be entirely lost.

For the sheep, mention has already been made of the relative nature of the anoestrous period in this species (see page 52). It may be that the fairly continuous action of the sheep ovary, especially in breeds such as the Merino (Robinson, 1950, 1951; Lamond, 1962), may mean that only a very small amount of photoperiodic stimulation is necessary to maintain a seasonal breeding activity. Yeates (1954) has shown that suitable light treatment can induce oestrus in ewes at any time of the year, and what is more, that the period of time from the beginning of the light treatment to the onset of oestrus does not vary significantly with season. This later observation suggests most strongly that the sheep has no inherent rhythm. Such a rhythm has been claimed to have been demonstrated recently by Wodzicka-Tomaszewska, Hutchinson, and Bennett (1967), who placed groups of 7 Southdown ewes and 7 Merino ewes on equatorial lighting and accentuated but reversed thermal seasons. The data for the first 18 months can be discounted for the reasons outlined above. After this time the workers claimed that experimental ewes showed the same season of breeding as 'controls' (on normal temperature but accentuated normal light). Their own data show that of the 7 Merino ewes used, 1 exhibited only a single oestrus in the last year of the experiment and another 3 were already cycling prior to the final breeding 'season'. For the Southdowns, although the seasons did coincide, the experimental ewes exhibited oestrus just following the decrease in ambient temperature below approximately 40° F – a temperature already known (Dutt and Bush, 1955) to induce oestrus anyhow. These points, together with the fact that the experiments were only run for 3 years plus another 2 when the authors themselves note 'that the rhythm may have been lost' (page 66), do not bear out the concept of an annual inherent rhythm in sheep breeding.

It is necessary, therefore, to reconsider whether or not such a rhythm actually exists in the light of reasonable counter-evidence to all the data which have been proposed as supporting its presence. There can be no doubt that the normal cycle of seasonal breeding does persist for a short period of time under altered environmental conditions, but the length of this time is so dependent on the timing of the commencement of the experimental regime that this merely suggests a seasonal pattern of refractoriness. Much of the remaining evidence purporting to demonstrate an inherent rhythm is based on small samples of animals with periods of oestrus whose distribution in time does not stand up to statistical analysis. It is therefore suggested that, to date, all the evidence presented can be more validly interpreted as demonstrating that the seasonal nature of breeding in mammals is entirely dependent on responses to external environmental stimuli and there is therefore no need to postulate any inherent annual rhythm in this group. The interested reader is referred to a discussion of the role of the inherent rhythm in breeding of ungulates in the recent book on their reproductive behaviour by Fraser (1968).

CHAPTER 10

The Effects of Temperature
on the Breeding Season

A. INTRODUCTION: THE SEPARATION OF CLIMATIC VARIABLES

The range over which any individual species of mammal occurs is governed by a large number of factors. However, if species with restricted ranges are excluded, the main factors which govern the geography of the limits of a species range are certainly climatic. These factors affect both the reproduction of many species of mammals and the recruitment of young animals into the adult breeding population, but each must be considered separately, as they are largely independent of each other. Climatic factors can alter the onset and cessation of breeding, either directly or by acting as modifying factors on a light-controlled breeding season. Secondly, climatic factors can have very considerable effects on the survival of young from the moment of parturition and indeed until they reach sexual maturity (see page 215). The selective action of climate will result in the evolution of breeding at the optimal time for survival and recruitment of young. Thus, climatic factors can regulate the breeding season both directly (i.e. as proximate factors, see Baker 1938a) through influencing rut and oestrus, and indirectly (as ultimate factors, see Baker 1938a) through survival of the young.

Climate, *per se*, cannot be evaluated as a single unit, so that it is measured by quantifying the various meteorological factors of which it consists. However, as far as the animals are concerned, all these factors act in a coordinated fashion, so that it is impossible, in most cases, to give the 'bio-climate' which affects individual mammals. We are thus forced to evaluate in an individual manner the effects (as far as reproduction is concerned) of individual factors such as temperature, rainfall, humidity, snow-cover, and wind, knowing full well that the animals do no such thing. For example, a musk ox in Greenland is subject to a combination of the cold temperature, wind velocity, and type of precipitation, and not by any one of these things divorced from each other. Therefore, many of the examples to be discussed below will cross-reference from one climatic factor to another and will be classified for purposes of discussion under that environmental factor which the writer considers the most crucial with regard to reproduction.

There is another very important point to make before the main part of this chapter is entered. Climatic factors control the food supply of all mammals by altering the availability of vegetation or of live prey. Also the level of past nutrition, as reflected in the present physiological condition of an individual, has a crucial role in determining the magnitude of effect climate will have on the individual. Therefore it is also somewhat academic to separate climate from nutrition, but again this must be done to systematize the presentation of information. The reader is warned!

To demonstrate the difficulties in separating the units of climatic phenomena,

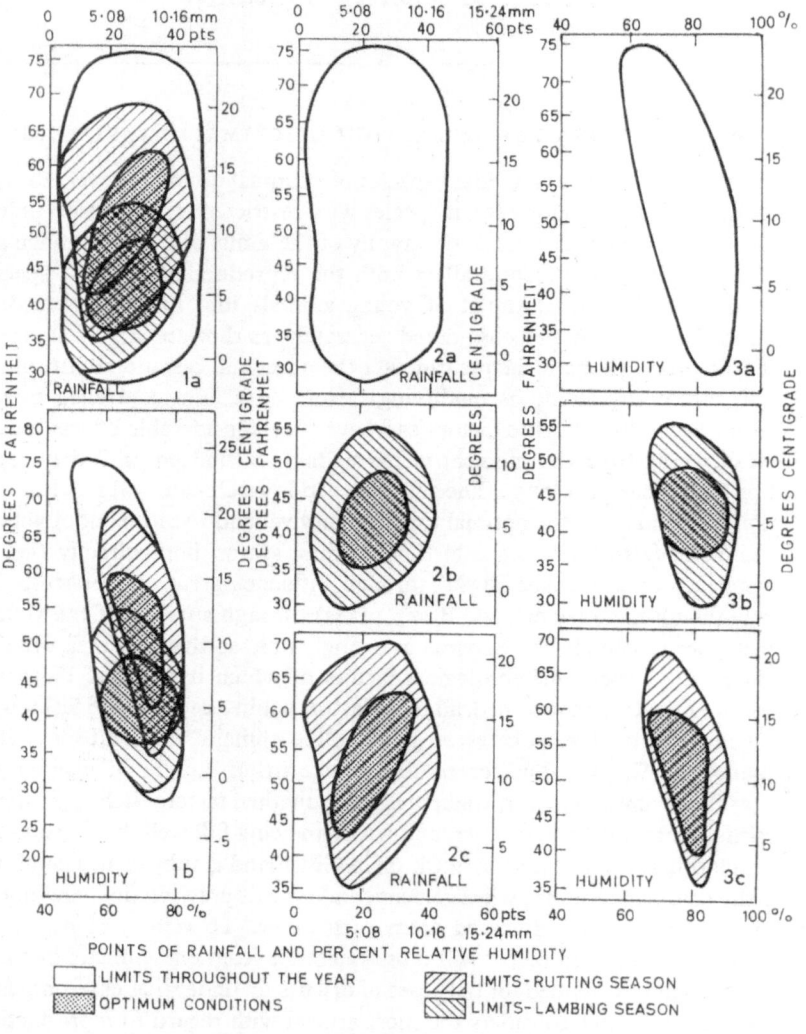

Fig. 25. (1a) and (1b), composite hythergraph and climograph of the dense areas of sheep population in the world; (2a), (2b), and (2c), hythergraphs of total range and range for critical periods: (3a), (3b), and (3c), climographs of total range and range for critical periods. From Johnson (1924) *J. agric. Res.* **29**: 491–500.

the classic paper by Johnson (1924) on the relationship between climate and sheep distribution will be discussed. Figure 25 shows that Johnson constructed hythergraphs (temperature compared with rainfall) and climographs (temperature compared with humidity) for those areas of the world where sheep commonly occur. The ovals on these graphs were made by joining the outer point of records from a number of meteorological stations where sheep occur. The graphs show that the two sensitive seasons, lambing and rutting, are relatively restricted when compared with the total yearly spread of the climatic conditions where sheep are found. The first few weeks are the most vulnerable period in the life history of a lamb, so that climatic conditions at this time are crucial (page 216). However,

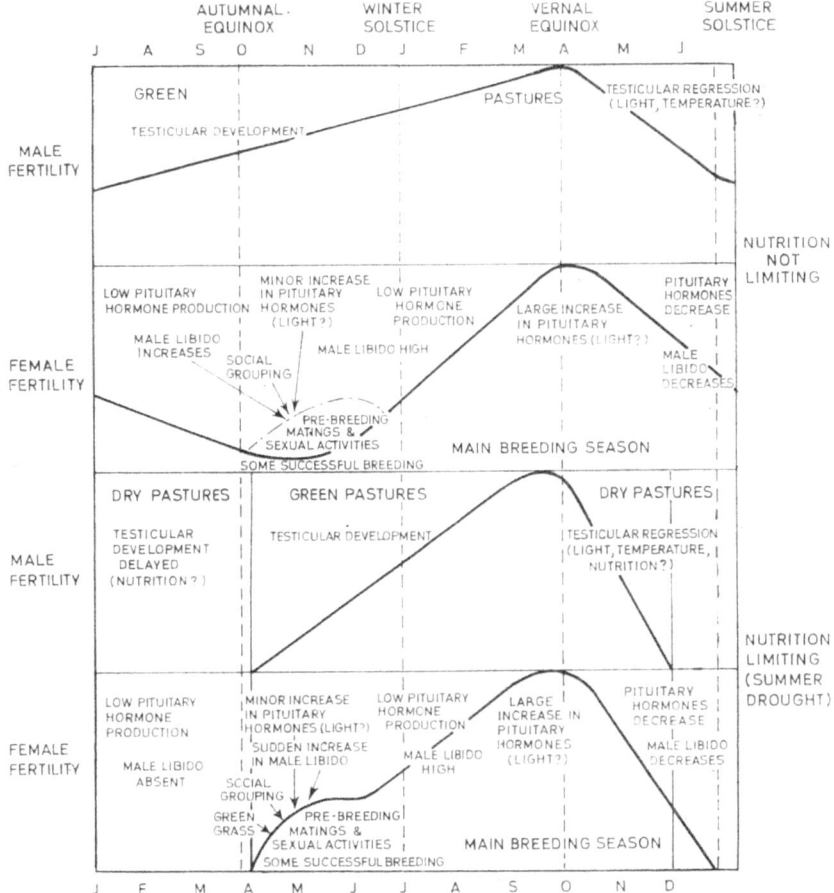

Fig. 26. Diagram of the main factors which appear to determine the breeding season of the wild rabbit (*Oryctolagus cuniculus*) in climates similar to England and parts of New Zealand, where green feed is present all the year and in Southern Australia with a well-defined summer drought climate. From Myers and Poole (1962) *Aust. J. Zool.* **10**: 225–67.

the main point from this example is that in this fairly typical domestic mammal the effects of two climatic variables are impossible to separate. The complexity of the interaction of climatic variables, nutritional changes, and the variation in light regimes is well illustrated in the discussion by Myers and Poole (1962) of the factors which control the breeding season and fertility of the wild rabbit (*Oryctolagus cuniculus*). They compare breeding in Australia, New Zealand, and Great Britain and conclude that the observations of workers in different parts of the world can be satisfied by postulating an annual cycle in male fertility correlated with the solar cycle (with a single peak in spring), a bimodal annual cycle in female fertility (with a minor peak in the autumn and a major peak in the spring), and the superimposed effects of nutritional changes in pastures under varying rainfall and climatic regimes. Figure 26 indicates Myers' and Poole's ideas on how these various factors interact and indicates that they consider nutritional changes of major importance. In this species, one of the most studied wild mammals, it has proved unrealistic to suppose that any single environmental variable is exhibiting complete control over the reproductive pattern, and it is better to consider that the general pattern of breeding is affected by a number of different factors acting at different times. The breeding seasons of hares (*Lepus europaeus*) also show a similar relationship to various environmental variables (Flux, 1965b).

Similar conclusions were reached by Lancaster and Lee (1965) when describing the environmental variables which affect the breeding of monkeys. They present data collected by Mizuhara from four populations of the Japanese macaque (*Macaca fuscata*) and compare environmental variables for the two months preceding the onset of conceptions:

Area	Date of onset of conceptions	Latitude	Daylength	Mean temperature (° F)	Monthly rainfall	Seasonal state of food
Shodoshima	Mid-Sept.	34° N	Decreasing	Down 5°	Steady	Beginning of fruiting season
Takago-yama	Mid-Oct.	35° N	Decreasing	Down 17°	Up 2 in	Fruiting season
Takasakiyama	Mid-Nov.	33° N	Decreasing	Down 21°	Down 5 in	Fruiting season
Koshima and Toino-misaki	Mid-Dec.	31° N	Decreasing	Down 21°	Down 2 in	End of fruiting season

A number of variables were common to the period preceding conception, but local differences in temperature and daylength are not enough to explain the fact that macaque populations on Honshu island (Shodoshima) start breeding up to four months earlier than populations on Kyushu (Koshima and Toino-misaki). The same authors then considered the environmental variations which occurred two months before the onset of conceptions in rhesus and langur monkeys in three areas in India. These species start breeding during and after a peak in rainfall, in the second half of the year, but the pattern was extremely variable with geographical locality. Kawai, Azuma, and Yoshiba (1967) analysed the birth dates of 25 troops of Japanese macaques and similarly concluded that 'the effect of environmental factors are complex and not always direct, and that the threshold value of each factor in causing the onset of the copulatory season might vary by troops' (page 35). All in all, the breeding pattern of the various species of *Macaca* were related to a number of environmental situations, and it would seem unrealistic to attempt to define any single variable as the primary cause of the timing of the onset of conceptions.

B. THE PRESENCE OF WINTER BREEDING

It has already been pointed out that before the importance of light changes was realized in the determination of the onset of breeding, it was generally considered that temperature was the main controlling factor in mammalian and bird reproduction. This was probably based on the many observations that, as the breeding commenced, there seems to be a correlation in temperate climates between the temperature immediately prior to breeding and the speed at which breeding commenced. Detailed work in the thirties, however, failed to produce any convincing experimental evidence. Baker (1930), studying the wood mouse, *Apodemus sylvaticus*, and the wood vole, *Clethrionomys glareolus*, in England between 1925 and 1928, failed to find any correlation between the presence of winter breeding in these species and the winter temperatures. Newson, R. (1963) found winter breeding occurred in two locations in Oxford over 1958–9 but at very different intensities. He could not explain this difference by temperature, or for that matter, nutritional or any other environmental variation, because the locations were only half a mile apart in similar vegetation. Similarly, Hamilton (1941) failed to find any correlation between winter breeding and average winter temperatures in *Microtus pennsylvanicus* in New York State.

It is now suspected that winter breeding in this type of rodent is actually dependent on other factors, including their stage in a population cycle (Krebs, 1963, 1966), the concurrent quality of the population (Chitty, 1962), and also the amount of food available during the late autumn and early winter (Smyth, 1966; Zejda, 1964). The absence of any correlation between breeding and temperature in these early studies may thus be fortuitous.

Other workers have suggested that the presence of winter breeding in non-cyclic rodents is governed by the temperature. During an outbreak of feral house-mice (*Mus musculus*) in California, Pearson (1963) postulated that the unusually large numbers of mice were due to the reproduction in winter and early spring. Whereas in previous years there was no breeding from December to April, between 1959 and 1960 many young were born inside this period. Normally the male mice are in breeding condition at this time, although females are not. Pearson has suggested that an exceptionally warm period during the latter part of 1959 may have been the cause of the extended breeding season. He presents information from three other house-mouse outbreaks in California: Kearn County, 1926; Davis County, 1926; and Davis County, 1941. On each occasion the mice reached a peak in autumn, and all were preceded by two years of un-usually favourable temperature conditions. Warm temperatures are thus ap-parently sufficient, but not necessary for winter breeding in the house mouse. There is no evidence that the house mouse has a regular cycle of abundance, but it is a species which is known to react very quickly to a favourable food supply, so it may be that the weather was only the primary agent in stimulating good crops, to which the mice then reacted. In a small study of the deermouse *Peromyscus maniculatus* in Kansas, Brown, H. L. (1945) found females pregnant right through the winter of 1944–5. Although the incidence of pregnancy was low: September, 54 per cent; October, 25 per cent; November, 14 per cent; December, 5 per cent; January, 4 per cent; February, 19 per cent, and March, 35 per cent, this had never been previously reported from Kansas. Brown attributes the breeding to a very mild winter where the minimum never dropped below $+4°$ F compared with the average lowest minimum for the previous five years of $-11°$ F. This conclusion suffers, however, from the usual difficulties of a single year's observations. Similarly, Zejda (1962) described how winter breed-ing was less intense during periods of extreme cold in *Clethrionomys glareolus* in Czecho-Slovakia. However, the direct factor affecting breeding intensity here was food supply, as there were local differences, depending upon the type of tree in the stand where the mice were captured. In this case temperature was not able to inhibit reproduction of these voles completely. These observations are of particular interest, because this species does not usually exhibit winter breeding. On the other hand, the extremely severe winter of 1959–60 in Oklahoma com-pletely stopped breeding in the autumn and winter in populations of the cotton rat *Sigmodon hispidus*, according to Goertz (1965). He found no pregnancies in 423 females examined at this time. In the following year when winter tempera-tures were more normal, pregnancies were found in all months except December and January. Although the density of cotton rats was very high in the former winter, Goertz suggests that extreme temperatures were the main inhibiting factor.

The effect of extreme cold was reported in a fascinating paper by Schiller (1956) on feral rats (*Rattus norvegicus*) living on garbage dumps outside Nome, Alaska. These animals were at 64° 30′ N and at the most northerly part of their range. This species normally breeds all the year (see Davis, 1953), but under these exceptional conditions reproduction ceased during the winter. During December and January there were very few or no pregnancies, and spermatogenesis stopped, probably due to the testis withdrawing into the body cavity. It is interesting that this population of rats maintains itself in this very extreme environment, the severity of which is indicated by the observation that over 60 per cent of rats have severe frostbite in the late winter.

Finally, temperature is reported by Rice (1957) to control the breeding of a bat (*Myotis austroriparius*). In Western Florida these bats hibernate from September until March in an area where the temperature is cold enough to result in fairly frequent freezing of water and where insect food is minimal during the winter. The epidydimes of male bats in Western Florida are enlarged from October to December, suggestive of mating at this time. However, a population of bats in peninsular Florida do not hibernate at all, as the winters are much milder and frosts do not occur in this area. These bats have enlarged epidydimes from February to April, which suggests that they are mating at a very different time of the year. Rice's observations are particularly interesting, as the two populations that he studied were at almost the same latitude.

C. THE EFFECTS OF TEMPERATURE ON BREEDING

Temperature has been suggested as having a modifying effect on the onset of breeding in a number of temperate-zone mammals. Generally it can be looked upon as giving the final extra stimulus to the onset of mating after photoperiodic stimuli have brought individuals up to breeding condition. For example, a number of workers on mammalian reproduction have faithfully and repeatedly referred to the colloquial evidence that the onset of roaring in the red deer (*Cervus elaphus*) stag depends on temperature (Marshall, F. H. A., 1937; Amoroso and Marshall, 1960; Bullough, 1951). It was supposed that a sudden fall in temperature was necessary to start male mating behaviour, but a modern red-deer worker (Lowe, *pers comm.*) has found no evidence for this during a long-term study on the 'strange island' of Rhum, Scotland, and Marshall, F. H. A. (1942) was unable to find any convincing data to support the suggestion. There is better evidence that a decrease in temperature may act to stimulate libido in the ram (Rice, 1942; Amoroso and Marshall, 1960; Yeates, 1965). A reverse effect of increasing temperature is reported (Einarson, 1956) to stimulate the onset of rut in mule deer (*Odocoileus hemionus*).

In other groups of mammals there is better field evidence for an effect of temperature on the onset of breeding. In a major paper on the ground squirrel (*Citellus tridecemlineatus*) Wells and Zalesky (1940) found that a decreasing temperature was correlated with an increased gonadotrophic potency of pituitaries from wild populations and that as the temperature rose this trend was reversed. This is an hibernating species in which the males undergo spermatogenesis during the winter and in early spring. Experimental work on the species is described on page 115. Temperature increases probably act to stimulate the onset of breeding in the Arctic ground squirrel (*C. undulatus*). This species, which has probably the shortest breeding season of any mammal, emerges from hibernation in Alaska in early May with its testes showing spermatogenesis. Sperm production and mating take place a week after emergence and last for precisely two weeks only, after which time the testes collapse, becoming abdominal by the end of June (Mitchell, O. G., 1959). Temperature increases are suggested by McKeever (1966, 1963) to be the cause of cessation of breeding in two species of ground squirrels in California. *C. beldingi* ceases reproduction when the high summer temperatures cause a drying of the vegetation and then immediately goes into aestivation followed by hibernation. *C. lateralis* reproduces for a longer period and changes from eating green vegetation to bulbs and tree seeds. When the temperature begins to drop this species enters hibernation.

In domestic mink (*Mustela vison*) kept in unheated farm pens the actual onset of first oestrus and mating was effected by temperature (Hanson, 1947). In the warm year of 1939 the mean onset of oestrus was 10 March, whereas in the colder 1940 it was delayed to 14 March. These dates are the means of very large samples. Increased temperatures can also affect the length of gestation in the mink by decreasing the period of delayed implantation (see also page 205). In Manitoba the initiation of breeding in muskrat (*Ondatra zibethicus*) depends on a rise in temperature (Olsen, 1959). Back-dating from birth dates to mating showed that many conceptions took place on the night of 7 May 1956. This night had been noted as one of unusual noise and movement by trappers and was also exceptionally warm with the temperature dropping only to 57° F, with fog and light rain. In 1955 there was a much warmer but earlier spring which did not result in an earlier onset of breeding, however; probably because the muskrats were not sufficiently primed by other stimuli (such as light increases) and by the onset of warm weather. Temperature still had an important effect because the early spring warmth was followed by a cold spell, which was in turn followed by a very warm night (min. tem., 56° F) on 1 May 1955. This again was the date on which mating started. McLeod and Bondar (1952) correlated the dates of conception in the same species with temperature variations as indicated by the break-up of pond ice. Detailed observations such as these have been carried out on very few species of wild animals, although there are a number of colloquial reports of this type.

There is some evidence that the breeding of grey seals, *Halichoerus grypus*, may be related to temperature. Davies (1957) has pointed out that the three populations of this species show a rather peculiar pattern of breeding. Those in the western Atlantic (around the Gulf of St Lawrence and the northern United States) and those in the Baltic Sea give birth and mate in the late winter and early spring, i.e. February and March (see also Cameron, 1967). The third population of the eastern Atlantic (Great Britain, Shetlands, Faeroes, Norway, and northern Russia) breed over a longer period, but the main pupping is in September and October. In the western Atlantic and Baltic populations breeding follows the extensive freeze-up of the seas at this time, a factor which may help pup survival. In Davies's own words:

'It is difficult to resist the conclusion that there is some connection between the late winter and early spring freeze-up in the Baltic and the Gulf of St Lawrence and the late winter and early spring breeding in those regions. The suggestion that the breeding season there is an adaption to life in contact with ice has been made but no explanation seems to have been offered. In neither region does ice prevent breeding on rocks or sandbanks during the autumn months. If there is any adaptive significance in the different breeding season, then it is probably connected with the mortality among the young in the first few months of life. Mortality during this time is normally heavy. It would be considerably increased if the young, born in autumn, had almost immediately to contend with an ice-covered sea. Calves born later on the top of the ice would be at an advantage and survival rates would be considerably increased if they took to the seas with the melting of the ice in spring' (page 305).

Similar suggestions have been made for four other species of seal in the Sea of Okhotsh (Fedoseev, 1965).

A possible effect of temperature on breeding termination was suggested by Ashby (1967) for *Clethrionomys glareolus* and *Apodemus sylvaticus* in northern England. In a twelve-year study he found that these species bred later in years when the summer temperatures stayed high well into late summer and autumn. In years with cooler summers breeding terminated earlier. This was a constant relationship with the exception of one 'crash' year in a *Clethrionomys* cycle. Ashby suggested that the effect of temperature was mediated through the food supply, as more food was available in the warmer summers and the mice and voles reflected this in their physical condition.

In a paper reporting the variation in the time of onset of breeding in Clun Forest ewes (*Ovis aries*), Lees (1966) found highly significant relations between the onset of oestrus and the mean ambient temperature during midsummer (for day numbering, see page 63):

		Mean day of first service		
	Mean temperature 16 July–5 Aug. (X) (° F)	7 ewes common to 1962–5 (Y_1)	18 ewes common to 1962–4 (Y_2)	13 ewes common to 1963–4 (Y_3)
1962	57·98	232·4	230·8	—
1963	60·43	247·1	246·2	245·5
1964	59·46	241·7	240·6	241·5
1965	56·60	227·3	—	225·5
		$rXY_1 = 0.9945$	$rXY_2 = 0.9993$	$rXY_3 = 0.9984$
		$P < 0.01$	$P < 0.02$	$P < 0.05$

Lees mentions the dangers inherent in selecting temperature data when looking for relationships such as this, but feels that selection is justified, as the period chosen is at the time when the ewes are just beginning to emerge from the anoestrous period, and would thus be most likely to respond to the effects of other important environmental variables.

During Brambell's (1944) study on the wild rabbit (*Oryctolagus cuniculus*) an interesting effect of temperature was noted. When breeding began in February 1942 there was a peak in pregnancies which declined to mid-March and then increased normally again. During a period of very severe cold from 5 February to 7 March many females failed to conceive at the *post partum* oestrus after the first pregnancy of the season, and were thus noted as lactating but non-pregnant in March. A similar situation was described in an excellent study by Wight and

Conaway (1961) of cottontail rabbits (*Sylvilagus floridanus*) in Missouri. These workers dated conception exactly for the first pregnancies in samples taken over three years. In 1958 and 1959 the median first conception dates were 26 and 27 February, but in 1960 the median conception date was delayed to 25 March. Temperature records showed that in the latter year there was an exceptionally cold spell, as the average daily temperature stayed well below the daily norm from 17 February to 27 March. Further experimental work on cottontails in large open pens gave more evidence of a temperature effect (Marsden and Conaway, 1963). The behaviour of these rabbits was closely observed during the onset of the 1962 season. Social interactions between individuals showed a 7-day cycle related to reproductive behaviour. The first peak in social activity was seen

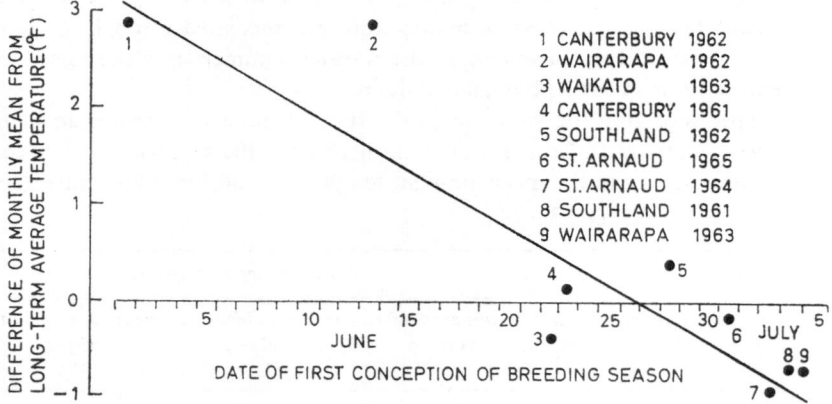

Fig. 27. The correlation between the date of first conception and the difference in temperature of the two preceding months from the long-term average for various populations of hares (*Lepus europaeus*) in New Zealand. The correlation is statistically significant ($r = 0.89$, $P < 0.01$). From Flux (1967), *N.Z. Jl Sci.* **10**: 357–401.

between 19 and 22 February. Severe cold weather with near blizzard conditions apparently suppressed the second peak, as social interactions were completely inhibited, the rabbits seeking shelter and feeding intensively. A third peak (5–8 March) in social interaction was characterized by the first mating activity. On two other occasions during subsequent peaks breeding was inhibited by severe weather and snowstorms. Finally, the effects of temperature on the onset of breeding of hares (*Lepus europaeus*) in New Zealand has been investigated by Flux (1967). He found a significant correlation, shown in figure 27, between the date of first conception and the temperature regimes for the two months prior to breeding. Flux emphasizes that the important variable is not the absolute temperature but the degree of deviation from the long-term average for the area being considered.

Human breeding may also be related to temperature. Cowgill (1966a, 1966b,

1966c) has shown that for most of the Northern hemisphere there are more births for the first six months of the year than the second six months. She suggests that this results from a peak in conceptions in spring and early summer. Patterns in the Southern hemisphere are exactly opposite, so that Cowgill postulates a meteorological effect on the pattern of human conception. As Vorhanger (1953) has found high significant negative correlations between summer temperatures and births the following spring, Cowgill implies that temperature is the most likely variable to affect human breeding (see also page 171).

D. THE EFFECTS OF THE EXPERIMENTAL ALTERATION OF TEMPERATURES ON BREEDING

The distinction between the effects of an environmental variable on the breeding season and on fertility during breeding is rather an artificial one. If the environmental factor is sufficiently detrimental to reduce the fertility to zero, then the species has stopped breeding. In the same way, as the environment becomes more favourable for breeding, fertility will also increase. Therefore the definition of a breeding season depends on a statistical concept of the proportion of animals breeding at any one time. This sort of difficulty becomes particularly apparent when trying to evaluate experimental manipulations of temperature and their subsequent effects on reproductive phenomena.

The alteration of temperature has been found by some workers to have possible effects on the onset of breeding in several rodent species, particularly those that live in underground colonies and hibernate during the winter. The ground squirrel (*Citellus tridecemlineatus*) comes out of hibernation in the third week of April around Chicago. Females appear about 10 days before males. There is a single oestrus soon after, and the males' testes are in peak reproductive condition at this time. Experimental alteration of temperature produced somewhat equivocal results (Moore, Simmons, Wells, Zalesky, and Nelson, 1934; Wells, 1935). For example, maintenance in constant warm, or in constant cool, temperatures or changing from cold to warm temperatures was not efficacious in inducing breeding in these animals. However, females kept in cold and darkness went into hibernation and showed the onset of sexual development at any time of the year. On the other hand, provided they were not kept in hibernation by cold and darkness, males came into sexual condition in the same way and timing as in the field. Later work (Wells and Zalesky, 1940) on the male altered these conclusions. They found that a constant temperature of $4°$ C would transform annually breeding male ground squirrels, with an annual aspermia, into animals capable of breeding continuously throughout the year. This occurred only when the treatment was begun during the mating season. Low temperatures during the normal aspermic interval caused only a slight speeding-up of the onset of breeding. These authors conclude that there is some temperature regulation of the gonad cycle of male ground squirrels in that low temperatures greatly favour reproductive activity and high temperatures are detrimental.

Modern work by Foreman (1962) on a closely related member of the Marmo-tinae, the black-tailed prairie dog (*Cynomys ludovicianus*), showed that the normal breeding period of late January to early March cannot be changed by altering the temperature. He maintains that the onset of breeding in this species is controlled by a genetic mechanism which produces hormonal changes regardless of the environment, but does not explain how such a mechanism can be timed in rela-tion to the environmental cycle. In a quite different type of rodent, the creeping vole (*Microtus oregoni*), alteration of temperature and light regime induced breed-ing during the winter (Cowan and Arsenault, 1955). This result is of particular interest, as this species is almost completely fossorial.

There is no clear picture from the above work on the role of temperature in inducing breeding in rodents. It is more than likely that this is due to complicat-ing factors of hibernation or to temperature sometimes acting in combination with the effect of light. There have been very few studies on mammals which have investigated the interrelationship of these two ecological variables particu-larly with regard to their relative importance immediately prior to breeding. The field observations suggest that temperature variations can alter within limits the actual onset of breeding, although photoperiod stimulates the mammal into a physiological state of readiness for breeding. It would be of interest to attempt to induce an out-of-season oestrus by altering the light regime but maintaining the experimental and control animals on either increasing as opposed to decreas-ing (or maintained low, as opposed to maintained high) temperatures. The rela-tive dates of mean onset of oestrus in experiments such as this would then give indications of the relative importance of the environmental temperature at the beginning of the breeding season.

Some work such as this has been done on a laboratory species, which hiber-nates, the golden hamster (*Mesocricetus auratus*). Hoffman, Hester, and Towns (1965) found highly significant differences in testes weights in the four treatments shown below:

	2 hrs light/24 hrs	14 hrs light/24 hrs
6° C	0·781 g/100 g body wt	1·539 g/100 g body wt
20° C	1·526 g/100 g body wt	2·030 g/100 g body wt

The additive effect of light and temperature are obvious from these figures, which were based on large groups of animals. These workers suggest that this species may have been a seasonal breeder in nature, although it breeds continuously under domestication. There is also evidence that temperature alone can alter breeding in hamsters. Grindeland and Folk (1962) found that prolonged cold exposure to 7·5° C for periods of 30, 60, or 90 days caused females to go into anoestrus. Removal from cold restored the oestrous cycles. These authors dis-cussed the distribution of cold-induced breeding cessation with regard to whether or not the species were hibernators or seasonal breeders. They concluded that 'the species differences in sexual responses to a cold environment are not explicable on the basis of the type of thermal regulation the animal has (homoio- or heterothermic), nor the number of oestrous cycles (mono- or polyoestrous) and appear to be species specific' (page 6).

In a cyclic species of rodent, the collared lemming, *Dicrostonyx torquatus*, the experimental alteration of light from conditions of continuous darkness to continuous light had no effect on the reproductive cycles in the laboratory (Quay, 1960). However, females which cycled fairly regularly when left at 70–80° F went into anoestrus when left under cold conditions (15–22° F).

The evidence for temperature affecting breeding is more substantial for the domestic sheep (*Ovis aries*). In an extremely interesting piece of research, Dutt and Bush (1955) put two groups of sheep (20 ewes and 3 rams per group) in rooms with natural lighting. One group was subjected to the normal range of temperature found in summer (average daily maximum 89° F) in Kentucky. The experimental group was subjected to a constant cool temperature of 45–48° F from 26 May to 8 October. The mean date of first oestrus of control ewes was 2 September $\pm$ 2·3 days, but the cooled ewes had a mean date of 10 July $\pm$ 4·0 days. Normal sexual activity was induced by the treatment, and the rams showed fertile semen. Conceptions were normal and resulted in lambs of control ewes being born around a mean date of 15 February and of cooled ewes, 10 December. The sheep-breeding season can thus be advanced by lowering the temperature, but the reader is reminded of the work of Robinson (1950, 1951, see page 52), which suggests that the sheep has a period of anoestrus where the ovary is still showing signs of activity. Unfortunately the taxonomic distribution of this pattern is unknown, so that it is possible that the sheep represents a special case. If this is so, the shallow anoestrus may be the reason why the sheep can be induced to breed by changing the temperature. It is a great pity that no other seasonally breeding mammal has been given this sort of experimental treatment. The effect of temperature in inducing sheep oestrus was confirmed by Godley, Wilson, and Hurst (1966) in South Carolina. They also varied the light regime and compared its effectiveness with that of temperature changes. In April, May, and June they put some ewes under October lighting conditions, or under October temperature conditions, or in rooms whose both temperatures and light were equivalent to October conditions:

	Mean % oestrous females	% ewes lambing	Mean number of lambs born/ewe
October light	93	87	1·13
October temperature	97	88	1·03
October light and temperature	97	91	1·06
Control ewes	58	54	0·68

The control ewes were those under normal conditions to the end of the breeding season.

I

E. THE EFFECTS OF TEMPERATURE AND ITS EXPERIMENTAL MANIPULATION ON MALE FERTILITY

Extremely high temperatures have detrimental effects on the fertility of both sexes. In the female this environmental variable acts after fertilization in most cases, and will thus be considered in Part Three (Pregnancy) (see page 167). The reader is referred at this point to the review by Bishop and Walton (1960), which details high-temperature effects on spermatogenesis.

High temperatures have been found to lower the semen qualities of many species of mammals. This was first demonstrated in any depth in the classic paper by Gunn, Sanders, and Granger (1942). These workers examined the semen of large numbers of rams in New South Wales and Queensland and showed that various seminal characters (morphology, density, longevity, etc.) were detrimentally affected in very hot weather. If rams were kept in areas where the maximum daily temperatures exceeded 90° F for more than 4 weeks the semen began to degenerate, despite good nutritional conditions. The change in seminal characters was reflected in the decline in the fertilizing ability of the rams. In some areas when the rams were exposed to prolonged daily maximum temperatures of over 100° F the semen showed marked degeneration to a degree that Gunn *et al.* described as being completely infertile. The morphology of the spermatozoa was a particularly good indicator of degeneration caused by high temperatures. The physiological reason for seminal degeneration is that normally in the ram, like other mammals with scrotal testes, the scrotum hangs well away from the body wall and the testis is maintained at a temperature of 5–7° C below the body temperature (Yeates, 1965). If the scrotal temperature is increased, spermatogenesis is interfered with and seminal degeneration sets in. As Yeates has pointed out, there are three mechanisms for preserving the temperature differential between the scrotum and main body mass: (*a*) reflex control of the dartos and cremaster muscles which can bring the scrotum up closer to the body in cold weather and which can also contract and expand the scrotum, (*b*) an arterial and venous pattern of blood vessels which ensure equality of temperature between the mass of testis tissue and the scrotum itself, and which can act as a thermal shunt in the region of the spermatic cord transferring heat from the descending arterial blood directly into the ascending venous blood without the extra heat getting to the testicular region; and (*c*) a tendency for excessive secretion and evaporation of sweat from the scrotum itself to cool its exterior. Once the environmental temperature is higher than the scrotal temperature, the only method which is effective is evaporative cooling.

The effects of high temperatures on ram fertility have been well documented by a very large number of workers, among whom are Moule (1950a), Smith (1964d), Yeates (1965), and Ulberg (1958), to mention a few.

As an example of the type of degeneration which has been seen, the table below shows the changes in spermatozoa, with the seasons, which were found by Smith (1962) in a group of rams in central western Queensland. It will be seen that the number of sperm abnormalities is high during the hot summer months, but in the cooler southern winter very few sperm abnormalities are seen. Similar effects have been documented for the bull. Yeates (1965) suggests that the upper level of ambient temperature for young Guernsey bulls is between 80° and 90° F. Ulberg (1958) has described how the semen of bulls in Louisiana

Sperm abnormalities	March	June	Sept.	Dec.
Detached tails (%)	4	0	2	5
Bent tails (%)	15	0	3	15
Protoplasmic droplets (%)	19	1	0	13
Proportion of live sperm (%)	61	87	80	70

declined in quality during the summer, and their fertility as based on the 60–90-day non-return rate percentage decreased during the same period. Schindler (1954) has described similar declines in bull semen in Israel. The decline in bull fertility is probably due to the same physiological causes as described for rams. The fertilization rates of cattle decline in hot areas in summer, probably due to combined female and male factors. For example, Holsteins in Arizona showed the following monthly percentages of pregnant females (examined 35 days after service) in a study by Stott and Williams (1962):

May	June	July	Aug.	Sept.	Oct.	Nov.	Dec.
62%	36%	28%	17%	31%	44%	48%	55%

In this region the hot weather lasts from June to September, and during this time the daily mean maximum dry-bulb temperature exceeds 95° F. This example has been chosen because it demonstrates the difficulty of separating male and female roles in the assessment of the effects of environment on domestic species.

As man has spread his agricultural development throughout the world he has taken many domestic species from the environments in which they originally evolved and tried to rear them in completely different types of environment. This applies to Merino sheep in the tropics of Australia (Smith, 1962) as well as to British breeds of sheep on the Equator in the Kenya highlands (Anderson, J., 1964) and to Holstein cattle in Arizona. Thus, reports of reproductive failure in animals such as these cannot be taken as representing a normal relationship between a breeding mammal and its environment. Nevertheless, the information gained from such observations can be of considerable use in interpreting the relationship between native animals and the environment and also in under-standing the ways in which native breeds of domestic animals have been more successful than the so-called 'improved breeds' which are introduced. For example, semen collected by the author from rams in the extremely hot Pilbara region of Western Australia showed the typical seminal degeneration during the summer. In this area the average daily maximum temperature exceeds 100° F from October to March. Semen from two species of kangaroo, the euro (*Macropus robustus*) and the red (*Megaleia rufa*), from exactly the same area showed no indication of any seasonal drop in spermatozoal characters. The locally evolved species of mammal has thus adapted its reproductive mechanisms to coping with this extreme heat. Method (c) (page 118) is used by the male to regulate testicular temperatures, as in the heat of the day the pendulous scrotum hangs well away from the body. Not only does it sweat profusely but extra evaporative cooling is achieved by the males salivating over the scrotum (Sadleir, 1965a). As an example of the second point, the higher reproductive efficiency of breeds of cattle such as Zebu in tropical areas, when compared with British breeds in these areas, could be quoted.

A number of agricultural scientists have investigated the use of native non-domestic mammals as farm animals in areas such as the tropics. For example, Skinner (1967) has appraised the eland (*Taurotragus oryx*) as a farm animal in central Africa, and concludes that this species has better physiological and be-havioural adaptations than domestic mammals to hot and semi-arid environments. A second biological advantage is its higher reproductive rate under these eco-logical conditions.

There are reports in the literature of indigenous mammals which show reduced fertility in hot weather. The Levant vole (*Microtus guentheri*) in Israel withdraws its testes into the body cavity during the hot months from May to July (Boden-heimer, 1949) and is infertile during this time. It is interesting that other species of *Microtus* show a winter decline in fertility, but this species shows an aestival effect. The cotton rat (*Sigmodon hispidus*) in Oklahoma shows a reduction in the number of pregnancies seen in midsummer which Goertz (1965) suggested was caused by summer temperatures. Even that classic of hot-climate mammals, the camel (*Camelus dromedarius*), shows a decline in gonad size and sperm production during the summer and is infertile during the extremely hot weather (Volcani, 1953).

The experimental effect of high temperature on male fertility has been investi-gated almost entirely in domestic animals, especially the sheep. Gunn, Sanders, and Granger (1942) reported on the effects of experimentally altering the environ-mental temperature. They exposed groups of Merino rams to varying tempera-ture regimes, usually about 90° F or about 100° F and showed that degeneration of spermatozoa was detectable after about 4 days of treatment. These responses depended also on the number of times the ram was electro-ejaculated and also to some degree on the individual animal. Some rams were placed under exactly similar diurnal variation in temperatures as had been recorded in one of the regions where they had investigated the seasonal change in fertility of non-experimental animals (see page 118). Seminal degeneration was also produced in these cases.

This and subsequent studies in Australia have been confirmed by similar studies carried out in the United States and reviewed by Ulberg (1958). Here the approach was to keep rams cool during the hot summer months and then investi-gate their fertility. It was found that cooled rams were considerably more fertile, both with regard to their semen characters and the number of ewes they successfully impregnated. It is interesting that shearing the rams during periods of high temperatures tended to reduce the infertility. Ulberg also mentions the reduction in fertility which occurs in dairy bulls in warm weather. Controlled-temperature chamber experiments with bulls have shown that exposure to temperatures of 85° F for 5 weeks can interfere detrimentally with spermatogenesis. Other reviews referring to the detrimental effects of high temperature on ram and bull fertility are those of Anderson (1945) and Maule (1962). The fertility of boars (*Sus scrofa*) is known also to be affected by temperature. Thibault *et al.* (1966) showed that boars kept in an air-conditioned piggery during the hot summer of southern France showed no decrease in fertility, whereas fertility declined in boars kept outside.

Quay (1960) has reported that male collared lemmings (*Dicrostonyx torquatus*) kept under temperatures of 70–80° F or 89–92° F showed reduced spermatogenesis. Basirov (1960) noted that libido and ejaculation were markedly inhibited even in a tropical species, the buffalo (*Bubalis bubalis*), when kept under temperatures of over 40° C.

There are a number of very real difficulties inherent in the interpretation of the experimental data quoted above. For example, the degree and method of heat exposure of the experimental animal is often not related closely enough to the sort of environmental conditions the animals are likely to meet. There is a natural tendency for the research worker to select temperatures at the extreme maximum or minimum end of the natural range so as to ensure a result in the duration of the experiment. Also, for simplicity, the animals may be maintained over long periods of time at these extreme selected temperatures when in nature the diurnal variation may act to alleviate the limited extremes. Behavioural changes related to the maintenance of an optimal body temperature are only just beginning to be appreciated by agricultural physiologists, and often the experimental manipulations themselves do not give the animals the chance to apply any behavioural selection of cooler environments. It would seem that there is a need for more precise monitoring of microclimatological data. The advent of radio transmitters which can be placed on free-ranging animals so that their temperature can be recorded by the experimenter, free from any environmental disturbance should lead to a closer experimental ability to duplicate the actual temperatures that mammals in the wild need to cope with. In this way also, the interesting data that have already accumulated from hot-room experiments could be checked to see if the temperatures used were realistic in terms of the non-experimental animal. These very small radios could also be used to follow the temperature of completely wild animals and to see how they regulate temperatures. Until the relationship of the true environmental temperature to the experimental temperature has been elucidated, much of the experimental effects of temperature on reproduction must be considered as giving insight as to the physiological effects of temperature extremes rather than reflecting conditions which the animal is likely to meet with in its normal existence. This is particularly true of domestic species such as the sheep, which, as has been pointed out before, is often maintained in areas with extremes of temperatures well above temperatures realized in its natural range before man transhipped it to other areas. Indeed, it is likely that as our knowledge of reproduction of wild animals grows, the relationship between environmental variables, such as temperature, and the maintenance of efficient reproduction, will be found to be one which has resulted from very long periods of selection, so that temperature in the natural range does not detrimentally affect reproduction. There is a fascinating open field in the study of the reproduction of populations on the edges of their ranges when compared with their reproduction in the centre of the range. It is most likely that environmental effects such as temperature may well delimit population ranges by their effects on the reproduction.

F. THE RELATIONSHIP OF HIBERNATION TO THE BREEDING SEASON

Hibernation in mammals is a physiological response to an as yet undetermined number of ecological variables. However, it is fair to say that in all probability the major variable is temperature, so that this may reasonably be considered in this chapter. Hibernation affects the reproduction of some hibernators so that breeding phenomena vary in response to the environment but are mediated through the physiological state of hibernation. Some mammals, such as bears and badgers, were once considered hibernators, but recent opinion (Harrison-Matthews, 1956) is that they are only species which undergo prolonged periods of sleep without undergoing the metabolic changes which are demonstrated by true hibernators, such as certain species of bats, rodents, and insectivores.

Wimsatt (1960) has pointed out that in some mammals hibernation really has little effect on the breeding season. Species such as the hedgehog, *Erinaceus europaeus*, and almost certainly the dormice, *Muscardinus avellanarius*, *Glis glis*, and *Dryomys nitedula*, spend the period of dormancy in a completely anoestrous state, so that their seasonal reproductive development occurs after they awaken from hibernation (Asdell, 1964; Nevo and Amir, 1964). In Wimsatt's words: 'There appears no reason to suppose that the factors which condition reproductive activity in these species are qualitatively any different from those operative in other spring-breeding mammals which do not hibernate' (page 249, 1960). He goes on to show that several bats of the families Vespertilionidae and Rhinolophidae show distinct patterns of reproductive phenomena which occur at the same time as hibernation but *are* modified by the hibernation process. This will be discussed below. In the light of evidence now available, it would seem that he was incorrect, however, to exclude certain other rodent species as not having their breeding patterns affected by hibernation.

The types of pattern of reproduction in the temperate living bats are very clearly shown in figure 28, which is taken from Wimsatt's excellent review of the subject (1960) and which is the basis for most of this paragraph. Leaving aside the special case of *Miniopterus schreibersii*, the temperate bat species show the following special adaptations to hibernation:

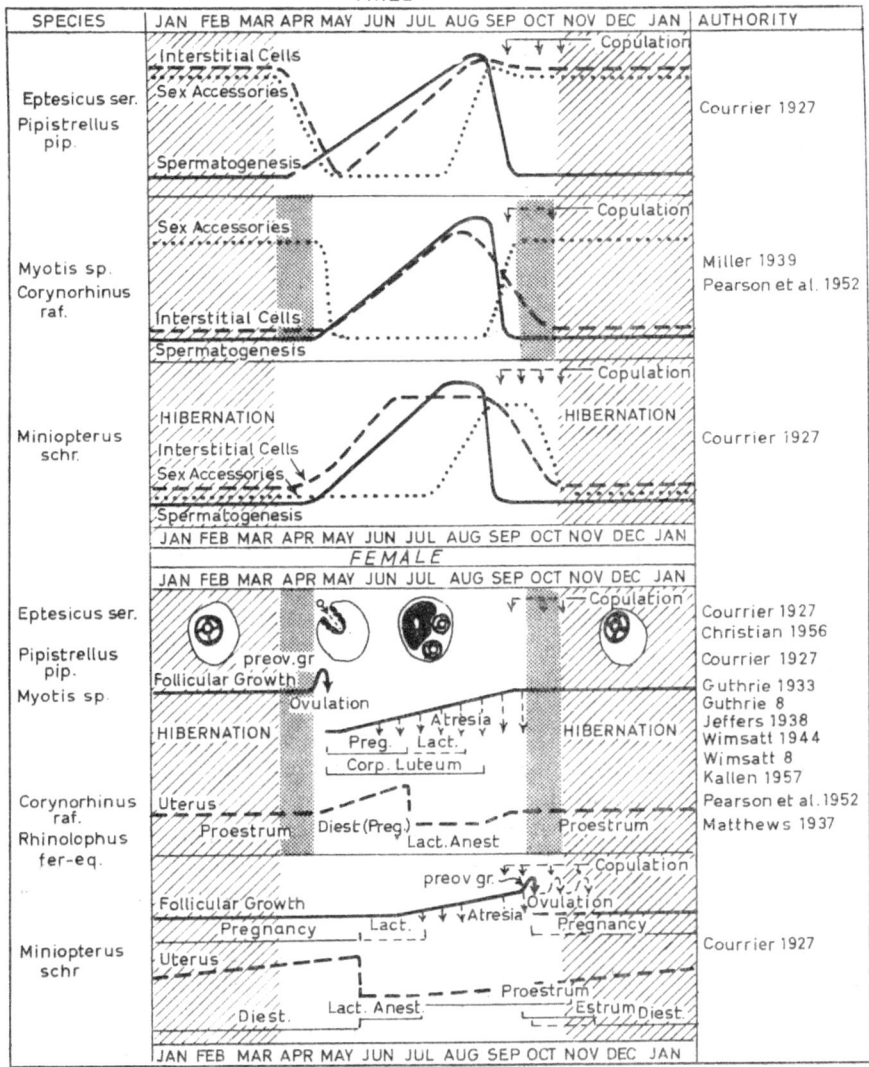

Fig. 28. Patterns of reproduction in various bats. From Wimsatt (1960) *Bull. Mus. comp. Zool. Harv.* **124**: 249–63.

In the male. (i) An exaggerated assynchrony between the spermatogenic and endocrinic cycles of the testis: (ii) the persistence of viable spermatozoa in the epidydimis for very long periods after the cessation of spermatogenesis: and (iii) the functional cycles of the Leydig cells of the testes, as judged by the usual histological criteria, do not coincide completely with periods of libido or hypertrophy of the accessory glands in certain American species (although they do in certain European species).

In the female. (i) A prolongation of the pro-oestrous stage of the cycle throughout hibernation: (ii) the maintenance of viable spermatozoa in the female tract throughout dormancy, and (iii) a marked delay in ovulation involving prolonged survival through hibernation of large follicles designed to rupture in the spring.

In the case of *Miniopterus schreibersii* the male seasonal cycle is like that of any normal seasonally breeding mammal, but the female cycle is unique in that after fertilization in autumn gestation is prolonged by slow embryonic growth during hibernation, so that the offspring are born in spring (but see page 206). Interestingly, the season of mating in this species is reversed from those of tropical members of the same genus who mate in spring.

It would seem, as Wimsatt suggests, that the breeding patterns of these several species of bat reflect a tropical origin on which the adaptation to cooler temperatures of hibernation has been superimposed. The most likely pattern is of an ancestral autumn breeding habit which when subjected to hibernation resulted in the main alteration in breeding being the delay in ovulation until the subsequent spring. There is a suggestion in one American species, *Myotis austroriparius* (Rice, 1957), that hibernation is dependent upon temperature in so far as it affects the food supply and that the breeding season can be either in spring or autumn in populations of this species, depending on their actual habitat. It would seem, then, that hibernation merely acts in bats to cause a delay period in the otherwise normal cycle. However, this does not explain problems such as Wimsatt has outlined of the endocrinic control of the reproductive processes during this delay, nor indeed is much known as to how the external stimuli affect the timing of breeding.

An approximately similar pattern is found in the small hibernating ground-dwelling rodents of the genus *Citellus*, although the endocrinological picture is not as well documented as for bats. The timing of breeding in this group is directly correlated with the time of emergence from hibernation. This is well demonstrated in figure 29, which shows the timing of these events in a Russian species, *C. pygmaeus*, from a paper by Kalabukov (1960). However, there is now evidence for a number of species (*C. undulatus*, Mitchell, O. G., 1959; Hock, 1960; *C. tereticaudatus*, Neal, 1965; and *C. lateralis* and *C. beldingi*, McKeever, 1966; *Cynomys leucurus*, Bakko and Brown, 1967) that the reproductive processes start during, and even before, the period of hibernation. The best documented of the changes are those by McKeever, who found that spermatogenesis started in autumn before (or while) these species were in hibernation and that there was a considerable enlargement of the male organs in the last few months of hibernation, so that mating activity began soon after emergence. The ovarian follicles started enlarging during the middle of hibernation. The involution of the gonads of both sexes, which occurs at the end of the summer, is followed by a degree of development of these organs in the late autumn before hibernation and in preparation for the subsequent breeding season the following spring. Thus hibernation in these rodents seems to have much smaller effect on the breeding patterns

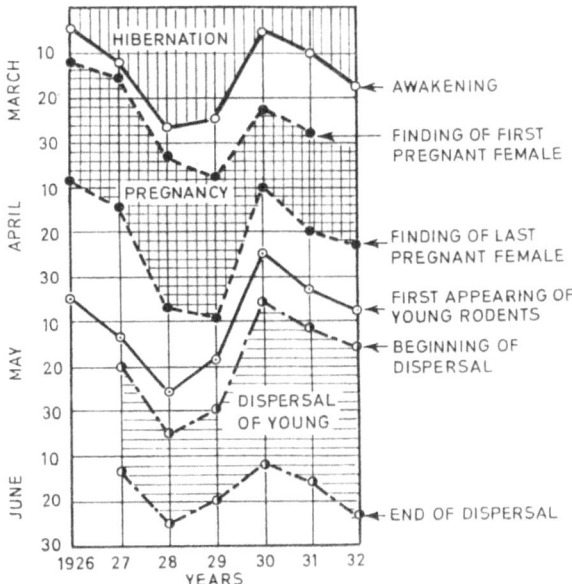

Fig. 29. The influence of the period of awakening on the breeding of ground squirrels (*Citellus pygmaeus*). From Kalabukhov (1960) *Bull. Mus. comp. Zool. Harv.* **124**: 45–74.

than was seen in the bats. There is, fortunately, one study of two species of *Citellus* living in the same area, one of which does and one of which does not hibernate (Neal, 1965):

	C. tereticaudatus	C. harrisii
Hibernation	Oct.–Dec.	Nil
Live sperm found	7 Jan.–8 April	4 Nov.–16 June
Embryos found	18 March–27 April	23 Feb.–12 May

Actually, due to the difficulty of collecting the hibernating species, no males could be sampled prior to 7 January, but in this sample spermatozoa were found in 3 of 10 males, indicating that some spermatogenesis had occurred during hibernation, but not to any great degree. The point here is that hibernation had resulted in a considerable delay in the onset of breeding, and thus limited the reproductive potential of this species as compared with the non-hibernator. Therefore it seems likely that hibernation in these group of rodents has a delaying effect, but not so marked as in the bats. This very lack of disruption to reproduction poses the interesting question as to the nature (especially in the Arctic species such as *C. undulatus*) of the stimuli causing the onset of gametogenic changes when the animals are in deep mid-winter hibernation.

CHAPTER 11

The Effects of Rainfall and Nutrition
on the Breeding Season

A. INTRODUCTION

The difficulty of separating the effects of rainfall and nutrition has been mentioned in the introduction to Chapter 10. There are many field studies which indicate the effects of changes in diet induced by rainfall conditions on reproduction in wild mammals. In most cases it is difficult to determine the relative roles of the possible direct stimulating effect of rainfall itself and its indirect effect causing growth in vegetation, increased nutritional supply, and thus improved physiological condition of the animals. Therefore, in the review which follows, studies in which separation of these roles is difficult have been considered under a joint heading. Only those studies, in which evidence is clearly pointing to one factor being much more important than the other, are considered as showing the effect of the separable variable.

The dangers of short-term studies should also be mentioned here. A causal relationship between two events, one ecological and the other physiological, is all too often postulated when the study has only encompassed a few breeding seasons. At this stage of a field study many workers have claimed that because an ecological event (i.e. exceptionally heavy rainfall) coincided with a physiological one (i.e. an exceptionally early onset of oestrus) that the two were causally related. As by far the most studies of mammals have not been continued for sufficient years really to prove or disprove such relationships, and indeed can never *disprove* them, this practice has led to a number of untestable hypotheses being suggested. Another danger is emphasis of obvious and easily observed ecological factors, without a true understanding of their relative causal importance. Many mammals start breeding in the spring in temperate regions. This coincides with a period of good vegetational growth and a consequent improving nutritional status at this time. However, it is demonstrably false to assume that this nutritional improvement has *caused* the onset of individual breeding at this time, although it may be the ultimate factor (Baker, 1938a) which has been selected for the individual genotype. If a species starts to breed each year at approximately the same time it is much more likely that photoperiodic changes are the physiological cause of the commencement of breeding. It is only when evidence is presented showing that the onset of oestrus happened constantly and consistently at times related to the more variable beginning of vegetation growth that a true causal relationship can be suggested.

B. THE EFFECTS OF PRECIPITATION ON BREEDING

There have been a few papers in which precipitation alone has been claimed to stimulate directly the onset of oestrus. Gooding and Long (1957), studying rabbits (*Oryctolagus cuniculus*) in Western Australia, found that young were captured barely 2 months after the first rains of the winter. From the timing of their appearance, in this 2-year study, these authors concluded that breeding occurred immediately rain fell and that there was not sufficient time for the grass to grow and thus rainfall to affect indirectly the onset of reproduction. In a more detailed study, on the same species in N.S.W., Poole (1960) attempted to explain how rabbits came into breeding condition so very quickly after rains fell:

'It is hard to see how this timetable of events could be satisfactorily explained except on the assumption that some factor which started to operate within a week or less of the autumn rain stimulated fertile matings among the local rabbits. With the temperatures occurring at that time, the appearance and growth of seedlings of the grass and herbage species that normally germinate in the autumn, would be very rapid; and this sudden availability of fresh high-quality feed seems to be the only factor that could have stimulated the reproductive activity' (page 41).

However, the same author describes how the gonads of such rabbits became fertile in early rains in August 1954 which broke a drought in Merricumbene, New South Wales. This was before the pasture had responded and still appeared 'extremely dry'. 'This response in fertility might have been due to a proximate stimulus from the rainfall or anticipation of the pasture response, or possibly to a combination of both factors' (page 33). Thus the picture is not entirely clear. It appears that after prolonged drought the rabbit can respond to rainfall alone, particularly with regard to male fertility.

In tropical climates and particularly jungle, the presence of clearly defined breeding seasons in some mammals has always presented a problem in explaining the ecological variables related to breeding at fixed times of the year. Rainfall is fairly heavy in these environments, so the suggestion has arisen that it is the reduction in precipitation which stimulates the beginning of breeding in some species. For example, Tamsitt and Mejia (1962) suggested that breeding in the bat *Artibeus jamaicansis* at Providencia, Puerto Rico (which starts in December), could be best explained as occurring at the end of the season of heaviest rainfall, October and November. Similarly, in a study of the breeding seasons of a number of genera of mammals in the lowland rain forest of North Borneo, Wade (1958) found that most species were seasonal breeders, and the onset of breeding coincided with the season of minimal precipitation.

At the other extreme there have been three reports of very heavy precipitation causing delays or interruptions in the onset of breeding. Lechleitner (1959) found that extremely heavy rains from December 1955 to February 1956 flooded the fields in Butte County, California, and this delayed the onset of breeding of the black-tailed jackrabbit (*Lepus californicus*) from the normal month of January until mid-February. Similarly, Birkenholz (1963) found that flooding due to heavy rains in March caused an interruption in pregnancies of the round-tailed muskrat (*Neofiber alleni*) in Florida. In a more northerly environment, Geist (1964) described the influence of snow on the behaviour at the beginning of rut in mountain goats (*Oreamnos americanus*) in British Columbia. Apparently in years of deep snow the males and females stayed in one small compact group, whereas in years of light snow cover there was more male movement from female group to female group and more formation of rutting pairs.

There have been suggestions that the presence of a snow cover during winter is the cause of breeding at that time in certain rodents. For example, Beer and MacLeod (1961) suggested that the presence of winter breeding in *Microtus pennsylvanicus* in Minnesota during 1954–5 was due to a continuous cover from December to March. These snow conditions were not seen in the other winters of this 8-year study, and no other winter breeding was observed. However, there are a number of alternative explanations for this winter breeding (pages 109, 244) including the changes described for microtines in their cycle of population density.

C. THE EFFECTS OF PRECIPITATION AND/OR NUTRITION ON BREEDING

The response of vegetation in tropical and arid regions to falls of rain is often an exceedingly rapid one. In many studies of animals living in these regions the onset of breeding has been observed to occur very soon after rain falls, but in most cases it is difficult to distinguish whether the rain itself or the general improvement in physical condition resulting from an improved nutritional regime is the factor involved. In some species the evidence that nutrition alone causes the start of breeding is incontrovertible, and there can be no doubt that in many of the studies reported below this is also the case. However, until further work is done on these species it is necessary to consider rainfall and nutritional improvement as factors acting in combination.

An example of the difficulty in separating out these factors is the study reported by Prakash (1960) of nineteen species of mammals in the Rajasthan desert in India. No common breeding season was detected, although the monsoons caused regular rains from July to September with a regular flush of vegetation. Species such as *Hemiechinus auritus*, *Antilope cervicapra*, *Gazella gazella*, *Tatera indica*, and *Meriones hurrianae* were observed to breed during or immediately after the rains, but four species of bats and two species of rodent bred in spring or early summer well prior to the rainy season. Prakash found that *Macaca mulatta* and *Presbytis entellus* had new-born young both in spring and after the rainy season. Although many of the species in this study were covered by rather small samples, the differences between species indicate the difficulty of environmental factor analysis. Field studies are required to determine the period of maximal stress during the reproductive processes, and thus to determine the selective effects on reproduction timing. It is more than possible that other factors, such as high temperatures, are those controlling the timing of breeding here.

(a) Rodents

In desert regions rodents tend to breed immediately vegetation improves after rain has fallen. The Merriam kangaroo rat (*Dipodomys merriami*) starts to breed in Mojave Desert of California about January when the annual plants which depend on winter rainfall start to grow (Chew and Butterworth, 1964). Other small rodent species (*Peromyscus eremicus*, *Onchomys torridus*, and *Perognathus longimembris*) also breed at this time, apparently in response to nutritional changes. *Dipodomys ordi*, by contrast, breeds over a longer period from September to March, on the southern great plains of Texas and Oklahoma. During the drought of 1956 very few females in this population came into reproductive condition (McCulloch and Inglis, 1961). In the next two years there were good rains and the onset of oestrus in 1957 occurred normally. In 1958 the onset of breeding was delayed, despite good rains, but the authors suggest that this was due to a density effect, as the populations were at a high level following the excellent previous breeding year. A feature of this study was that McCulloch and Inglis actually measured the seed crop and showed that drought did significantly reduce its magnitude. In the cold desert of the Urals the steppe lemming, *Lagurus lagurus*, reproduces continually under favourable conditions, but drought and consequent poor nutrition results in a complete inhibition of reproduction (Shevchenko, 1963). Similarly, the presence of continual or interrupted breeding in *Meriones hurrianae* in the Rajasthan desert depends on the rainfall incidence (Prakash, *pers. comm.*).

A very interesting relationship has been suggested by Orlova (1955) between the occurrence of rain in summer and the breeding of the Siberian suslik (*Citellus pygmaeus*). He concluded that the nutritional conditions of the *preceding* year determine the onset of breeding the following spring. If the summer is a dry one breeding starts later and carries on longer, whereas if the summer is a favourable one there is an earlier season the following spring, and breeding is more intense and synchronized. This could be justified from more recent work on sheep breeding, which is discussed on page 147.

In more tropical regions there seems to be a tendency for some rodents to breed not at the beginning but at some other time relative to the rainy season. For example, the multimammate mouse (*Mastomys erythroleucus*) in Sierra Leone actually breeds throughout most of the year (Brambell and Davis, 1941). However, by far the greatest number of females are pregnant in October and

November during the later part of the rainy season and immediately after it has finished. The lowest levels of reproduction occur at the very beginning of the rainy period. A similar picture is reported by Chapman, Chapman, and Robertson (1959) for *Rattus* (*Mastomys*) *natalensis* in the Rukwa valley of Tanzania. Again breeding occurred mainly at the end of the rainy season and the beginning of the dry period. Similar breeding is reported by Southern and Hook (1963) in *Paraomys morio* and *Lophuromys flavopunctatus*. In contrast, another species of the same genus, *Mastomys natalensis* in Malawi, has a true breeding season from November to May. The period of oestrus here coincides with the duration of the rainy season, which is also the warmer part of the year (Hanney, 1965). A very rare species of rat in Nyasaland, *Beamys major*, is reported by Hanney and Morris (1962) as reproducing only in the rainy season, whereas the harsh furred rat, (*Lophurus flavopunctatus*) also breeds during the warm wet rainy season in that country (Hanney, 1964).

The genus *Tatera* has been studied in this respect in Southern Africa and India. Measroch (1954) found pregnancies in all months in *Tatera brantsii* in the Transvaal, but most adult females went into anoestrus from October to December during the wet. Similar results were reported by Allanson (1958). These workers also studied *Tatera afra*, which is in complete anoestrus from April to July in Cape Province during the rainy cold season. These two species both stop, or markedly reduce, their breeding at the beginning of the rains, but reverse this when the rains cease. This pattern occurs independently of the temperature. Similar responses were reported by Prasad (1956) for *Tatera indica* in South India. Breeding occurs there from August to February, and is less closely related to rainfall and more to the growth of sprouted seeds which start from the end of July.

An interrelationship between nutrition, precipitation, and temperature has been suggested by Ashby (1967) as affecting the prolongation of the breeding season of two rodent species *Clethrionomys glareolus* and *Apodemus sylvaticus* in northern England. Breeding occurs later than usual in years when the summer is warm and dry and prolonged into autumn. This results in higher levels of available plant and invertebrate nutrition for the mice and voles. In years where the summer is wet and cold during the latter months breeding ceases early.

The majority of studies reported above show no constant relationship between the presence or absence of breeding and the timing of rainfall and consequent good nutritional conditions. It would seem that only in desert environments is the relationship positive. Unfortunately there have been no intensive long-term studies in this type of habitat, so that the coincidence of breeding with a regular season of rainfall and good nutrition may reflect not an immediate relationship but rather indicate that breeding is under photoperiodic control, the seasons of rainfall being constant. Until there is a study showing that breeding in desert rodents can occur at different times of the year (and thus the light cycle), but always related to rainfall and nutrition, the timing of the onset of breeding relative to a good nutritional regime must be considered as secondary.

(b) Lagomorphs

Perhaps the best understood mammal from the point of view of reproduction in the wild is the European rabbit (*Oryctolagus cuniculus*). The papers on its reproduction in Wales by Brambell (1944) and in New Zealand by Watson (1957) are models for the study of the physiology of reproduction under natural condi-

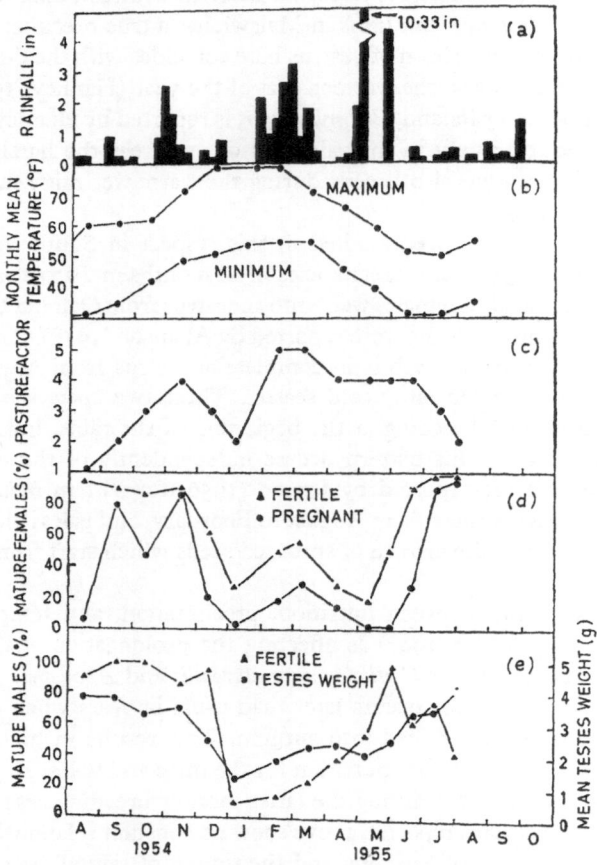

Fig. 30. Patterns of breeding in wild rabbits (*Oryctolagus cuniculus*) in Merricumbene, N.S.W., Australia. From Poole (1960) *C.S.I.R.O. Wildl. Res.* 5: 21–41.

(a) Seasonal distribution of rainfall in ¼ monthly periods.

(b) Seasonal changes in extreme temperatures.

(c) Seasonal changes in pasture conditions: 1, dry; 2, slight green tint; 3, green; 4, green sward; 5, lush green sward.

(d) Seasonal changes in percentages of fertile mature females and percentages of pregnant mature females.

(e) Seasonal changes in percentages of fertile mature males and mean testes weights of mature males.

tions. The paper by Myers and Poole (1962) on the same species has already been
referred to (page 108). There have also been a number of studies of rabbit breeding in relation to environment, the most important being that of Poole (1960) in
Australia. He studied two populations in New South Wales, at Lake Urana and
Merricumbene Creek, during 1954 and 1955. Figure 30 shows that at Merricumbene the fertility dropped so low that breeding ceased entirely at one stage.
Examination of Poole's data demonstrates the arbitrary nature of separating a
discussion of the onset and cessation of oestrus from a discussion of fertility in
females, one being only an extreme of the other.

   In this study Poole describes three dominant features of rabbit reproduction
in Australia: first, a very strong tendency to be reproductively inactive or quiescent in midsummer; secondly, a main flush of breeding in the spring; and lastly
occasional breeding activity in the autumn (particularly where the summer is
a dry season, as in the south of the continent). On the basis of those trends
figure 30 will be discussed in detail, because it is an exceedingly good example
of the difficulties inherent in trying to distinguish between various environmental factors which usually act in concert on the physiology of reproduction.
The break in drought conditions in August 1954 was followed by only a gradual
increase in the condition of the pasture, and yet there was a marked increase in
the numbers of pregnant females. Fertility then dropped temporarily, but again
increased in November, when there were good rains and the pasture conditions
reached a high level. In December two factors became important, namely the
slightly delayed effect of increased temperatures (mean monthly maximum of
61° F in October–79° F in December), and also the lack of any effective rain, so
that the pasture began to dessicate. This resulted in a sudden decrease in the
proportion of males fertile and a considerable drop in the proportion of does
pregnant.

   Rain which fell from January to March 1955 caused a response in pasture
conditions, but little response in breeding was seen. Although about 56 per cent
of mature does were now fertile, only half of them conceived. Poole suggests that
this was due to high summer temperatures. Although temperatures were falling
in April, the heavy rains which occurred then and in May again did not stimulate
breeding, because they flooded practically all burrows and breeding stopped.
By June, however, breeding was again increasing, as pasture conditions remained
good and the temperature was not yet low enough to interfere with breeding.

   This example has been detailed to emphasize the interrelated nature of the
environmental variables. It has also shown that any variable has a different
effectiveness in stimulating reproduction, depending on the time of the year
and on the relative intensity of other variables. Despite the complexity of the
reproductive patterns described, the responses of a lagomorph to sudden improvement in pasture conditions following rainfall are well demonstrated. In another
population at Gunbower in Victoria there were good rains in February, but the
response was negligible, due either to the lack of pasture improvement (which was
adapted to the normal winter rainfall and did not sprout) or due to the does

K

being 'physiologically unprepared to come out of anoestrus'. There were heavy falls of rain in the second half of March over a period of 4 days starting 16 March. Samples taken in April showed that there was a mass period of conceptions from 26 to 28 March. Poole suggested that this was due to a very rapid germination of fresh grass and its immediate effect on the rabbits' physiology. Indeed, the rapidity of this response suggests that a general improvement in condition could not have been the cause, but rather that some specific pasture factor which stimulated breeding must have become included in the diet of the rabbits. This idea is of interest in view of the experimental results of Friedman and Friedman (1939, 1940), who showed that green-plant extracts had gonadotrophic qualities when injected into rabbits.

Similar responses by rabbits to green feed have been reported for the Riverina and Southern Tableland areas by Hughes and Rowley (1966). These workers also found that the availability of green food was a necessary prerequisite to breeding in *Oryctolagus cuniculus* and for the maintenance of maximum fertility. Drying pastures caused reduced breeding and later resulted in extremely low breeding or its cessation. However, in the Southern Tableland regions restoration of pasture conditions by autumn rains resulted in only abortive breeding success.

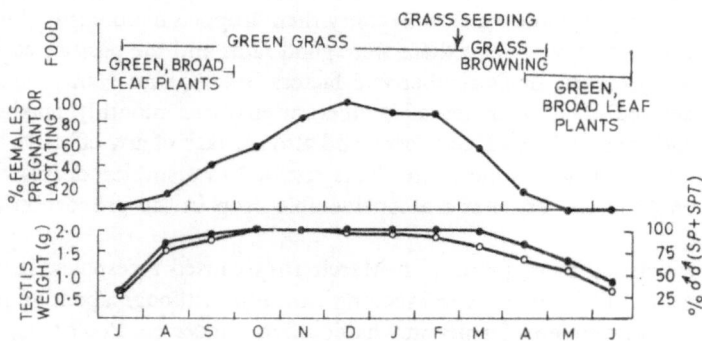

Fig. 31. The breeding season of rabbits (*Oryctolagus cuniculus*) on Macquarie Island. In the lower graph the small open dots indicate the mean testis weight in grams and the large closed dots the percentage of males in which the testis was producing either sperm (SP) or spermatids (SPT) or both. From Shipp *et al.* (1963) *Nature, Lond.* **200**: 858–60.

The authors suggested that this may have been due to some type of reproductive exhaustion which was associated with the exceedingly brief anoestrous period following the usual intense spring–summer breeding.

In other areas the reproduction of the rabbit does not seem to be so markedly affected by variations in food supply. Lloyd (1964) has suggested that rabbits in Wales are not as affected by food because there is a relatively higher level of nutrition all the year in Wales (and seemingly New Zealand) when compared with Australia. It is fortunate that there is one small study on rabbits living in a very uniform environment which throws light on to the importance of nutrition.

Strictly speaking, this paper should be discussed in the following section, as it shows the effect of nutritional changes on the onset and cessation of breeding independently of rainfall, but it is included here for comparison with the previous paper. Macquarie Island is found in the sub-Antarctic at 54° 30' S between New Zealand and the Antarctic Continent and is characterized by little variation in temperature from a mean of 40° F and by uniform annual precipitation. Rabbits were studied there in 1950–51 and 1956 by Shipp, Keith, Myers, and Hughes (1963). Figure 31 shows that breeding was related to plant growth. The short period of anoestrus from May to July occurred when grass was not available,

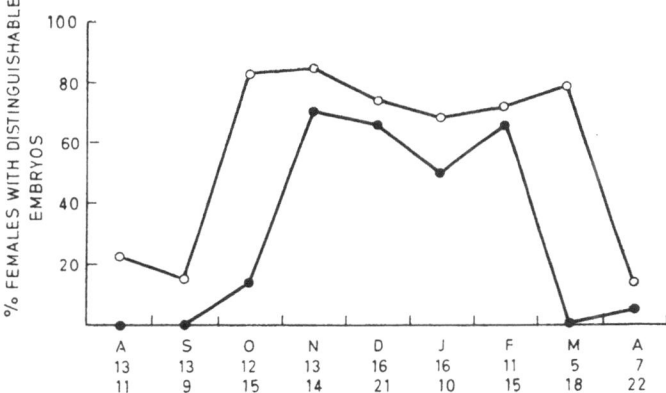

Fig. 32. The effect of habitat of the breeding of wild rabbits (*Oryctolagus cuniculus*) on Macquarie Island. Open circles refer to a well-vegetated area, the sample size being given in the uppermost of the two lower lines of numerals and the closed circles referring to an area denuded of vegetation with the sample size being on the lower line. From Shipp *et al.* (1963) *Nature, Lond.* **200**: 858–60.

there being considerable weight loss at that time. The importance of nutrition is further demonstrated in figure 32 (from the same paper), which shows that breeding began early and ceased late, at one locality on the island where there was luxuriant grass growth. On an area which was overgrazed (to the extent that rabbits were occasionally seen digging for roots), breeding started later and ceased earlier, and the average level of fertility was lower too. In the light of these papers and those showing that social factors can have a marked effect on the breeding of rabbits (page 263), it is interesting to consider the phenomenal success of rabbit breeding in so many different types of environment. Temperature, rainfall, nutrition, and social factors may all play a role in initiating, or inhibiting, the reproduction of this species. It would seem possible that this is a case where the introduction of the rabbit to areas with very different climatic conditions to its original home in Spain and Mediterranean Europe (Zeuner, 1963) has resulted in the expression of some susceptibility to environmental conditions not met with in the original range. It would be fascinating to compare the reproduction of this mammal in its original homeland with its reproductive cycle in other areas. The selective advantages of responding to different environmental variables could then be assessed, and it might thus be possible to gain a clearer picture of why certain species such as the rabbit have been such successful colonizers of habitats outside their original range, whereas other species have been relatively unsuccessful.

Other lagomorphs show breeding responses to precipitation and nutritional conditions. Studies on the European hare (*Lepus europaeus*) in Poland have shown that under severe winter conditions the snow cover reduces food availability and breeding ends earlier than usual (Raczynski, 1964). In mild winters reduced snow improves the food supply and breeding starts earlier in the year. On the Caucasian steppes, in a more arid environment, this same species will curtail the end of its breeding season in drought years when the good supply is limited by insufficient rainfall.

There is evidence for North American lagomorphs also having breeding seasons which are related to rainfall and nutrition, although it is most likely that nutrition is the primary factor. The black-tailed jackrabbit (*Lepus californicus*), for example, breeds in California when rain falls and vegetation is at its greenest (Lechleitner, 1959). Comparison of different populations in different parts of the same state showed that the length of the breeding season is closely correlated with the length of the rainy season. In Arizona there are two peaks in breeding corresponding to two peaks in rainfall. In this species nutrition is almost certainly the major factor because individual jackrabbits, with home ranges which included irrigated fields, continued to breed, whereas individuals without this extra nutrition did not. Similarly, Fitch (1947) described how cottontail rabbits (*Sylvilagus auduboni*) breed only in late autumn, winter, and spring in central California at the time when green forage is abundant. In one exceptional case he found young cottontails born in the dry season, but was able to show that these were born to parents feeding on artificially watered lawns, which kept the grass green and fresh. It is strange that Sowls (1957) was unable to find any relationship in Arizona between the breeding of this species with rainfall incidence or nutrition.

(c) Larger Mammals

Among the artiodactyls the onset and cessation of breeding is only closely related to nutritional conditions in areas where these factors show great variability. There are very few studies where the effects of rainfall and nutrition have been clearly separated. Some early workers have apparently not recognized that nutrition availability plays only a secondary role in timing breeding compared to the primary role of photoperiodic stimuli. A case in point is the oft-quoted example of the breeding of gazelles (*Gazella* spp.) in the Cairo Zoo (Flower, 1932, also quoted in Bodenheimer, 1938). Under zoo management, with optimal and even feeding regimes throughout the year, gazelles gave birth in seasons which coincided with the seasons of rainfall in their natural habitats in Arabia. This must be explained as resulting from photoperiodic control of breeding, otherwise the pattern of breeding in the zoo would have changed. Unfortunately the breeding season of these gazelles in the wild is not known. In many desert areas a feature of the environment is the complete failure of rains in occasional years. Under these conditions it would seem that severe undernutrition could cause a delay or complete inhibition of breeding which would otherwise have occurred as the result of photoperiodic stimulation. It is a great pity that there are so few studies on the biology of the later mammals which live in arid or semi-arid areas. Evidence of the responses of these species to rainfall and nutritional variations is almost non-existent. There is one report by Lavauden (1927) that during an exceptionally severe drought in the Tunisian Sahara various species of gazelles and mouflon (*Ovis musimon*) completely ceased breeding at the normal time. When winter rains fell, this sterility proved temporary and normal breeding occurred again.

In tropical regions the relationship between rainfall, nutrition, and the onset of breeding is again not entirely clear. Dasmann and Mossman (1962a) found that impala (*Aepyceros melampus*) in Southern Rhodesia rutted at the beginning of the dry season, so that the period of births was concentrated at the beginning of the rains. They suggest that the timing of impala breeding (which differs in different parts of southern Africa) is regulated in each locality by the onset of rainfall and plant growth. However, in a later paper (Dasmann and Mossman, 1962b) dealing with the breeding season of a number of large ungulates in a single area they could find no obvious infraspecies mechanism, as different species produced their young at different times relative to the rainy and dry seasons. Although the seasons of birth were different in different species, they were often concentrated at a fixed time. Thus it is possible that rainfall and nutritional changes were acting as ultimate factors through varying stages of the reproductive cycle (i.e. late pregnancy or lactation or conception, etc., see page 48).

In one study of a primate the rhesus monkey (*Macaca mulatta*), Koford (1965, 1966) has suggested that delays in the onset of mating in summer on Cayo Santiago, Puerto Rico, are related to delays in the start of the heavy spring rains and a consequent delay in the increase in nutritional quality of the natural vegetation. This example is particularly interesting in that this is an introduced population from India, which is artificially fed at hoppers, as well as utilizing the natural vegetation. In the four years of the study mating began when the cumulative rainfall for the year was between 25 and 30 inches. In North India, Southwick, Beq, and Siddiqi (1965) found that mating mainly occurred shortly after the wet monsoon season – a similar relationship to rainfall as on Cayo Santiago.

The seasonal breeding of rhesus monkeys has been the subject of investigation by Vandenbergh and Vessey (1968) in another introduced colony near La Parguera, Puerto Rico. Detailed behavioural observations over more than four years resulted in the data shown in figure 33. Outside the season of births, an increase in rainfall and growth of new leaves was always followed by an increase in mating incidence. Vandenbergh and Vessey compared their data with those of other workers in different localities (including those mentioned in the previous paragraph), and from this comparison it would seem that the rhesus monkey in all areas so far studied starts breeding soon after there is an increase in rainfall. The difference in times of mating between Cayo Santiago (August to December) and La Parguera (October to February) is particularly interesting, because the two areas are at practically the same latitude, and rainfall incidence is the only major environmental variable that appears different.

In the arid regions of Australia the breeding of the larger marsupials is very

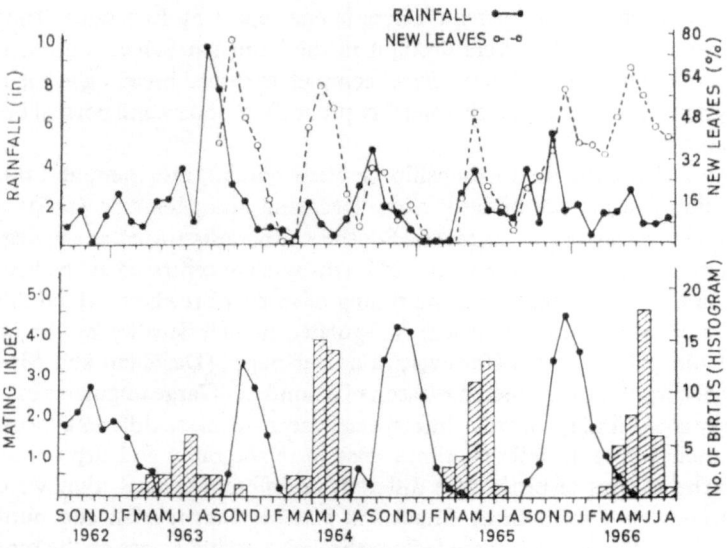

Fig. 33. Month-to-month variation in rainfall, vegetation, mating activities, and incidence of births in the free-ranging population of Rhesus monkeys (*Macaca mulatta*) at La Parguega, Puerto Rico. From Vandenbergh and Vessey (1968). *J. Reprod. Fert.* **15**: 71–9.

closely related to rainfall and nutrition. The most detailed study in this respect has been made by Newsome (1965) on the red kangaroo (*Megaleia rufa*) in two areas near Alice Springs, Northern Territory. Newsome has developed the concept of an *index of aridity* to quantify the intensity of drought conditions and the changes found in the vegetation at different times of the year:

'The index was built up as follows. When enough rain fell to cause a general response in pastures, the estimated evaporation from the soil was subtracted day by day until the remainder was nil, and throughout that time the index of aridity was set at zero. Thereafter, the estimated evaporation was accumulated to provide a measure of the aridity on a given date. Small falls of rain were merely subtracted from the index on the day it fell, but large falls sufficient to promote a general response in pastures caused the index to be put at zero once more. In effect, the index represents the period during which drought was experienced, with each month or part thereof being weighted according to its drying power. As such, it should be a good indicator of the physiological stress imposed on plants by a drought and also on the animals grazing on them' (pages 738, 739).

Using this index, Newsome showed very clearly that the proportion of individual females which were in anoestrus increased as the drought proceeded. The red kangaroos breed all the year round, in that even after severe droughts there are still *some* females in breeding condition, even though the proportion may be

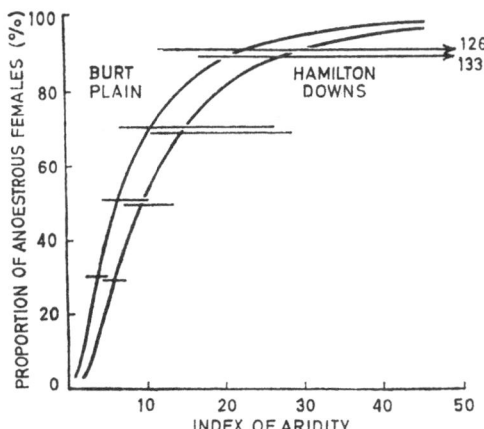

Fig. 34. The theoretical regressions of the proportion of anoestrous female red kangaroos (*Megaleia rufa*) on the index of aridity (see text) for two study areas in central Australia. The horizontal lines indicate the 5 per cent confidence limits of the index of aridity required to induce the 30, 50, 70, and 90 per cent of females to go into anoestrus. From Newsome (1965) *Aust. J. Zool.* **13**: 735–59.

very low. This is true of a number of the large kangaroo species. The presence of breeding females in every month of the year has been reported from the euro (*Macropus robustus*) in Western Australia by Ealey (1963) and Sadleir (1965a) and in other populations of the red kangaroo in New South Wales and Queensland by Frith and Sharman (1964) and in Western Australia by Sadleir (1965a). However, Newsome has shown clearly that individual animals react to drought by reduced breeding. Indeed, he was able to find a statistical correlation (figure 34) between the proportion of anoestrous females at any one time in the two populations he studied and the index of aridity for the area. If $X =$ the probit transformation of the proportion of anoestrous animals and $Y =$ the logarithm of the index of aridity, then:

$$\text{for Burt Plain} \quad Y = 2 \cdot 340997X + 3 \cdot 177184$$
$$\text{for Hamilton Downs} \quad Y = 2 \cdot 646574X + 2 \cdot 467400$$

Both these regressions were highly significant ($P < 0 \cdot 01$).

Although the two populations showed the same sort of general relationship, drought was 1·57 times more effective on Burt Plain than on Hamilton Downs in inhibiting breeding. The former station was more intensely grazed by cattle, and so carried less food for the kangaroos. The relationship shown indicates that when rains fell and the index of aridity dropped to zero, then most females ovulated as green feed became available. The advantage of using the index of aridity is that it enables an estimate to be made of the amount of rainfall necessary at any time of the year to stimulate breeding in the red kangaroo. In a later paper Newsome (1966) found a direct relationship between food supply and the percentage of breeding females on Burt Plain and gave a significant regression expressing this relationship. However, after the breaking of a drought the females came into heat so very quickly after rains fell that this immediate response could not have been due to nutrition *per se*. When rains fell on one occasion in January 1960 there was a detectable increase in the percentage incidence of oestrous females only 5 days later. Newsome suggests that this type of response was due to oestrogenic substances in the growing grass shoots and quotes as evidence enlarged reproductive tracts and enlarged Graafian follicles in immature females and an increased incidence of twinning at this time. Sharman and Clark (1967) found female red kangaroos in populations in New South Wales which, although they had not undergone a *post partum* oestrus during a preceding drought period, came into oestrus and ovulated after drought-breaking rains while still carrying joeys in the pouch up to 163 days of age. Ovulation did not occur at this time if these females already had functional or resting corpora lutea. Kirkpatrick and McEvoy (1966) noted that a prolonged drought in southern Queensland, from September 1964 to December 1965, inhibited the reproduction of grey kangaroos (*Macropus giganteus*). Eight months after the drought period had commenced, all breeding had ceased and there was drastic mortality of the pouch joeys (see page 224).

In the case of the hill kangaroo or euro (*Macropus robustus*) in Western Australia, Ealey (1963) suggested that females anticipated the onset of the wet season and started to come into breeding condition *before* the rains fell. However, Newsome (1965) and Sadleir (1965a) have both shown that Ealey's data can be interpreted in another way connected with the development of delayed blastocysts (see page 209). The termination of breeding in two smaller marsupials, the bandicoots *Perameles gunni* and *Isoodon obesulus*, was found by Heinshohn (1966) to be related to drought in Tasmania. Breeding extended for two months longer in a year of adequate rainfall than in a year of drought conditions when food availability was reduced.

D. THE EFFECTS OF NUTRITION ALONE ON BREEDING

In those cases where nutritional changes in the environment alter breeding but no rain actually falls, it is possible to demonstrate that nutrition alone affects reproduction. In some field studies situations have been described where without deliberate experimentation (see page 143) definite comparisons can be made between the breeding of populations on different levels of nutrition. Common examples of this are found where wild mammals utilize domestic crops which may be either fertilized or irrigated.

(a) Rodents

The breeding pattern of rodents living in arid areas has been investigated only in a few studies. Bodenheimer (1949) found that breeding of the Levant vole (*Microtus guentheri*) in Israel varied with location. In some months breeding had ceased completely or was at very low levels in non-irrigated fields but continued at high levels in irrigated fields. Bodenheimer relates these differences to differences in the diet as a result of the higher nutrition available under irrigation. There is evidence that alfalfa roots are a greatly preferred gopher diet. In a study of the mountain pocket gopher (*Thomomys talpoides*) in Colorado, Hansen (1960) found that the development of sex organs of this seasonally breeding species started in December when there was minimal green feed available. By March, however, Hansen collected gophers with stomachs full of alfalfa roots at the time that breeding was just beginning. It is not clear whether this change in diet was the cause.

It has been suggested by various workers that the beginning of breeding in some small mammals is related to the nutrition available in late winter and spring. Adamczewska (1961) concluded that the main environmental factor controlling the onset of oestrus in *Apodemus flavicollis* is the fruiting of various species of deciduous trees. From a study over 5 years in Poland, he found that in years with considerable fruiting there was autumn breeding and a mass appearance of young field mice in the spring of the next year. In poor fruiting years spring and summer breeding was reduced. The relationship was complicated by the role of winter weather in the survival and general condition of the animals each spring.

The extension of the end of the breeding season in another species, *A. sylvaticus*, and the bank vole (*Clethrionomys glareolus*) has been related to food supply by Smyth (1966) for populations in woods outside Oxford. These two species usually stop breeding in October, although the acorn crop falls during this month. On the basis of body weight distribution, Smyth showed that for sixteen breeding seasons this was extended through to December only in those years when there were good acorn crops. He also attempted to show that breeding at the end of the season on individual plots correlated with different falls of acorns. The general relationship was suggestive, but the information on acorn fall was limited, and in his examination of relatively small plots, Smyth does not seem to have realized the importance of determining the feeding range of animals sampled. The relationship between extended winter breeding and good acorn fall suggests that nutrition regulates the cessation of breeding in these species. Similar conclusions were reached by Zejda (1962) in his study of winter breeding of the bank vole in Czechoslovakia. Winter breeding only occurred in vole populations in stands of oaks, whereas populations in stands such as alder showed no winter reproduction. The presence of winter breeding depended on the type of vegetation and thus presumably the level of nutrition for old-field mice (*Peromyscus polionotus*) in South Carolina (Caldwell and Gentry, 1965). The incidence of pregnancy varied with vegetation, and in some areas, i.e. the *Lespedeza* stage of the old-field succession, no winter pregnancies were found. Jameson (1955) also noted the effect of food conditions on the cessation of breeding in *Peromyscus maniculatus*.

Detailed studies of the muskrat (*Ondatra zibethicus*) by Errington have shown

	Percentage of litters born by months		
	Iowa mean (1935–57)	Little Wall Lake (1961)	Goose Lake (1961)
Number of litters examined	3,209	70	88
	(%)	(%)	(%)
March	0·2	0·0	0·0
April	14·0	2·9	0·0
May	30·6	17·1	14·6
June	28·6	21·4	16·8
July	17·4	30·0	29·2
August	8·5	21·4	33·7
September	0·6	7·1	5·6

that its breeding can be affected by the nutrition available. The length and intensity of breeding in Iowa is shown in the table above, which has been taken from a paper by Errington, Siglin, and Clark (1963). The Goose Lake population started to breed later because there was a markedly reduced food supply available in 1961. This was partly due to there being very great numbers of muskrats present, so that density effects cannot be ruled out, especially as many fights were seen there in early June. However, the degree of deterioration of the habitat in 1961 can be seen from the following estimates of the area of emergent vegetation available to the muskrats above the lake surface: 1959, 82 per cent; 1960, 66 per cent; 1961, 17 per cent. The effect of nutrition on breeding (independent of population density) can be seen from the surge of litters which occurred later in the season. At this time, although the numbers stayed high, the muskrats in Goose Lake moved into cultivated tracts of land surrounding the habitat and started feeding very intensively on corn. The later breeding in Little Wall Lake was most likely caused by density effects alone, as here food did not appear to be at all short, even though muskrats were excessively crowded together as a result of the artificial raising of the water level.

(b) Other mammals

The effect of nutrition on breeding of rabbits of the genus *Sylvilagus* has been investigated several times. Ingles (1941) found that *S. auduboni* bred all year in the Sacramento Valley of California and attributed this to the continuous supply of fresh green feed available due to irrigation. The same species in the foothills of the Sierra Nevada limits its breeding to periods when green forage is abundant (Fitch, 1947). In *S. aquaticus* in Texas breeding continues throughout the year, but this has also been attributed (Hunt, 1959) to the peculiar availability of fresh feed through the normally sparse period at the end of summer and fall.

The breeding of a small marsupial, the quokka (*Setonix brachyurus*), in Western Australia has been shown to be affected by nutritional status (Shield and Woolley, 1963; Shield, 1964). On Rottnest Island births take place from February to April, and occasional young can be found in the pouch until August. Seasonal anoestrus occurs from then until the following January. The island experiences a Mediterranean climate with a dry and hot summer from October to April and a cool wet winter from May to August. Although the onset of breeding occurs in the hot dry summer, it is probably still related to nutrition due to the following factors. Populations of quokkas on the mainland have been found to breed continually in swampy areas which maintain fresh green feed all the year. Quokkas brought into captive pens on a good diet and under identical conditions of light and temperature, as on Rottnest Island, breed continuously. Finally, a small subpopulation of quokkas on the island, which live in the area of a municipal rubbish dump, have been found to be carrying pouch young well outside the normal breeding season due to the better nutrition available from food scraps. This species is a good example of the complexity of the effects of nutritional levels on the onset and cessation of breeding. The breeding season begins when nutritional levels are at their lowest, so that unlike some of the examples already mentioned, good nutrition *at the time* cannot be considered as the stimulus causing the animals to come into reproductive condition. Despite this, breeding in February means that the young will be leaving the pouch at a time of good nutritional availability, namely at the end of the winter in August and September. Therefore, whereas the *ultimate* cause of timing of the breeding season is in all probability the nutrition available (Baker, 1938a), the *proximate* cause is really unknown. Evidence from pen-kept animals rules out light and temperature variations.

The dates of birth of four Swiss ibex (*Capra ibex*) populations from separate management areas ('gerhegen') have been documented by Nievergelt (1966). Over a number of years the nannies of the Langenberg gerhegen bred significantly earlier (and had a higher rate of twinning) than the other populations studied, and this difference was attributed to undocumented differences in available nutrition. Transport of ibex from one gerhegen to another showed that environmental and not genetic factors maintained the different onset of breeding. Contrastingly, Cheatum and Morton (1946) reported that white-tailed deer (*Odocoileus virginianus*) started breeding significantly earlier in northern areas of New York State, where nutritional conditions were worse (Morton and Cheatum, 1946) than in the south.

E. EXPERIMENTAL INVESTIGATIONS INTO THE EFFECTS OF
NUTRITION ON BREEDING

There is only little experimental information on the effect of varying nutrition on breeding. In domestic species there has been much research into the effects on fertility of the level of nutriment and its quality, and this has resulted in the fairly obvious observation that if animals are fed on reduced diets (with low calorie intake) or diets deficient in any essential factor (such as proteins, minerals, or vitamins), then reproduction can be interfered with to such an extent that breeding may cease. This is really an expression of minimum fertility, but as

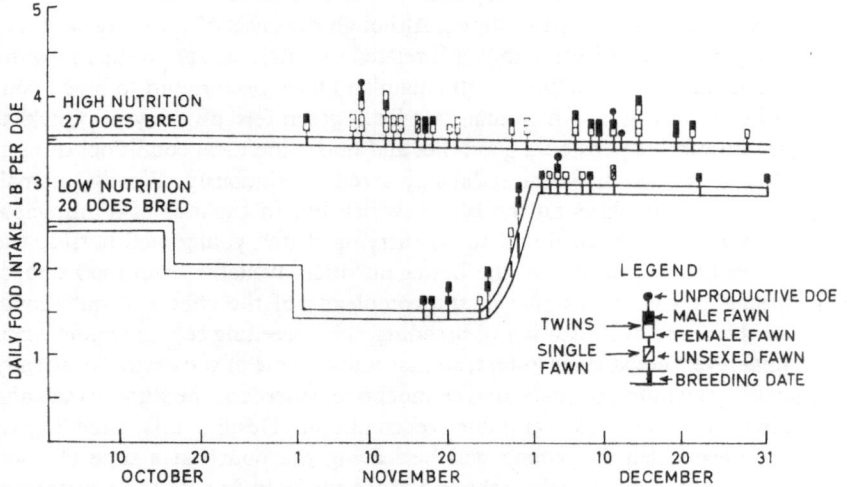

Fig. 35. Breeding dates of penned female white-tailed deer (*Odocoileus virginianus*) and the subsequent fawns produced in relation to nutritional levels. From Verme (1965) *J. Wildl. Mgmt.* **29**: 74–9.

most domestic species are continuous breeders, it is reasonable to consider these results as representing a pathological extreme. Starvation as such is a rare phenomenon in wild mammals, although it is more than possible that seasonal deficiencies in certain dietary items may affect their reproduction.

The little information that is available on non-domestic mammals relates to artiodactyls and lagomorphs. Studies of white-tailed deer (*Odocoileus virginianus*) have shown that the level of dietary intake can slightly alter the onset of oestrus (Verme, 1965). The daily food intake in one group of deer was reduced just before the normal breeding season began. It was later increased after a control group had commenced breeding. The onset of oestrus of individual does is shown in figure 35, and it will be seen that low-nutrition does started breeding later. It has been postulated by Jaczewski (1958) that the cessation of breeding in the bison (*Bison bonasus*) has been altered by the feeding regime under management conditions in Poland. The extra food given is of a relatively poor quality, and more importantly is given over a long period of the year. The result has been, according to Jaczewski, to extend the period of breeding so that approximately one-quarter of the calves are born after June and strung out into the winter months (original birth season was May and June). It is difficult to see exactly how the nutritional alteration could have had this effect in the absence of changes in any other environmental stimuli. The impression is gained from this paper that the original breeding season of bison is not known with accuracy, as it has been under conditions of semi-domestication for many hundreds of years in Poland.

A study of the effects of altering the nutrition of penned populations of wild rabbits (*Oryctolagus cuniculus*) in Australia showed that the length of breeding was changed by the addition of supplements of fresh green feed or oat grain (Stodart and Myers, 1966). The length of breeding seasons (being the mean number of days from first conception to date of last birth) after different treatments was as follows:

	Breeding season (days)
Pasture only (1962–3)	106
Pasture and fresh green feed supplement (1962–3)	139
Pasture and excess oat grain (1962–3)	138
Pasture only (1963–4)	106
Pasture and fresh green feed in excess (1963–4)	141
No pasture or green feed, oaten hay in excess (1963–4)	77

These results emphasize the absolute necessity for the presence of some green food before rabbits can reproduce. The very poor reproduction in the last treatment is considered to have resulted from the sprouting of the grain fed after rain.

For domestic animals the effect of nutrition on the onset of breeding can be really tested only on those species, such as the sheep, which exhibit seasonal breeding. It was thought many years ago that the practice of flushing, that is suddenly increasing the quality of food available to the ewe before the breeding season, would hasten the onset of oestrus, but modern work has shown that this is not so (Hafez, 1952b; for further references see Thomson and Aitken, 1959). In contrast to this has come the exceedingly interesting results reported by Smith working with Peppin Merinos in central Western Queensland. In this area the rains fall in summer (December to March), and this is the period of good feed, while the winter is a period of reduced nutrition. Smith (1964c) first noted that the duration of the period of anoestrus was different in ewes lambing at different times of the year, and also that the duration of anoestrus varied if the ewes were or were not lactating:

| | Mean duration of anoestrous period | |
	Lactating (days)	Non-lactating (days)
Ewes lambing November 1961	97	81
Ewes lambing in March 1962	41	39
Ewes lambing in July–August 1962	195	176

Now most of this variation in the period of next oestrus can be accounted for in terms of the influence of photoperiodic control, in that ewes lambing July–August are under the influence of increasing light (Southern hemisphere) for the next 6 months, whereas those in November are subjected to a shorter period of increasing light, and those in March are under decreasing light. Smith, however, interprets these results in a different way, and suggests the nutritional regime met with by the ewes at the time of lambing governs the period to next oestrus. His experimental evidence supports this idea. He put one group of sheep on extra lucerne feeding from October 1962 until the rains fell in December of the same year. The experimental and control groups of ewes were then mated in February 1963, with the following results:

	Extra feed	Controls
Oestrus 1–14 days	229	26
Oestrus 15–21 days	53	37
Oestrus 22–28 days	28	25
Anoestrus	9	19
Total ewes	319	107

This result showed that the fed ewes came into heat significantly faster than unfed ewes, the important point being that this feeding was carried out some 2 months before mating occurred.

In a later paper Smith (1965b) reported the results of a more complex experiment. In September 1962 he constituted three groups of ewes and put them on maintenance, submaintenance, and flushing diets until mid-December 1962, when the flocks were combined together until May 1963. From this time onwards the three original flocks were subdivided into two groups each and maintained for 12 weeks on either a low- or high-plane diet during the first half of the southern winter. The incidence of oestrus in each group was tested by putting them with vasectomized rams in January, March, and again in July. The results of this experiment are shown in figure 36. The important point is that the level of nutrition in the spring of 1962 (September to December) affected in a statistically significant manner the incidence of oestrus not only in January 1963 but in March and even July of the subsequent year. There was evidence of an effect of spring nutrition in the previous year when the ewes were put with the rams in December 1963.

The table below gives the mean time to onset of oestrus at that time, depending upon previous nutritional regime:

	Level of nutrition in spring 1962		
	Maintenance (days)	Submaintenance (days)	Flushed (days)
Level of nutrition in winter 1963			
High	32	45	37
Low	52	51	53

'Despite the fact that there were no significant differences since February 13, 1963, in the body weights of the three subgroups on a high level of nutrition during winter 1963, there remained a significant effect of the level of nutrition during spring 1962 upon the onset of oestrous activity during summer, 1963–64' (page 98).

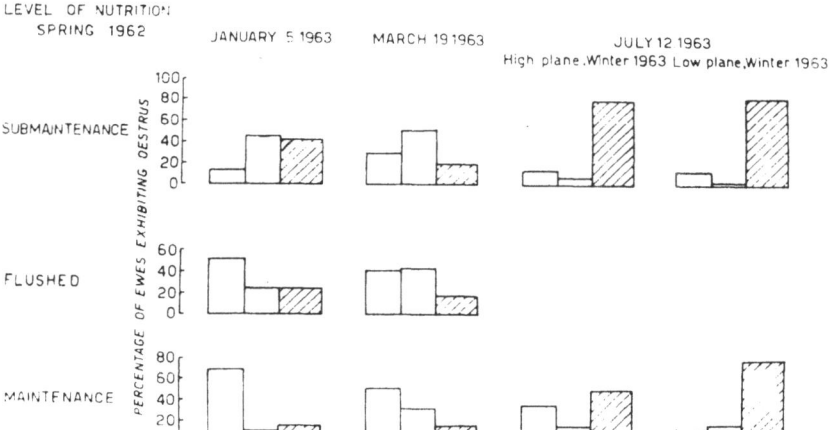

Fig. 36. The effects of level of nutrition during the spring of 1962 and winter of 1963 upon the incidence of oestrus at various times of Merino ewes. From Smith (1965b) *Wld. Rev. Anim. Prod.* **4**: 95–102.

These results are of extreme significance with regard to the role of nutrition in breeding, in that they indicate that there can be a considerable delay in the effect of nutrition on regimes which can last for at least 12 months. Smith suggests that nutrition may act on pituitary sensitivity to light response and be related also in some way to the concept of an annual rhythm.

'The cessation of the breeding season is influenced by the level of nutrition during the autumn and winter (i.e. during the later part of the synchronized phase of activity) which, if low, may cause the onset of anoestrus before the winter solstice, probably due to exhaustion of asynchronous suppression of adenohypophyseal activity. A definite period of anoestrus appears to occur only in Merino ewes in those areas where ewes are on senescent pastures during the late summer, autumn or winter months, resulting in reduced body-weights during the autumn and winter months. The onset of oestrous activity is delayed until after the summer solstice in those areas where pastoral conditions during the winter and spring are poor, otherwise it usually occurs well before the summer solstice' (page 99).

Be this interpretation as it may, the significance of these results should be remembered in wild mammals. The demonstration of a long time period between altering an environmental variable and the reproductive response has already been met with in Thorpe's (1967) work on the effect of light on the ferret. Smith's demonstration of long-term nutritional effects suggests that other environmental effects can also have delayed results. This should therefore be remembered in any consideration of the timing relationships between an observed environmental change and its effect on the reproduction of a mammal.

F. THE EFFECTS OF NUTRITION ON MALE FERTILITY

The most common method of assessment of male fertility in wild mammals has been determination of testes position (abdominal or scrotal) in the live animal or testis weight in sacrificed samples. The fertility of males of domestic species is evaluated by their ability to achieve conception (although this involves another variable, the fertility of the females used) or by examination of semen characters. The later method has also been used in one study of wild mammals (Sadleir, 1965a). The same problems of studying the animal's relationship to its food supplies, as were considered earlier (see page 14), affect the assessment of male nutrition, but the interrelationship between nutrition and reproduction is further complicated by the variation in the means of evaluating male fertility. Thus the literature on this subject is very unbalanced, much more information being available for domestic stock than for wild mammals.

(a) Rodents

It is best to commence with the comprehensive review by Russell (1948) on nutrition in the laboratory rodents, and similar useful reviews by Moustgaard (1959) and Moule (1963). The reader is referred to these papers for the original references on which the following summary is based. Reduction of total feed intake in rats and mice results in a loss of libido and decrease in the number of motile spermatozoa in the semen. This can occur when the voluntary energy intake is reduced by from 15 to 30 per cent. The general evidence is that inanition in rats results primarily in a lowered hormone production which secondarily leads to a disturbance in accessory glands and testes. The pituitaries of underfed rats have a lower gonadotrophic potency than those from well-fed controls.

Reduced protein intake is followed by a decrease in the size of the testes and by their disfunction, especially if the protein concerned is lysine. On the other hand, excess protein causes an increase in frequency of spermatozoa abnormalities. There appears to be little information on the role of minerals in male rodent fertility, except that there are reports that managanese-deficient rats show testicular degeneration and no libido. Deficiencies of various vitamins have been investigated in more detail. Vitamin A deficiency results in atrophy of the testes in mice, rats, and guinea-pigs. It is interesting that if very low quantities of this vitamin are given to deficient animals testicular function can return, even though the amounts were not sufficient to have any effect on body weight. Apparently this deficiency acts independently of inanition. In rats and hamsters Vitamin E deficiency results in a decline in sperm numbers and motility, and leads into complete aspermia and finally an irreversible testicular degeneration. This effect has not been found in other mammals. For example, Vitamin E deficiency in rabbits can be so severe that muscular dystrophy develops but no effect is seen on the testis. Finally, Vitamin C deficiency is reported to reduce spermatogenesis in guinea-pigs, and a deficiency of Vitamin B_6 results in sterility in male rats.

There appears to have been little research into the effect of nutrition on male fertility of wild rodents, the only two papers found in the literature referring to one genus. Bodenheimer (1949) described variations in the seasonal fertility of voles (*Microtus guentheri*) as shown below:

	Jan.	Feb.	Mar.	Apr.	May	June	July	Aug.	Sept.	Oct.	Nov.	Dec.
Testis weight in non-irrigated fields (mg)	322	418	435	222	162	22	20	30	60	130	255	364
Testis weight in irrigated fields (mg)	481	678	—	—	—	31	—	—	—	300	344	495

Although the data are not complete, the effect of nutrition can be easily seen, as voles in irrigated fields had a much higher level of available nutrition. From May to July vole testes were abdominal, and the animals therefore infertile.

An indirect effect of nutrition was reported by Hinkley (1966), who fed plant extracts from sprouting wheat to male *Microtus montanus*. This resulted in a 41 per cent increase in the frequency of the gonadotrophs in the anterior pituitary, the increase being largely in the delta-type basophils, which are primarily responsible for stimulating the gonads. This result was almost certainly due to the hormonal nature of the extract, and not to its nutritional value (see page 134).

L

(b) Domestic mammals

Information on the effects of nutrition on male fertility is widely scattered over the agricultural literature. Reviews by Moustgaard (1959), Moule (1963), and Duncan and Lodge (1960) have been used as the basis for the brief summary which follows. The interested reader is also referred to abstracting journals, such as *Animal Breeding Abstracts* and *Bibliography of Reproduction*, for more recent papers.

(i) *Gross feed intake.* Decreases in intake lower spermatozoa concentrations and spermatozoa motility in the bull, and also result in a decrease in the amount of secretions from the seminal vesicles, lowering the fructose and citric acid content of the semen (Moustgaard). Contrary to this, Moule commented that experiments on small numbers of bulls failed to demonstrate a depression of semen quality, or libido, as a result of underfeeding. Lower nutritional intakes cause seminal degeneration in rams (see also Gunn, Sanders, and Granger, 1942), but this is often complicated by deficiencies in Vitamin A.

(ii) *Protein intake.* In bulls intakes lowered to 1·6 per cent protein (when compared with 8 and 14 per cent) resulted in reduced spermatozoa concentrations in the semen, and overfeeding of protein did not improve normal fertility. Whether the protein supplied in the diet is of plant or animal origin has no effect in bulls. Rams kept on a nitrogen-free diet showed a decrease in semen volume some 100 days after treatment commenced, and there were indications of a decrease in number of spermatozoa. As in both sets of experiments the low-protein animals died as a result of the feeding regime, the effects of chronic inanition cannot be ruled out. Very little effect on fertility due to underfeeding has been reported for boars, although semen volume was decreased.

(iii) *Vitamins.* Deficiencies of Vitamin A result in lowered male fertility in most domestic species. Libido is reduced in young bulls, and the numbers of morphologically abnormal spermatozoa increase. Carotine deficiencies in the ram have the same effect, but both rams and bulls can respond remarkably quickly to Vitamin A supplements. Volcani (1953) suggested that seasonally lowered fertility in the domestic camel (*Camelus dromedarius*) may be due to Vitamin A deficiency. No effects are known in domestic males of Vitamin E deficiency. As ruminants can synthesize Vitamin C, it is extremely unlikely that deficiencies of this vitamin could affect reproduction in domestic animals, as suggested by some workers.

(iv) *Minerals.* Calcium and phosphorus levels do not apparently affect farm animals, and although there is good evidence for manganese deficiencies causing male infertility in rabbits and rats, there are no reports of any effects in farm animals.

(c) Man

In his review, Moule (1963) mentioned that the evidence from famine areas or in prisoner-of-war camps was equivocal with regard to the effects of nutrition on male fertility. The gross size of testes are not reduced in proportion to the gross reduction in body weight. During the Ukranian famine of 1922 and 1923 testicular tubule degeneration was recorded, although the interstitial cells remained normal. The exhaustive study of the effects of starvation on humans by Keys, Brozek, Henschel, Mickelsen, and Taylor (1950) reported that prisoners in prisoner-of-war camps had oligospermia and hyalinization of the seminiferous tubules. The sperm count and sperm motility were reduced by chronic starvation in the camps, and these still remained low after 4 months of intensive rehabilitation. A common feature in American males in prisoner-of-war camps was enlargement of the breasts (gynecomastia), although this has not been reported elsewhere. Under the controlled conditions of semi-starvation carried out in the Minnesota experiment as directed by these workers, the total number of spermatozoa in the ejaculate actually increased. However, the percentage of motile spermatozoa found 3 hours after ejaculation was reduced by over 50 per cent, and the longevity was greatly decreased. There was a slight tendency for an increase in the proportion of morphologically abnormal spermatozoa during starvation.

(d) Rabbits

The detailed studies on rabbits in Australia by the C.S.I.R.O. Division of Wildlife Research have also included investigations as to the effect of environmental conditions on male fertility. In his study of a wild population, Poole (1960) showed that the state of the pasture combined with the temperatures controlled the level of male fertility (see figure 30, p. 132). Male fertility in this study was assessed by testis weight and the presence of epidydimal spermatozoa. The results of this study suggest that temperature effects were more important than pasture in causing infertility. On the other hand, unless green pasture was available, Myers and Poole (1962, see figure 26, page 107) concluded that males would remain at a low level of fertility. In one of their pens high densities had caused feed deficiencies, so that starvation was prevalent. Males in this pen, although subjected to the same external temperatures as those in other pens with better feed, remained infertile much longer.

(e) Kangaroos

In the arid deserts of Northwestern Australia two species of kangaroo (*Megaliea rufa* and *Macropus robustus*) live in an area of highly erratic summer rainfall. Immediately after cyclonic rains fall, the protein content of the feed available rises to its highest level, only to drop to very low levels for most of the year. This is reflected in the physical condition of *M. robustus*, whose blood values are highest in late summer and lowest in spring (Ealey, 1963). Despite this fluctuation in nutrition, the level of male fertility in the two species does not apparently alter. Analysis of semen obtained by electroejaculation has shown (Sadleir, 1965a) that neither the total number of spermatozoa in the ejaculate nor the spermatozoa density varied significantly. It would seem that this species is well adapted to its environment, in that, despite rather extreme fluctuations in condition, male fertility is unaffected to any great degree. This may be due to the importance of maintaining a high frequency of conceptions in every month of the year in these kangaroos (see page 225).

(f) Discussion

In domestic species experimental deficiencies of various nutritive elements result in remarkably small effects on male fertility unless the deficiencies are practically total or maintained over long periods of time. There appear to be very few reports of gross nutritive deficiencies (unconnected with other factors, such as temperature) resulting in male infertility in managed stock, even in otherwise suboptimal environments. With the exception of the rabbit (a species almost entirely dependent on green grass for reproduction, see page 107) the very small amount of information available in the literature for other mammals would suggest nutrition is very rarely limiting to male fertility of wild mammals. Little or nothing is known of the specific energy requirements for spermatogenesis, and other male reproductive processes, although an increase can be inferred in the case of some mammals, such as artiodactyls, who markedly increase activity during the rut (see page 158). It can be inferred that the nutritional intake required for males to maintain fertility forms only a low proportion of the total intake, so that the males' reproductive processes can be maintained for long periods of poor nutrition, perhaps at the expense, in extreme cases, of other bodily functions. Acute starvation is of such rare incidence in wild mammals that there appears to have been no study of male fertility under these conditions. It would be of interest to compare the type of reproductive response seen under these rare conditions to those found experimentally in domestic stock.

Of the mechanisms whereby nutritional intake affects the male reproductive processes little is known. Moule (1963) has suggested that they could be of three types namely: (a) a decrease in hormone synthesis; (b) a failure of hormone release mechanisms; or (c) an insufficiency of substrates for the target organs to function normally. He concludes that 'a reduction in the level of circulating hypophyseal gonadotrophin appears to be the key to reproductive malfunction among laboratory animals subject to protein deprivation, although adequate substrate must also be available for the target organs to produce their own secretions' (page 302). In farm animals the hypophysis must be of major importance, as injection of gonadotrophins can restore or maintain fertility in males under conditions of inanition. A purely subjective comparison of the levels of inanition reported in wild and domestic mammals would suggest that in the former process (c) above will rarely, if ever, occur.

CHAPTER 12

The Effects of Minor Ecological Factors on the Breeding Season

A. THE EFFECTS OF ALTITUDE

The actual height at which an animal lives does not in itself cause variation in the breeding season, but a number of other ecological factors vary considerably with height, and thus can affect breeding phenomena. It is a pity that so little is known about the exact effect of increasing height on other environmental variables. At extreme heights the effect of reduced oxygen concentration in the atmosphere is well known, but we know little of the changes in variables other than temperature or oxygen. The altitude itself may not be so important as the aspect of the region under consideration, as this will alter the light regime (if the aspect is other than due east–west) and will also affect the rainfall and possibly nutrition available to mountain species. A common feature of many mountain-dwelling large mammals is a vertical seasonal migration pattern, so that the effects of the altitude itself will need to be considered in the light of these movements. Thus it will be appreciated that the breeding pattern of mammals will vary according to the altitude, but this is due to a combination of individual ecological factors which are usually not separable as to their individual action. The interested reader is referred to a paper by Roberts, Hock, and Smith (1966) which analyses the metabolic responses of deermice (*Peromycus maniculatus*) to temperature and altitude.

The onset of breeding in species with wide altitudinal limits occurs later in the year in high-altitude populations. Dunmire (1960) found that *Peromyscus maniculatus* in California started breeding in late summer at altitudes of 12,400 ft, whereas in the same region populations at 9,800 ft and below started in the normal spring. Long (1964) found his first litters in the same species in June at high altitude, and Zejda (1966) reported that *Clethrionomys glareolus* in Czechoslovakia started breeding in June at high altitudes, but in April at lower altitudes. Shubin (1964) reported that chipmunks (*Tamias sp.*) started breeding a month later at higher than lower altitudes in Siberia. It would seem that the total length of the breeding season is reduced with increased height, as the evidence from these studies is that breeding ceases at approximately the same time of the year at all altitudes or possibly even earlier in the higher regions. The effect is not confined to rodents, as Wodzicki and Darwin (1962) found that the rabbit

(*Oryctolagus cuniculus*) in New Zealand also had a shorter breeding season at high altitude. One of the effects of this shortened season is that the total number of young produced is reduced, as shown by Dunmire (1960) for *Peromyscus maniculatus*: Numbers of young per adult female per year, 4,500 ft, 19·2; 7,100 ft, 15·2; 9,800 ft, 5·5; 12,400 ft, 4·8.

The actual ecological variable which causes the delayed breeding season is often the lower temperatures which prevail at higher altitudes. This affects the time of melt of the snow cover and thus the time of growth of the vegetation. McKeever (1963) found that the Belding ground squirrel (*Citellus beldingi*) became active and started breeding in early March at altitudes of 4,500 ft in the Great Basin of the Western United States, but at 5,600 ft and above it did not commence activity until early April. The time of emergence from hibernation and subsequent breeding was correlated with the growth of grass and sedges, and less with the actual snow melt, as in certain years the animals emerged when the snow was still 2–3 ft deep. Similarly, Hoffman (1958) compared the breeding season of a lowland species of vole (*Microtus californicus*) with a mountain species (*M. montanus*). The latter started breeding in June, compared with the almost-year-round breeding of the former. This was directly correlated with the protein content of the grass available, which was also reflected in litter size (see page 178).

Altitudinal effects parallel the effects of breeding seen as latitude increases. The shorter breeding season (page 64) and increased litter size (page 193) at higher latitudes have been fully described. The litter size of mammals does not seem to increase as altitude increases so that the shorter breeding season results in fewer litters, which are not compensated for by increased numbers. Therefore the total production of young (see example above) would seem to be smaller in populations of species living at higher altitudes. This poses a very interesting problem with regard to population control when recruitment is reduced.

Apparently altitude does not affect the breeding season of some species. For example, Dunmire (1961) could find no reduction in the length of breeding in *Reithrodontomys megalotis* at high altitudes in California.

Exceedingly high altitudes have certain traumatic effects on reproduction in mammals. Testis function is greatly impaired when humans, cats, rabbits, or rats are transferred above 10,000 ft. For example, Altland and Highman (1968) have shown that exposures of male rats to stimulated altitudes of over 20,000 ft have induced considerable degenerative changes in testis structure. Bishop and Walton (1960) have reviewed the literature on this phenomenon and show that the reduced testicular function is due to hypoxaemia. A fascinating historical result of this effect was that the capital of Peru was transferred in 1535 to Lima (sea-level) from Jauja (11,500 ft) because horses, pigs, and fowls would not reproduce at the latter locality.

B. THE PHENOMENON OF INTERRUPTED BREEDING SEASONS

For most mammal species the seasonal changes in the environment which result in conditions being optimal for reproduction occur only once in the annual cycle but last for a continuous period once they have begun. Therefore the reports from the literature of mammal species with either two marked peaks in their continuous breeding season or two completely separate periods in the year during which mating occurs should be considered in the light of the environmental conditions causing such interruptions of breeding. There are also species which normally breed continually but can have their continuous breeding interrupted. Some species which mate at two separate periods during the year probably do so because their relatively long gestation and lactation periods only allow two litters per year. This cannot be considered as interrupted breeding, because, in effect, the species is continually reproducing. However, in the smaller mammals with short periods of time from conception to weaning (or to postpartum oestrus) there are cases where both sexes cease reproduction during one period of the year and then restart inside the same calendar year. *Peromyscus boylii* and *P. leucopus* in Missouri have two definite breeding peaks in February and April, and later in September and November (Brown and Conaway, 1964; Brown, L. N., 1964). Although the testes do not regress entirely in some summers, females do not become pregnant during the period of interruption. This biphasic reproduction cannot be easily explained by light or temperature variations, and while a lack of water may be the cause of cessation of breeding in midsummer, this does not explain why females start breeding in October, before the first winter rain falls.

Aestival interruptions to breeding have been reported or other species of mammals. Prakash (1958) found that the rhesus monkey (*Macaca mulatta*) did not breed from June to August at Jaipur, India, and Hoffmeister (1963) found no pregnancies in midsummer in the cotton rat, *Sigmodon ochrognathus* in Arizona. Dunnet (1956) reported that the brush-tailed possum (*Trichosurus vulpecula*) did not breed during the Southern hemisphere summer from November to February. These summer interruptions would seem to be due either to hot weather or to lack of water during the middle of the normal breeding season, but there is no available evidence to determine which of these factors is more important. The effect of high temperatures on fertility was considered on page 118.

The patterns of interrupted breeding in certain mammals would seem worthy of more detailed investigation. The environment variations causing the interruption may have traumatic effects on reproduction, which in other years and in other localities is not interrupted by environmental stimuli. Documentation of these patterns would thus give an insight as to the nature of the physiological response of the species to relatively traumatic effects but under natural conditions.

C. MIGRATION AND BREEDING

The effect of environmental stimuli on the reproduction of many species of birds and some species of fish is closely related to the long-distance seasonal return movements known as migrations. Migrations are not nearly so common phenomena among mammals, but it is appropriate to consider here the very little information available which relates migratory activity to breeding seasons in this group.

Perhaps the best-known migration of a mammal is that of the caribou (*Rangifer tarandus*) in the northern parts of Canada. This species lives in the Barren-grounds during the summer and migrates (usually south) to the timber line in late autumn, remaining there until spring, when it migrates north again. The timing of mating is not fixed in relation to the variable autumnal migration, but the calves are dropped mainly in June. In a recent monograph Kelsall (1968) has made it clear that parturition takes place in special calving grounds and not during migration as had been suggested by some authors (Bouliére, 1954). The calving grounds are always at the highest altitude available, where the weather conditions are often exceedingly harsh. There appears to be little information as to the effect of environmental variables on the reproduction and migration of this species, although it would be most peculiar if a photoperiodic effect was not involved. The high mortality of the caribou calves due to environmental factors is discussed on page 217. Other species of ungulate, such as bison, elk, and mule deer, also undergo migrations of greater or lesser length, but there appears to be little information linking reproduction to this behaviour.

As they have been more and more subjected to intensive study, it is becoming evident that migration is a fairly common feature of the seasonal activity of those aerial mammals, the bats. Bouliére (1954) shows maps of recoveries of European species and speculates on their migratory paths. In the United States Findley and Jones (1964) have reported that females of the hoary bat (*Lasiurus cinereus*) migrate north in May and June, earlier than males. The young are born and raised in summering grounds in the eastern United States, while apparently the males are concentrated in the western part of the continent. The return migration south is in the autumn. There is no available information on the environmental factors involved.

The aquatic mammals have definite migration patterns, which are especially evident in the Cetacea. The effect of migration on the attainment of puberty in the southern fin whale (*Balaenoptera physalus*) has already been mentioned (page 10). This and other species, migrate regularly from the Antarctic to tropical waters and back again each year. Dawbin (1966) has studied the breeding and migrations of humpback whales (*Megaptera nodosa*). In this species the calves are born in the tropics at latitudes below 15° S in water of high surface temperature. Pregnant females and lactating females migrate north from the feeding grounds at two slightly different times of the year. Dawbin has suggested that the non-pregnant lactating females are more sensitive to photoperiodic changes than their pregnant sisters, so that they migrate north earlier. This species would thus seem to be adapted to giving birth to young in warmer waters, but the environmental stimulus which sets off this migration is possibly light change.

Some of the pinnipeds (harp seal, *Pagophilus groenlandicus*; Alaska fur seal, *Callorhinus ursinus*) are known to undergo extensive migrations. As these species, among others, give birth and copulate in only particular localities, migrations to the breeding beaches are related to reproductive phenomena. The ecological factors involved here have been little studied, except that the nutritional strain of cows during the severe lactation, and on the territory-defending bulls, must necessitate the animals being in areas where they can lay down stores of blubber before moving to the locations where they reproduce. The breeding beaches and ice-shelves themselves, as well as having a strong traditional attraction for the species, would also seem to have peculiar physiographical, and thus environmental, features which are attractive to the seals. Migration is thus closely interwoven with breeding in a number of pinnipeds. Studies on migratory birds (Farner, 1954) have shown very clearly that the migration itself is closely related to basic metabolic changes concerned with the laying down of fat reserves and their subsequent utilization, a whole interrelated picture being built up between the animal's activity, its fat reserves, and its state of reproduction. In birds all this is controlled by photoperiodic phenomena, so that it is not unreasonable to suggest that further investigation of the migratory pinnipeds would demonstrate similar mechanisms operating in these species.

D. BREEDING AS AFFECTED BY LUNAR VARIATIONS

Probably as a result of the human menstrual cycle being the same length as the lunar cycle, the suggestion has often been made that there is some causal relationship between them. Other aspects of human reproduction have been suggested as being affected by the lunar cycle, but in most cases the statistical analysis has not indicated any real significance, but merely a possible relationship. For example, McDonald (1966) analysed the time of human births in South Carolina inside the lunar month and suggested, on data which showed no statistical evidence of significant variation, that more births occur during the full moon and the new moon phases than in the quarter phases.

In other mammals Cowgill, Bishop, Andrew, and Hutchinson (1962) suggested that there was a higher incidence of mating in species of *Lemur* and *Galago* during the full moon. Harrison, J. L. (1952, 1954) investigated the breeding of several species of forest-dwelling *Rattus* in the Malayan jungle, and found a strong tendency for conceptions to be more frequent in the period immediately before the new moon. This deduction is somewhat questionable as to statistical methods used, and also because he combined data from several different species in the same analysis.

Although the possibility exists that the extra light available during the full-moon phases has a stimulating effect on reproduction in nocturnal species, to date no evidence has been presented to support this idea. It is also possible that crepuscular species mate more often during the full moon simply because the extra light results in longer periods of activity at this time, so that an increased incidence of mating is a secondary effect.

E. THE ROLE OF THE MALE IN BREEDING

In previous sections breeding ecology has been considered largely in terms of the reaction of the female. We will see, however, that the male presence can have important effects in stimulating the female at the onset of breeding and can synchronize her heats with those of adjacent females (page 253). Thus the male himself acts as an environmental factor in the reproduction of the female, and can be very important in his effects on the fertility of the female population. This section will attempt to collate and consider other aspects of the male's role in the ecology of reproduction.

I. The male rut

In many species of mammal the breeding season of the male is characterized by a considerable amount of sexual display, which is closely associated with a high level of inter-sexual aggression. This type of behaviour has led to the development of different types of social system exhibited during the period of mating. These systems have been extensively and expertly reviewed by Eisenberg (1965), and the reader is referred to his monograph for a more detailed consideration of the ethological aspects of the subject. A few examples will be given here to demonstrate how common is male inter-sexual aggression in mammals.

The phenomenon of rut is best documented in the Artiodactyla (in particular the Cervidae), where the development of antlers as a secondary sexual characteristic is associated with much fighting. This has been described for many species, including red deer (*Cervus elaphus*) (Lowe, 1966), roe deer (*Capreolus capreolus*) (Short and Mann, 1966), lechwe (*Kobus leche*) (Allen, 1963; Robinette and Child, 1964), puku (*K. vardoni*) and waterbuck (*K. ellipsiprymnus*) (de Vos and Dowsett, 1966), kob (*Adenota kob*) (Buechner, 1961a, 1963; Leuthold, 1966); wildebeest (*Connochaetes taurinus*) (Estes, 1966), mountain goat (*Oreamnos americanus*) (Geist, 1964), bighorn sheep (*Ovis canadensis*) (Blood, 1963), reindeer (*Rangifer tarandus*) (Meschaks and Nordkvist, 1962), whitetailed deer (*Odocoileus virginianus*) (Severinghaus and Cheatum, 1956), blacktailed deer (*O. hemionus*) (Dasmann and Taber, 1956), moose (*Alces alces*) (Dodds, 1958), saiga antelope (*Saiga tatarica*) (Rashek, 1963), American bison (*Bison bison*) (Soper, 1941), and muskox (*Ovibos moschatus*) (Tener, 1965). This by no means exhaustive list indicates how common aggressive fighting at the time of rut is among this order. The interested reader is referred to the review of the behaviour of North American Cervidae by de Vos, Brokx, and Geist (1967) and to the excellent coverage of the phenomena of rutting in ungulates generally by Fraser (1968).

In many cases (lechwe, puku, waterbuck, mountain goat) this aggression results in considerable wounding and occasional death from wounds as the result of prolonged fighting.

In some species there is also a considerable loss in weight and overall physiological condition, which, though noted, has unfortunately been little quantified. Severinghaus and Cheatum have reported a whitetailed deer buck which dropped from 225 to 207 lb during 8 days of the rut. Much of the energy expressed at this time is also taken up with activities such as roaring, pawing the ground, shaking antlers in undergrowth, and rubbing them on the ground and on trees, as well as actual fighting.

Similar behaviour has been documented for other mammals. Among the Perissodactyla fighting during the breeding season has been described for a number of species of horses, including the kulan (*Equus hemionus*) (Rasek, 1964), and also in the black rhinoceros (*Diceros bicornis*) (Goddard, 1966). The territorial behaviour of seals and the very considerable amount of between-male aggression and wounding which occurs in these species is well known: grey seals, (*Halichoerus grypus*) (Hewer, 1960; Hewer and Backhouse, 1960); Weddell seals (*Leptonychotes weddelli*) (Mansfield, 1958); elephant seal (*Mirounga leonina*) (Laws, 1956). Aggression specifically associated with breeding has also been reported in the lagomorphs: cottontail rabbit (*Sylvilagus floridelanus*) (Marsden and Conway, 1963), rodents: ground squirrel (*Citellus beldingi*) (McKeever, 1966) and primates: Rhesus monkey (*Macaca mulatta*) (Vandenbergh and Vessey, 1968).

The rut in all the above species coincides with the time of mating, so that it forms part of the environment in which conception is taking place. Probably the main ecological factor affecting the rut itself is the level of nutrition available. During this time males use up a very considerable amount of stored energy in the form of fat, as in most cases the short intensive rut is a period when they do not feed. There is little information available as to the amount of energy required, and the necessary alteration in dietary requirements to store this energy which precedes the time of male breeding.

The examples given above are mainly of species in which the rut takes place over a limited period of time during the year. The majority of male mammals have a more prolonged breeding season, but during this time their breeding activity is not of such an intense nature. This activity almost certainly results in very considerable alteration in the nutritional and metabolic picture when compared with the non-breeding season.

The work of Ewing, Green, and Stebler (1965) has shown that major biochemical and metabolic changes occur, relative to the histological changes in the testis of the cotton rat (*Sigmodon hispidus*) at different seasons of the year. An indication of the nature of the changes found is given in the table on page 160 (from their paper).

During the period of redevelopment of the cotton rat's testes the metabolism is characterized by a high utilization of oxygen, whereas during the stage of

degeneration glycolysis predominates. These changes would probably be associated in the wild with marked changes in dietary preferences, which would thus
be affected by the nutrition available to the species at the time. There is a very
real need for an extension of this sort of approach, combining laboratory metabolic studies with field studies of nutritional changes, to other male mammals.
Laboratory work in mice has already shown that in the smaller species (with

	Period of redevelopment, Feb.–May	Active breeding season, June–Sept.	Period of degeneration, Oct.–Jan.
Testis weight (mg)	569 ± 134	2,268 ± 94	750 ± 155
Seminal vesicles (mg/100 g body wt)	87 ± 25	979 ± 106	95 ± 31
Ventral prostate (mg/100 g body wt)	0	219 ± 32	29 ± 9
Oxygen uptake (μg/mg dry wt/hr):			
Endogenous	8·22 ± 0·65	6·28 ± 0·37	6·66 ± 0·36
Substrate added	7·53 ± 0·42	8·96 ± 0·45	6·12 ± 0·39
Glucose uptake (μg/mg dry wt, hr)	90·16 ± 11·70	80·08 ± 5·39	95·41 ± 10·45
Lactic acid production (μg/mg dry wt/hr):			
Endogenous	1·66 ± 0·99	1·77 ± 0·50	6·62 ± 1·89
Substrate added	17·78 ± 2·78	26·55 ± 2·07	43·92 ± 5·95
Protein content (μg/g testis)	83·21 ± 3·86	76·17 ± 1·63	74·81 ± 2·66
RNA/DNA ratio	0·710	0·484	0·435

faster metabolic rates and thus a need to feed continually) short periods of nutritional depletion can cause infertility. McClure (1966) found that if mice were
fasted for periods of 36 or 48 hours their libido was reduced for the next 2 days.
He suggested that this effect was due to an acute energy deficiency. Little is
known of this in higher mammals, although chronic starvation is known to
reduce libido in humans under conditions of starvation in prison camps (Keys,
Brozek, Henschel, Mickelson, and Taylor, 1950).

II. Male breeding season relative to female breeding season

The necessary prerequisite for fertile conception is the presence and co-operation
of both a male and female of the same species. In many field studies of the reproduction of mammals the continual fertility of the males of the species is too often
assumed without documentation. The timing of male breeding relative to female
reproduction is basically maintained in one of two ways. In the first, the males
are fertile and capable of mating throughout the year, and the females either
breed continuously or have a discrete breeding season, and in the second the male
also has a breeding season which coincides with or overlaps, that of the female.

Probably the commonest arrangement of these two is the latter. Generally the
male breeding season is slightly longer than that of the female's, in that fertile
males are found before and after the dates of conception in the female. A few
of the many examples of this pattern are: field vole (*Microtus agrestis*) (Brambell
and Hall, 1939); Indian gerbil (*Tatera indica*) (Prasad, 1956); and snowshoe

hare (*Lepus americanus*), (Bookhout, 1965). The later paper gives dates which show the relative timing of male and female season rather exactly:

	1959	1960	1961
Earliest known spermatogenesis	10 Feb.	9 Feb.	22 Feb.
Earliest conception date	10 Apr.	10 Apr.	3 Apr.
Earliest parturition date	17 May	17 May	10 May
Latest known spermatogenesis	13 Aug.	15 July	17 Aug.
Latest conception date	30 July	6 July	25 July
Latest parturition date	7 Sept.	13 Aug.	31 Aug.

In species with rather long breeding seasons such as this one the onset of female breeding would appear to be more dependent than males' on actual weather conditions. Kline (1962) found that male cottontails in Iowa (*Sylvilagus floridanus*) all attained breeding condition in February, but in four different years the first pregnancies were found in weeks varying from that beginning 14 February to 21 March. In more northern regions the timing of the male season is much more closely tied to that of the short female season. Moore, Simmons, Wells, Zalesky, and Nelson (1934) found that male *Citellus tridecemlineatus* emerged from hibernation 10 days before females in Illinois, and mating took place only in April. Farther north this breeding period is even more reduced, as Mitchell, O. G. (1959) found that the male Arctic ground squirrel (*C. undulatus*) started producing spermatozoa 1 week after emergence from hibernation and would mate only in the subsequent 2 weeks in early June. By late June the testes have completely degenerated (see also Hock, 1960).

A slightly peculiar variant of this pattern is seen in the dik-dik (*Rhynchotragus kirkii*), which has two breeding seasons per year in the region of Old Shinyanga, Tanganyika (Kellas, 1955). In this case the male's testes and accessory glands show a peak in May and another in November, and the female shows two groups of conceptions in June and July and in December and January.

The former pattern of relationships, in which the males are fertile all the year round and the females have a discrete breeding season, is a much more relative arrangement, because male fertility can drop to low levels on a population basis without having actually interrupted the sequence of conceptions. However, there are a few clear cases where males are continually fertile and yet the females have a breeding season. In two species of Australian marsupials, the possum (*Trichosurus vulpecula*) and the potoroo (*Potorous tridactylus*), the females give birth during two discrete seasons of the year and yet the males are apparently continuously fertile (Dunnet, 1956; Hughes, 1962, 1964). The female breeding season in these species may be due, however, to interrupted breeding (page 155) and may be more continuous than previously reported (Sharman, Calaby, and Poole, 1966). In other normally continually breeding species the female season may be interrupted and yet the males may continually be capable of conception. Perry (1945) found fluctuations in the incidence of pregnancy in the wild brown rat (*Rattus norvegicus*), but maintained strongly that this could not be due to the male breeding pattern, as there were always breeding males present. Similarly, Pearson (1963) maintained that a *Mus musculus* outbreak in California was due to the unusual breeding of females between December and April. In other years females had not bred during this time, though males were always in breeding condition. However, their breeding condition dropped over the winter, according to Breakey (1963). Without data on the actual number and distribution of the males causing conceptions it is hard to accept the conclusions of workers such as DeCoursey (1957), who suggested that the reason that he found no pregnancies in *Microtus ochrogaster* in January in Ohio was due to the absence of fertile males in December. This worker presented data indicating that in any month from October to December less than 20 per cent of the males sampled were fertile, so that the actual size and representivity of his sample is crucial. At the same time, data such as are presented by Hanney (1965) for *Mastomys natalensis* in Malawi would strongly suggest an effect of male fertility on the female season:

	Feb.–Mar.	Apr.–May	June–July	Aug.–Sept.	Oct.–Nov.	Dec.–Jan.	Feb.–Mar.	Apr.–May	Jan.
Males fecund (%)	100	100	36	54	100	100	100	86	60
Females fecund (%)	—	81	53	43	0	14	90	60	67
Females perforate (%)	54	100	18	0	0	79	100	100	67

However, this and similar data can be biased by the number of juveniles sampled, which alters the proportions of reproducing animals per month. There can be little doubt from the data presented by Myers and Poole (1962) (see also figure 26, page 107) that reduced fertility in the male rabbit (*Oryctolagus cuniculus*) can cause interruptions to the breeding of females.

The presence of breeding males at seasons of the year when female breeding does not occur raises some interesting questions. The continual maintenance of a physiological state whose original function cannot be fulfilled for part of the year would suggest that some other function is involved. For example, there is evidence connecting the continual production of testosterone with aggressive and often territorial behaviour in birds, so that in mammals it may be necessary to maintain the male of a species in a certain behavioural state for ethological or population dynamical reasons. Buss and Johnson (1967) have suggested that the aggressive behaviour of male elephants (*Loxodonta africana*) is related directly to the testosterone content of their testes. Although in most mammals the male physiology is such that spermatogenesis and interstitial cell development proceed in conjunction with the development of the accessory glands (i.e. in the roebuck (*Capreolus capreolus*) (Short and Mann, 1966)) there are species in which the different stages of the male's reproductive cycle occur at different times of the year. For example, the seasonal cycle of the interstitial tissue in a number of species is separate from the spermatogenic cycle (eastern mole (*Scalopus aquaticus*) (Conaway, 1959); several bats (Wimsatt, 1960)). In these species nutritional or other environmental factors may necessitate the occurrence of the different stages at separate times of the year.

F. THE EFFECTS OF LACTATION

Lactation is known to affect a number of other stages in reproduction of mammals, particularly in terms of the implantation of the blastocyst (page 205). There is one report, however, which presents clear evidence that lactational incidence can alter the timing of breeding in the subsequent season. Koford (1966), in his study of rhesus monkeys (*Macaca mulatta*) on Cayo Santiago, Puerto Rico, noted that the birth distribution in any year was affected by the survival of infants from the previous year because multiparous females which were not lactating (due to post-natal mortality or a failure to reproduce) came into heat earlier than lactating females. It is possible that this effect was due to the increased nutritional strain caused by lactation in this primate, but as the population was fed from hoppers in addition to the natural food supply, this seems unlikely. Lactation may also affect other species in the timing of their breeding. A single observation by Erickson, Nellor, and Petrides (1964) suggested that oestrus in the black bear (*Euarctos americanus*) is inhibited by lactation. A trapped female which was held away from her cubs for 3 days during July of one year was noted as giving birth in the subsequent year. This is unusual in this species, in which individuals usually breed only every second year. From the date of birth it was calculated that the interrupted lactation caused the female to come into oestrus and conceive.

PART THREE

Pregnancy

CHAPTER 13

The Effects of Temperature on Pregnancy

A survey of the literature has revealed that almost nothing is known in wild species of the effects of temperature on mammalian pregnancy once conception has taken place. In wild mammals this period of the reproductive cycle appears relatively unaffected by temperature effects, as homeothermy renders females independent of this environmental variable. In domestic mammals, other than sheep, the same would apparently apply, so that this particular species may be a special individual case.

(a) Sheep (*Ovis aries*)

To commence at the beginning of pregnancy, three papers by Dutt and his associates (Dutt, Ellington, and Carlton, 1959; Dutt, 1963; and Dutt, 1964) have reported investigations of the effects of relatively high air temperatures on early pregnancy. These workers exposed groups of ewes to temperatures of 32° C (90° F) for 24 hours on either the 12th day of the oestrous cycle or 8 days after artificial insemination at oestrus. Only 1 of 20 in the former group lambed, compared with 17 of the later 20. Comparisons of the ewes heated before breeding with unheated controls showed the following:

	Heated ewes	Control ewes
Cleavage rate of ova (%)	51·9	92·6
Proportion of abnormal ova (%)	44·2	3·7
Proportion of fertilized ova which died (%)	91·7	4·0

Later work narrowed down the actual time when heat most affected pregnancy, as ewes were heated either on the day of insemination or 1, 3, or 5 days after. The percentage of abnormal ova under each of these treatments showed that the ova were most sensitive at the initial stage of cleavage. In further experiments ewes were heated up to 8 days after breeding, but the results still confirmed the sensitivity of ova. It is possible that extreme cold at the same time may have a similar effect; Averill (1955) has reported that an 11-day period of very adverse weather with heavy frosts and snow resulted in a decrease in the percentage of ewes conceiving at first mating during a study of seasonal conception rates at Cambridge, England. Aircooling of Rambouillet ewes in Kansas during the summer increased the percentage of fertilized ova, whereas forced exercise reduced it (Spies, Menzies, Scott, Coon, and Kiracofe, 1965).

The remaining papers are concerned with results from ewes which have been subject to longer periods of high temperatures right through pregnancy. Moule (1950a), writing before the work above was carried out, postulated detrimental effects of heat on pregnancy. He compared the neonatal mortality of lambs born in Queensland to ewes mated at different times, and thus subject to varying lengths and intensities of heat during the Southern hemisphere summer (May–August):

Season of mating	Neonatal mortality (%) (proportion of lambs born)
October–November	22
February–March	14
April–May	9
May–June	12

Despite the excellent nutrition available to the earliest mated group of ewes, high temperatures had such a detrimental effect that lambs were weak at birth and were often unable to suck the plug of colostrum out of the teat. Yeates (1965), in a review of climate and reproduction in farm animals, makes the general comment that the effects of high temperatures on ewes in northern Australia are often increased by the poor nutrition available at the same time. The general vulnerability of the ewes to hot weather during pregnancy has also been noted in South Africa and in more southern areas of Australia, which are cooler but occasionally subject to unusually hot summers, such as that of 1951–2, which resulted in a very high incidence of prenatal mortality (Morley, 1954). Experimental work by Yeates (1956a) compared the effect of prolonged heating of Merino ewes at either 105° F dry-bulb, 92° F wet-bulb or at 112° F dry-bulb, 98° F wet-bulb. He found that the birth weights of lambs in the former group were not reduced from controls, but the higher-temperature group was lighter. This confirmed Moule's suggestions and demonstrated that Merinos are relatively less affected by heat than Romney Marsh, as previous experiments had shown that no ewes of the latter breed lambed at the lower of the two temperature regimes used.

The very high temperatures used in these experiments or the sudden 24-hour maintenance of temperature used in the first group suggest that the effects seen may be of a pathological nature. Even though sheep over their present distribution in Australia and South Africa are kept in areas where these temperatures repeatedly occur under range conditions, the important question is whether these are temperatures that the species would be subjected to in its native pre-domestication environment. It should be remembered that, if this is not so, then these sheep are being kept in an environment to which their genetic and physiological constitution is not adapted. Therefore field and laboratory experiments only emphasize that sheep reproduction is being detrimentally affected because the species is being maintained outside its natural environment. There appears to be a very real need for climatological investigations of the areas where domestic species live, so that comparisons can be made with the climates to which they are subjected in different areas of the world to which they have been 'exported'.

(b) Rabbits

A study of reproduction in caged domestic rabbits (*Oryctolagus cuniculus*) in Southern California showed that some of the documented parameters varied inversely with the maximum temperature (Sittman, Rollins, Sittman, and Casady, 1964). Figures 37 and 38 indicate that conception rate and percentage mortality (consisting of still-born rabbits from litters with at least one rabbit born alive) showed a particularly close relationship with the temperature. These graphs are based on nearly 18,000 births from 1943 to 1960 in colonies kept on a continually high standard of nutrition, and in cages maintained in sheds cooled by sprinklers during the summer months. It will be seen that the maximally detrimental effects were noted in September at the end of the summer.

A single observation by Kline (1962) indicates that extreme cold may also be detrimental to pregnancy in the cottontail rabbit (*Sylvilagus floridanus*). Two of seven females collected on 11 March 1959 after a very severe blizzard which lasted from 4 to 6 March showed complete resorption of their litters, a very unusual phenomenon in this species. These litters would have been conceived about the day before the blizzard struck.

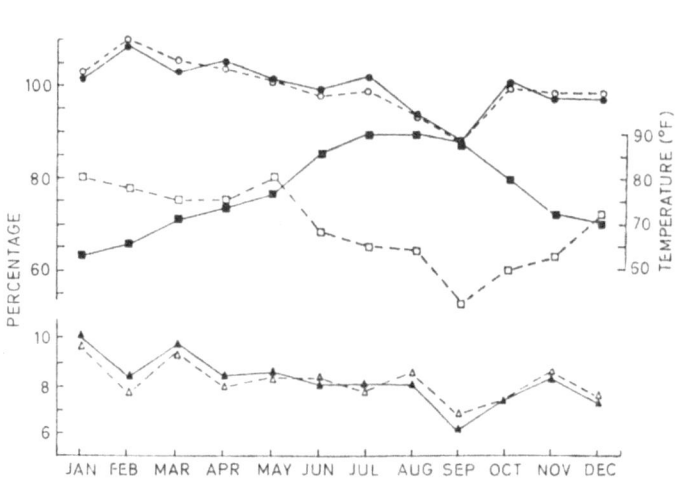

Fig. 37. The effect of season on reproduction in the caged rabbit (*Oryctolagus cuniculus*). ○: average born alive per litter per month as percentage of overall average; ●: average total litter size per month as percentage of overall average; □: percentage conception; ■: average maximum temperature; △: number of litters born per month as percentage of overall total; ▲: number of rabbits born per month as percentage of overall total. From Sittman, Rollins, Sittman, and Casady (1964) *J. Reprod. Fert.* **8**: 29–37.

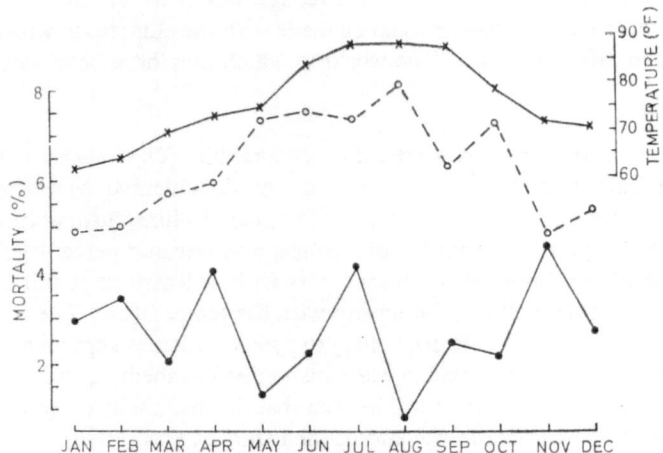

Fig. 38. The effect of season on still-births in the caged rabbit (*Oryctolagus cuniculus*). × : average maximum temperature; ○ : still-born rabbits from litters with at least one born alive as a percentage of all rabbits born per month; ● : still-born rabbits from litters with no rabbits born alive as a percentage of all rabbits born per month. From Sittman, Rollins, Sittman, and Casady (1964) *J. Reprod. Fert.* **8**: 29–37.

(c) Rodents

The effects of temperature on breeding of rodents was first experimentally investigated by Parkes (1924), who noted that the litter frequency of mice (*Mus musculus*) maintained out of doors was less than when the mice were maintained in a heated laboratory. Baker and Ransom (1933a) found that if *Microtus agrestis* were kept at 5° F they bred less frequently than normal temperature controls, although Knudsen (1962) reported that the litter size in mice did not vary when they were kept at 18°, 25°, and 32° C.

The breeding of laboratory mice at colder temperatures has been investigated in very great detail by Barnett (Barnett and Coleman, 1959, and Barnett, 1962, will give a bibliographic coverage of this work), but he has concentrated more on the frequency of breeding (i.e. length and distribution of the oestrous cycles and frequency between litters). Generally mice kept under cold conditions produced less young per year, but there were very considerable differences between strains in the effects of cold.

At the other end of the temperature scale the effect of severe heating of laboratory rats has been investigated by MacFarlane, Pennycuik, Yeates, and Thrift (1959). These workers halved the litter size by exposing females to 35° C between the 6th and 12th days of gestation. This decrease was found to be due to fetal resorption, which could be somewhat reduced by increasing the protein and vitamin content of the diet, and also by the administration of either thyroxine or progesterone. If animals were acclimatized for periods of between 14 and 79 days the foetal loss rate dropped till approximately equal with controls, but the ovulation rate dropped and there was a greater loss at the stage of implantation.

In subsequent experiments the same workers maintained a strain of rats over their whole lives at 32–35° C. Some of the differences which were seen are shown in the table below:

	Rats at room temperature	Rats at 32–35°
Proportion possible oestrous females served (%)	81·9	51·4
Proportion females pregnant (%)	65·3	34·7
Corpora lutea	9·1 ± 1·0	7·4 ± 2·5
Implants	8·4 ± 1·2	4·7 ± 1·9
Viable 16-day foetuses	8·3 ± 1·2	1·2 ± 1·2
Viable young	8·1 ± 1·3	0
Gestation (days)	22·1 ± 0·5	24·6 ± 1·7

The addition of vitamin supplements to the rats maintained at the higher temperatures considerably reduced the numbers of implants lost in the heat. Depending on the stage at which the supplement was implemented, the number of viable young was raised from 2·2 to 6·2 per female.

(d) Humans

The conception rates of humans show seasonal changes which seem to be most closely related to temperature. Figure 39, which is taken from a very interesting

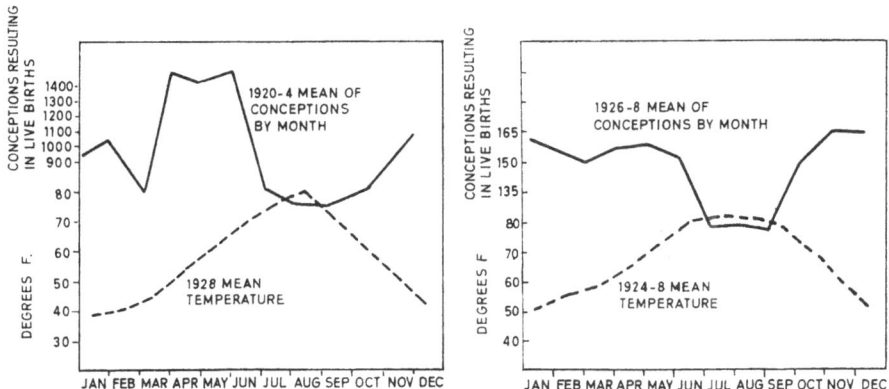

Fig. 39. The effect of temperature on conception rate in humans in Japan and Charleston, South Carolina. From Mills, C. A., *Medical Climatology*, 1939. Courtesy of Charles C. Thomas, Publisher, Springfield, Illinois.

and apparently frequently overlooked book by Mills (1939), shows the conception rates per month in South Carolina and Japan, and compares them with the mean monthly temperatures at these two localities. From this and other data, Mills concludes that human conception rate declines above 70° F and below

40° F. This is particularly evident in the graph of data from Japan. What is more, there is excellent evidence that for Japan this is a real shift in fertility and not due to a drop in the incidence of copulation. The methodical Japanese police have data (unfortunately not presented), according to Mills, on the monthly incidence of the use of brothels, and this shows no decline in hot weather.

More recent work by MacFarlane and Spalding (1960) has borne out Mills' deductions. The conception rates in tropic Queensland were reduced in the hot summer months of December and January by about 20 per cent when compared with the maximum winter rates. The drop in conception in coastal towns occurred 2 months later, in February and March, but again this was the period of the hottest temperatures. A comparison of conception rates in town at high altitudes in the same state with towns at lower altitudes also confirmed that the main environmental factor affecting conception rate was temperature. At the other end of the scale (and continent) the conception rate was highest in the summer in Tasmania, and showed a distinct drop during the winter months.

Other work has confirmed these findings. MacFarlane, Pennycuik, Yeates, and Thrift (1959) showed that conceptions in summer (December to February) were significantly lower in frequency than in winter (June to August), but at lower latitudes:

Locality	Latitude	Mean temperatures (° C)		Mean population over 8 years	Average conceptions	
		Summer	Winter		Summer	Winter
Townsville	19° S	27·2	20·6	34,280	1,438	1,548
Rockhampton	23° S	27·2	17·2	36,950	1,488	1,608
Toowoomba	27° S	21·7	11·1	34,250	1,554	1,527

The conceptions between summer and winter are significantly different ($P < 0.001$) for the first two localities. Unfortunately, little is known of the relative roles of male and female in this reduction of fertility.

The reader is referred at this point to page 121, where the difficulties of interpreting much of the experimental data relating the effect of temperature to breeding were discussed. The points made there are just as relevant to a consideration of the effects of temperature on pregnancy.

CHAPTER 14

The Effects of Nutrition on Pregnancy

The phenomenon of mammalian gestation, involving as it does the prolonged maintenance of rapidly growing uterine young, alters the requirements of the female for nutrition. The developing embryo needs energy to maintain its own metabolism and also has requirements for nutritive elements to increase its body mass. In addition, the pregnant female will have extra demands on her own metabolism – for example, increased requirements for respiratory processes, and increased requirements in late pregnancy, related to the work of carrying the extra weight of uterine contents.

As was mentioned in the section on puberty, the nature of the data available relating nutritional intake to reproduction processes is very different in laboratory and field studies. In the laboratory, fairly exact monitoring of the level and quality of the various nutritive elements is possible, but there are sometimes practical difficulties in obtaining the maximum data about the reproductive state of individuals if this involves examination of the tract. In the field, however, there are few studies where the actual food intake has been monitored in wild mammals (let alone analysed for protein, mineral, or vitamin content), although more exact information about the state of the reproductive tract is often obtainable from wild samples. Most studies of nutrition and reproduction in the wild have consisted of documenting differences in levels of fertility or fecundity, and then, *a posteriori*, attempting to explain these differences in terms of a general description, usually without quantification, of habitat or seasonal differences in available nutrition. The writer knows of no published paper which describes the setting up of a hypothesis relating nutritive intake to reproductive efficiency in a wild population and the subsequent testing of this hypothesis.

(a) Rodents

(i) *Rats.* The laboratory rat (*Rattus norvegicus*) is perhaps the best known non-domestic species with regard to the effects of nutrition on reproduction. That which follows is based on the excellent review by Russell, F. C. (1948), to which the reader is referred for further details. The gross intake of food increases considerably during pregnancy:

	Mean intake per day (g)
Cycling, non-pregnant	6·0
First week of pregnancy	13·4
Second week of pregnancy	15·5
Third week of pregnancy	16·9

and the calorific value of the intake also increases. Fewer females became pregnant if the gross intake was reduced, and starvation in pregnancy resulted in resorption of embryos or in still-births. The crude dietary component of most importance was protein, and here Russell quotes in length work by Slonaker (1931) from which the data below are taken:

Group	Protein level (%)	Proportion of fertile matings (%)	Average no. of litters	Average litter size	Proportion of young weaned (%)
I	10	96	5·30	5·26	60
II	14	100	5·76	5·15	72
III	18	81	3·94	4·68	51
IV	22	61	3·00	4·38	60
V	26	50	3·92	3·90	49

Later workers have confirmed that excess protein in the diet can lower fertility, but modern research has shown that 17–19 per cent protein is the optimal level – not 14 per cent as suggested by Slonaker. The importance of mixed as opposed to single-protein diets is also relevant in terms of the protein concentration.

The most important mineral element in the diet of the pregnant rat is calcium. Females fed on reduced calcium will drain their own body reserves, and a reduction in the number of viable embryos is found throughout. The calcium content of the individual embryos borne is maintained throughout at a high level. Other minerals seem to have little effect on rat pregnancy as long as general inanition is avoided. Levels of carotine below 5 ppm decrease the litter size. The effect of deprivation of Vitamin E is quite specific in the pregnant rat, as ovulation, conception, and implantation occur successfully, but uterine death of the embryos follows.

There have been very few studies on the effect of nutrition on pregnancy in wild *Rattus norvegicus*, and dietary differences can be only strongly inferred in the next two papers mentioned. Perry (1945) found that during the winter samples of adult female rats from Liverpool sewers showed a 35·5 per cent incidence of pregnancy, while samples from a boneworks during the same season showed an incidence of 24·0 per cent. In a major paper, Leslie, Venables, and Venables (1952) have pointed out the difficulties inherent in comparing simple fertility data from large samples. For example, the number of embryos varies with the weight of the female, so that all samples must be corrected for the bias of weight distribution. This paper cannot be too strongly recommended as a detailed evaluation of the potential statistical and biological pitfalls inherent in this type of analysis. These authors found that rats captured in corn ricks in and around Oxford during the winter were significantly more fertile (higher proportion pregnant, higher embryo rates) than samples captured outside ricks. Dietary differences were certainly involved in this disparity, but as the corn ricks also form a different microhabitat as far as temperature is concerned, this factor may also have added to the differences observed.

In tropical rats the incidence of pregnancy fluctuates seasonally, but in studies of what is now known as the same species (*Rattus coucha*) in three different latitudes, the highest incidence of pregnancy was found at the end of the wet season when nutritional conditions were best (Brambell and Davis, 1941, *Mastomys erythroleucus*, Sierra Leone; Chapman, Chapman, and Robertson, 1959, *Rattus (Mastomys) natalensis*, Tanganyika, now Tanzania; Coetzee, 1965, *Praomys (Mastomys) natalensis*, South Africa). In tropical-living *Rattus rattus* and *R. norvegicus*, the incidence of pregnancy among females also increases during the rainy season, and has been related by Buxton (1936) to some subtle and unmeasured effect through the food supply. Similarly, McDougall (1946) described the relationship of rainfall to nutrition and its correlation to the periods of greatest incidence of pregnancy of *Rattus conatus* in Queensland cane fields.

(ii) *Mice*. During pregnancy the laboratory mouse (*Mus musculus*) reacts in much the same way as the laboratory rat to dietary differences (Russell, F. C., 1948). An interesting aspect of the effect of diet on pregnancy has been reported in a series of papers by McClure (1961b, 1961c, 1962, 1966, 1967). Forty-eight-hour periods of fasting in early pregnancy (i.e., from the third to the fifth day after finding copulation plugs) resulted in the failure of the mated mouse to litter. Other periods during pregnancy were relatively less susceptible to the effect of fasting. The starvation resulted in death and subsequent degeneration of the deciduoma followed by embryonic death, and evidence was found that carbohydrate was the crucial factor. In experiments where starvation was accompanied by hormone treatments results indicated that the cause of embryonic death was probably 'due to depression in production or liberation of gonadotrophic hormone by the anterior pituitary . . .' (1962, p. 241). Early pregnancy and conception are also affected by short periods of food depletion, in that 30-hour fasts resulted in a failure to ovulate and an increased incidence of the death of ova before implantation.

For wild mice, Laurie (1946) described differences in the incidence of pregnancy found in the following environments: cold stores – wartime meat stores with temperatures averaging 15° F below freezing; buffer depots, unheated, uncooled flour or sugar stores; corn ricks – samples obtained at threshing; urban shopyards or houses. Some of the data from this study are tabled below and demonstrate the importance of considering the weight of the sampled mice.

Weight group (g)	Proportion of female mice pregnant (%)			
	Buffer depots	Urban	Ricks	Cold stores
7·5—	2·2	6·5	8·6	0·0
12·5—	29·5	25·1	42·4	9·4
17·5—	39·7	24·6	50·6	23·3
22·5—	49·2	31·9	66·7	32·1
27·5—	80·0	100·0	—	45·7
32·5—	—	—	—	83·3
37·5—	—	—	—	100·0
Total number of females	499	433	468	524
No. pregnant	158	95	190	139
Proportion pregnant (%)	31·6	21·9	40·6	26·5

There were no significant monthly differences inside each environment, but when the non-rick samples were combined the rick mice had significantly more pregnancies from February to April and in September and October. When standardizations such as are outlined by Leslie et al. (1952) were applied by Laurie to his data the pregnancy rate for the rick mice was still significantly higher than the non-rick mice over these periods. An interesting comparison is shown below, in which the annual productivity is estimated by multiplying the mean proportion of females pregnant by a factor of 365 divided by the length of time between litters, 14·5 days:

	Maximum number of litters per year
Urban	5·52 ± 0·50
Buffer depots	7·97 ± 0·52
Ricks	10·22 ± 0·57
Cold stores	6·68 ± 0·49

It will be seen that the mice in corn ricks are breeding at close to the maximal rate possible (13–14/season). The litter sizes of the non-rick groups were not significantly different from each other, but the rick mice did have significantly larger litters, (non-rick 5·60 ± 0·08 embryos per female, rick 6·37 ± 0·20 embryos per female).

Once again it is difficult to separate out the effects in this study of poor nutrition from the effects of temperature. For example, the cold-store mice bred remarkably well at the low temperatures found there, but at the same time they were feeding on an exceedingly nutritious diet of frozen meat. As for rick rats (see above), the rick formed a continuous food source and a microhabitat with reduced temperature range.

(iii) *Other rodents.* Little is known experimentally of the effects of nutrition on pregnancy in any other rodents. Pinter and Negus (1965) investigated this effect in conjunction with the effects of light. *Microtus montanus* was kept on two light regimes and fed with dietary supplements of sprouted wheat at two frequencies. The results are given below:

	18 hrs light/day		6 hrs light/day	
Sprouted wheat germ:	*Every 3 days*	*Every 15 days*	*Every 3 days*	*Every 15 days*
Mean no. litters/ female/100 days	1·73 ± 0·26	0·96 ± 0·23	1·30 ± 0·23	0·99 ± 0·19
Mean no. offspring/ female/100 days	6·19 ± 0·99	3·36 ± 0·85	3·90 ± 0·73	2·94 ± 0·61
Mean litter size	3·57 ± 0·16	3·49 ± 0·21	2·80 ± 0·19	2·93 ± 0·14

Although the effect of the light regime was more marked than that of the nutritive regime, those voles fed more sprouted wheat produced significantly more litters and young, although the litter size was not significantly increased.

Negus and Pinter (1966) also found that feeding sprouted wheat or an extract containing plant oestrogens produced more litters than controls, but again this result seemed to be the result of a higher incidence of *post-partum* matings than to an increase in the size of litters. It would seem from these results that the diets fed had a hormonal, rather than a nutritive, effect.

A paper with more direct bearing on the effects of nutrition in relation to pregnancy (Kaczmarski, 1966) reports the bioenergetics of pregnancy and lactation in the bank vole (*Clethrionomys glareolus*). Figure 46 on page 220 shows that during pregnancy the calorific intake of food increases considerably. Non-reproductive adult female voles required on the average 7·5 kcal/day, whereas pregnant females required 23·7 kcal/day at the end of pregnancy (an increase of 24 per cent for the total period of gestation). Allowing for normal metabolism, the female requires a total of 75 extra kilocalories to produce a litter of five young. In penned populations of nutria (*Myocastor coypus*) there is evidence that if the nutritive intake is not increased during pregnancy there is considerable pre-natal mortality (Ehrlich, 1966). Sealander (1964) reported that during pregnancy the blood values of *Onychomys leucogaster* and *Sigmodon hispidus* were lower than in non-gravid females, suggesting a metabolic shift in this species.

Agricultural practice has artificially altered feed availability to a number of wild mammals, and papers described below show the effects of crop irrigation on two species of rodent with regard to pregnancy. Bodenheimer (1949) presented the following data comparing fertility of *Microtus guentheri* populations in Israel.

	Percentage of females pregnant											
				1947						1948		
	M.	J.	J.	A.	S.	O.	N.	D.	J.	F.	M.	A.
Non-irrigated fields	41	0	0	0	0	0	18	58	17	7	72	47
Litter size	8·8	—	—	—	—	—	5·5	5·2	4·6	10·0	9·2	8·8
Irrigated fields	—	31	—	—	—	0	25	59	58	75	—	—
Litter size	—	5·0	—	—	—	—	4·5	8·5	7·2	10·0	—	—

Despite the incompleteness of this information, it demonstrates that the incidence of pregnancy was consistently higher in voles feeding on irrigated fields with therefore better nutrition and that the litter size was variable. Similar data were presented by Miller, M. A. (1946) for pocket gophers (*Thomomys bottae*) in Southern California, which were in some cases feeding on irrigated alfalfa fields. In non-irrigated fields the level of available protein is much lower. Note that the

		N.	Proportion of adult females (% and S.E.)	
			Pregnant (%)	Post-partum (%)
March to	Alfalfa fields	64	17 ± 4·7	34 ± 5·9
May	Other fields	40	25 ± 6·8	28 ± 7·1
June	Alfalfa fields	62	27 ± 5·7	10 ± 3·8
July	Other fields	15	0	13 ± 8·7

only major difference between these two groups occurred during midsummer when nutritional deficiencies would exert a maximal effect.

There appears to be only one field study on a wild rodent species in which the

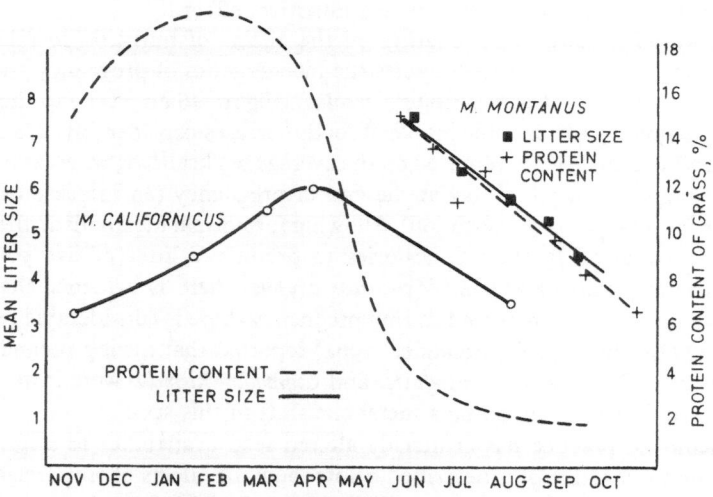

Fig. 40. The seasonal variation in litter size and food quality in *Microtus montanus* and *M. californicus*, showing direct correlation between litter size and crude protein content of common grasses. From Hoffman (1958) *Ecol. Monogr.* **28**: 79–109.

nutritive value of the available food was analysed and then related to pregnancy. In a very detailed study of *Microtus montanus* and *M. californicus*, Hoffman (1958) documented a seasonal change in litter size. As shown in figure 40, the litter of *M. californicus* is smaller at the beginning of the season, increases to a peak, and then declines, whereas that of *M. montanus* declines steadily. This difference is explained by Hoffman in terms of the protein content of the diet.

Unfortunately the curve in the figure for protein available to *M. californicus* is based only on a general evaluation of growth, but that for *M. montanus* is based on actual analysis of the protein content of one of its major foods, *Carex*. The agreement between litter-size trends and the changes in protein content strongly suggests a causal relationship. However, it should be remembered that both species are cyclic microtines, so that it is possible that in other years this sort of relationship may not be exhibited. A peculiar case was described by Hamilton (1962) for the red tree mouse (*Phenacomys longicaudus*), a species whose sole diet is Douglas fir needles. This is very low nutritionally and is possibly the cause of the relatively small litter size (2–3) and its relatively long gestation (28 days or 42 days if implantation is delayed) when compared to other rodents of similar size.

There are a number of other reports in the literature on rodents in which relationships between nutrition and pregnancy are suggested rather than directly substantiated from the data presented. For example, Orlova (1955) suggests that the intensity of breeding the Siberian suslik (*Citellus pygmaeus*) was affected by the level of nutrition available in summer. Caldwell and Gentry (1965) reported consistent but non-significant differences in the litter sizes and proportion of females pregnant in populations of *Peromyscus polionotus* found in different stages of habitat succession, where different nutritional regimes were suspected. Similarly, Errington, Siglin, and Clark (1963) reported lower litter sizes (6·4, 7·1) in two muskrat (*Ondatra zibethicus*) populations who had low food availability compared with the normal average litter size (8·0+) for Iowa. However, the low food availability coincided, or was the result, of a population high, so that social factors may also have been responsible for the differences noted. Peaks of pregnancies were found in *Dipodomys merriami* during times of new growth of vegetation (Reynolds, 1960).

The suggestion has been made twice in the literature that the ability of some rodents to live in desert environments is related to nutrition during pregnancy. Bodenheimer (1954) postulated that *Microtus guentheri* cannot maintain itself in the Israel desert because it cannot maintain pregnancies on dry food alone, whereas desert-living *Meriones* and *Acomys* can do just that. Similarly, Eisenberg and Isaac (1963) suggested that the reduced litter size in some desert-living *Dipodomys* species may be adaptive, as this reduces the maternal nutritional strain. It would appear that a rodent-like marsupial (*Sminthopsis crassicaudata*) which (though a carnivore more than a seed eater like the heteromyids) lives in a similar habitat in the central Australian desert can also alter its reproductive pattern in relation to food availability. Spencer (1896) reported that this species had litters of four to five in drought years instead of the usual ten.

(b) Artiodactyls

(i) *Sheep*. An agricultural practice of considerable antiquity indicates that practical sheep farmers have long realized the importance of dietary effects on pregnancy in sheep. When sheep are put on a rising scale of nutrition ('flushing') just prior to tupping it is well known that this results in an increased lamb crop. Agricultural biologists (Hammond, 1949; Moustgaard, 1959; Robinson, 1951; Schinkel, 1963; Thomson and Aitken, 1959, to list a few) have often commented on this phenomenon and there has been much research into its causes. It is now known that the main effect of the treatment is to increase the ovulation rate so that the proportion of ewes bearing twins increases. However, it is apparent from the literature that many comparisons made are not strictly valid due to differences in the nutritive regimes. Schinkel (1963) has pointed out that variations in fertility have been reported in the following three types of situations:

(a) when flushed and control groups have both been losing weight but at different rates;

(b) when flushed and control groups have both been gaining weight but at different rates;

(c) when flushed groups have gained and control groups have either maintained or lost weight.

These three situations cannot be compared, as controls are not standard. Another factor of importance is the duration and onset of the feeding regime relative to the stage of pregnancy of the experimental ewes. The discussion which follows will briefly consider some of the effects of nutrition on the various stages in reproduction from ovulation to parturition in the sheep.

To consider first the effect of nutrition on ovulation rate. Hammond, writing in 1949, reported that the injection of FSH did not cause follicular growth or ovulation in sheep on a low plane of nutrition, and suggested therefore that the lowered nutrition had a direct effect on the ovary itself. He commented, however, that if the nutritional deficiency was of energy or protein, ovulation could be induced by hormonal injections in deficient ewes. It is not clear from this paper as to exactly what type of nutritional deficiency induced ovaries unresponsive to hormonal dosage. Indeed, all later work has suggested that the effects of undernutrition on ovulation do not act directly on the ovary. For example, Allen and Lamming (1961) reported the following mean ovulation rates in ewes under different nutritional regimes. Five other groups of ewes were put on similar

Store ewes	1·33 ovulations/ewe
Ewes on experimental submaintenance diet	1·00 ovulations/ewe
Ewes flushed for 5–8 days	1·50 ovulations/ewe
Ewes flushed for 1 oestrous cycle	1·83 ovulations/ewe
Ewes flushed for 2 oestrous cycles	2·17 ovulations/ewe

feeds but also injected with PMS, and no significant differences in ovulation rates could be detected after hormonal treatments. Semi-quantitative assays of gonadotrophic potency of the anterior pituitary indicated that the unfavourable nutritive regimes resulted in incomplete release of gonadotrophins, and the workers suggested 'that nutrition affects ovarian activity not by any direct effect on the responsiveness of the ovaries to gonadotrophin, but by regulating the rate of release of gonadotrophin from the pituitary gland' (page 79). As with other workers, Allen and Lamming found no advance in the actual onset of oestrus, but they did find a longer duration of oestrus and higher incidence of 'silent heats' in the submaintenance group.

In the above work the ewes were sacrificed soon after treatment to determine ovulation. Hunter (1961, 1964) maintained groups of Merino ewes in South Africa with vasectomized rams for nearly a year to investigate the effect of diet on the incidence of oestrus uninterrupted by pregnancy. Four groups in two pairs on either high- or low-plane diets were set up in September 1959, and in March 1960 two groups under each treatment were reversed. The results indicated that dietary effects did not become evident till about six months later. This suggests that photoperiodic effects (in this Southern hemisphere experiment) may mediate nutritional differences – a field worthy of more investigation.

There appears to have been only one paper investigating the components which are found in the weight changes related to flushing. Coop (1966) postulated that there were two: (a) a static live-weight effect not specifically related in time to the mating period, and (b) a dynamic effect of the rising or falling body condition specific to that period. He reported that the twinning rate was altered by 6 per cent per 10-lb body weight difference in static live weight in the post-weaning preflushing period. The truly dynamic effect was shown by the mean response to flushing of 12 per cent twinning when differential nutrition for 3 weeks premating (plus 3 weeks mating, at fairly extreme nutritional levels) was maintained. These effects can be presumed to be due directly to ovulation-rate differences, and not to differential mortality after ovulation.

Before finishing the consideration of the effect of nutrition on ovulation a phenomenon of a completely different nature is here relevant. The oestrogenic action of certain subterranian clovers, which was first discovered in Western Australia, has been found not to affect the oestrous cycle but to reduce the ability of the ewe to conceive at oestrus. The substance concerned (gestinin) has also been found in soya-bean oil, which is known to reduce the fertility of laboratory mice (Moule, Braden, and Lammond, 1963).

Nutrition also affects the survival of the fertilized egg. Although, in early pregnancy, there is a steady increase of voluntary food intake from conception onwards (Hadjipieris and Holmes, 1966), only one paper was found which presented evidence of a detrimental effect of poor nutrition. Edey (1966) reported that if ewes were drastically restricted to 15 per cent of their normal maintenance for 0–1, 6–13, or 13–20 days after mating, the pooled treated groups had a higher loss (46·9 per cent) of ova than in untreated controls (31·0 per cent). Significantly the loss of ova from those ewes who had twin ovulations was higher than in ewes with a single ovulation.

During later pregnancy there is much more evidence connecting nutrition to fertility. A number of authors (Hammond, 1949; Schinkel, 1963, for example) have reported that reduced dietary intake results in the restricted development of the lamb, which often dies at birth. Moustgaard (1959) commented that

N

reduced gross food intake from the 28th to the 91st day of pregnancy had no effect on foetuses but that underfeeding during the last 2 months caused gross emaciation of the ewe and the foetal growth rate slowed down. In his review, Moustgaard listed the effects of individual nutrient deficiencies on the pregnant ewe. Lowered protein intakes resulted in still-births, reduced birth weights, and higher rates of post-parturent mortality. Reduction in calcium intake resulted in greatly debilitated lambs and reduced copper induced a spastic paralysis ('sway-back') in the new-born lamb. Vitamin A intake seems less important, as sheep can store it in considerable quantities, so that deficient ewes can apparently undergo at least two successful pregnancies before lambing is affected. Although Vitamin E deficiencies result in abortion in laboratory rats, in sheep and all other domestic stock investigated Vitamin E deficiency has no effect during pregnancy. A common feature of sheep on lowered nutritional intakes during late pregnancy is often fatal pregnancy toxemia (Hammond, 1949; Schinkel, 1963). This is known to be especially prevalent in ewes carrying twins who are losing body-weight. The importance of nutrition at this time relative to the well-being of the foetuses is also emphasized by the differences in increased food intake – ewes carrying one lamb increasing about 50 per cent over normal, ewes carrying two lambs increasing about 75 per cent. The reader is referred to the detailed review by Thomson and Aiken (1959) for other information about the effect of diet on sheep pregnancy.

Two further papers will be mentioned here. Thomson and Thomson (1949) placed ewes on a low-protein diet during the second half of pregnancy, with the following results:

	High-plane diet		Low-plane diet	
	Singles	Twins	Singles	Twins
Survival to 4 days (%)	100	82	64	14
Number which never stood	0/13	0/17	1/22	6/27

Ewes carrying twins on a poor diet showed reduced maternal interest, and milk production was lowered, but the main effect was an extreme weakness of the lambs. An interesting and somewhat unusual effect of lowered nutrition was reported by Alexander (1956), who lengthened the period of gestation by putting ewes on low-plane diets. The heavier the uterine contents at the time of commencement of the experimental diet, the more parturition was advanced.

The importance of lamb production in sheep husbandry has therefore resulted in considerable experimental investigation into the effect of diet on pregnancy. The review above indicates, however, that only more recent research has looked into the effect of nutrition at the various stages from ovulation to parturition, as opposed simply to monitoring the end-product, i.e. the lamb crop.

(ii) *Cattle*. Although it is perhaps a function of the literature that was available to this author, there seems to be much less information on the effect of nutrition on pregnancy in the cow. General reviews on nutrition and reproduction in domestic animals, such as those by Hammond (1949) and Moustgaard (1959), cover the subject adequately, and generally cattle seem to respond similarly to sheep as far as nutrition during pregnancy is concerned. Special features mentioned by these two reviewers are: deficiency of phosphate delays the ripening of ovarian follicles if cows are also suckling (due to the extra drain on the phosphate reserves); pregnant cattle in copper-deficient areas give birth to debilitated calves; unlike sheep, a Vitamin A deficiency affects the processes of organogenesis in the uterine calf which often dies *in utero* (sometimes with placental retention).

More recent work by McClure (1961a) in New Zealand suggested that cattle on pastures which were low in calcium though high in phosphorus showed low fertility in their first service after parturition because of the nutritional strain resulting from lactation (see also Alderman, 1963). Later experimental work in Australia consisted of supplementing the diet of cows from parturition to 3 weeks after mating (McClure, 1965). This resulted in a reduction of the normally seen weight loss and significant increase in conception rate. It is interesting that the manifestation of behavioural oestrus is not affected by this weight loss after birth, even though the conception rate is reduced if supplemental food is not given.

A rather more indirect effect of nutrition on early pregnancy was reported by Fallon (1961) for Hereford cattle in Queensland. Despite the known detrimental effect of high temperatures on cattle, Fallon reported the following interesting data relating conception rates to seasonal rainfall incidence:

	Percentage total annual rainfall	*Percentage total annual conceptions*
Spring	22	21
Summer	41	50
Autumn	22	19
Winter	15	10

These figures (based on 9,800 births) show that the best conception rates occur during the wet hot season, so that Fallon suggests that the differences are due to this being a period of better nutrition. It would seem that detrimental temperature effects are minimized in this species if nutrition is good.

(iii) *Pigs*. As was mentioned in the discussion of the effects of nutrition on the attainment of puberty (page 16), it is often difficult to interpret the available literature on nutrition in pig reproduction because of the disagreement between various workers as to what levels of intake can be considered 'normal' or 'underfeeding'. There have been a number of investigations as to the effect of the level of dietary intake prior to mating on the ovulation rate, and it seems in general that pigs exhibit a 'flushing' response in a similar manner to sheep. Sorensen,

Thomas, and Gossett (1961) reported increased ovulation rates after flushing, and Zimmerman, Spies, Self, and Casida (1960) obtained increases from 2·8 to 3·9 ova per sow when placed on high-level diets from 6 to 14 days prior to service. The picture is somewhat complicated by an apparently reverse effect of nutritional intake on embryonic survival after conception, as the former workers reported that reduced feed intake after service increased embryo survival.

Recent work has confirmed the increased ovulation-rate response to increased feed intake. Schultz, Speer, Hays, and Melampy (1966) increased the mean number of corpora lutea per gilt by 3 when they increased the intake from 1·81 kg per day to 3·63 kg per day, but noted a significant decrease in embryo survival at 27 days in comparing gilts fed at 1·81 kg per day with those on 2·72 kg per day. Lodge and Hardy (1968) reported that doubling the intake from 1·8 to 3·6 kg per day for the first feed following service increased the subsequent litter size from 8·95 ± 0·53 to 10·80 ± 0·40. The same workers increased the intake from 1·4 to 4·1 kg at a feed given 12 hours after the onset of oestrus and noted that the number of corpora lutea per sow increased from a mean of 12·14 ± 0·86 to 14·71 ± 0·42. However, they comment that both of these results may be due to bringing the sows up to normal intakes from subnormal levels, as the larger litter sizes and corpora lutea counts are at levels found in gilts on normal intakes.

Two recent papers have investigated the effects of short-term and long-term differences in dietary intake (after service) on embryo survival and demonstrated little effect. Heap, Lodge, and Lamming (1967) placed groups of sows on 3, 6, and 9 lb of feed per day from 1 day after service to slaughter at 28 days. When adjustments were made in the data analysis to compensate for the differences in embryo numbers due to the weight of the sows at service no significant effects of the treatments were found. Ray and McCarty (1965) fasted gilts immediately after mating for 24, 48, and 72 hours and found no significant differences in ovulation rate or embryonic survival.

It would thus seem that while the role of nutrition prior to mating is well established in the pig, there are considerable differences in interpretation of the effects of nutrition on early pregnancy. Part of this is apparently the result of a lack of establishment of norms for levels of feed intake by the various workers concerned. Perhaps the most realistic approach was that of Lodge, Elsley, and Macpherson (1966) who maintained groups of sows on either 6 lb of meat meal per day, or 3 lb per day, or 3 lb per day for the first 76 days of pregnancy with a subsequent increase to 6 lb per day; but who continued these three treatments for 3 consecutive pregnancies. They reported no significant differences in litter size or in post-natal survival, but the mean piglet weight at birth was significantly altered.

(iv) *Moose*. Examination of over 200 uteri from moose (*Alces alces*) in British Columbia by Edwards and Ritcey (1958) showed that from autumn onwards an increasing proportion of tracts dissected exhibited twinning. Samples were taken at low altitudes from cows which had recently migrated down from higher pastures where the feed was of a better standard than on the valley floors. It was suggested that differences in nutrition resulted in higher twinning rates of higher altitude, better-fed moose populations than for females which stayed on the poorer valley-floor pasture.

Comparison of uteri and field observations of cow/calf ratios over various areas of Newfoundland also demonstrated an effect of nutrition on twinning rates in moose. In the Central area, where nutrition was poor, only 5 per cent of 151 tracts showed twins, compared with 21 per cent of 132 tracts from the remainder of the island. Pimlott (1959) felt that this was a true difference in the ovulation rate of adults and was not due to differences in yearling rates (see page 22).

(v) *Deer*. The importance of the fawn crop to recruitment in North American deer and their interrelationship with man as competitors for cattle and sheep feed has led to considerable investigation into the effect of the quality of deer range on their reproductive processes. Unfortunately, there has been little actual measurement of nutrition available (i.e. protein content of browse), as opposed to reliance on visual range estimates in these reproductive studies, but there are some recent papers giving details of experiments on penned deer.

A series of three papers (Morton and Cheatum, 1946; Cheatum and Severinghaus, 1950, and Severinghaus and Cheatum, 1956) describes investigations into the effect of range conditions on pregnancy in the white-tailed deer (*Odocoileus virginianus*) in New York State. In the western region, where the food supply was very good during the winter, and there was little winter snow, an average of over 90 per cent of adult does and over 30 per cent of fawns were pregnant, compared with 78 per cent of adults and 3 per cent of fawns pregnant in the central Adirondack mountains, where the range was very poor and malnutrition was common in winter. The Western-region adult does were carrying an average of 1·7 embryos/female, while the Adirondack does were carrying 1·1–1·2 embryos. These rates were significantly different at the 1 per cent level, whereas comparisons with rates within three other areas in the Catskill mountains showed no significant differences. Unfortunately the data presented in these papers do not appear to have been corrected for the age of does in each sample, as the 1956 paper clearly showed that the average number of fawns seen per doe altered with the age of the mother. Another variable recognized by the workers concerned was the density of the population, which was much lower in the Western region, due to antlerless deer hunting seasons. It was suggested that this sort of intense hunting reduced the numbers and thus increased the relative amounts of food available. In each case where increased hunting was documented for a small subregion it was followed by an increase in the proportion of does pregnant and in the numbers of embryos per doe. Little is known about the effect of social factors and density on deer reproduction, so that the most probable explanation of these fertility changes is increased food supply.

A 'natural experiment' was reported by Ransom (1966, 1967) in his study of three populations of white-tailed deer in Manitoba. The does on Turtle Mountain had significantly higher ovulation rates than females from two other areas, and this seemed to be due to their habits of feeding during the breeding season on crops of alfalfa, coarse grains, and flax, which resulted in an unusually high protein intake at this period, so that the does in this population were effectively 'flushed' like sheep. A result of this was that those does with three ova ovulated and showed very high rates of ova loss.

More recently the effect of nutrition on reproduction in white-tailed deer has been investigated by Verme (1962, 1965) in experimental pens. Using experimental regimes shown in figure 35, page 144 he reduced the body weight of one group of does by 15 per cent, with the following effect on fertility:

	High-plane group	Low-plane group
Number of does	27	22
Fawns born per doe	1·74	0·95
Fawns born per pregnant doe	1·96	1·11

In another experiment, groups of does were kept on five sustenance levels ranging from excellent (level I) to near starvation (level V), with the following results:

	Excellent diet (levels I and II)	Medium diet (level III)	Poor diet (level IV)
Number of does	9	5	10
Fawns carried	18	7	18
Litter size (1:2:3 fawns)	0:9:0	3:2:0	3:6:1
Gestation length in days (mean ± standard error)	199 ± 2·8	203 ± 3·5	205 ± 5·0
Fawn birth weight (lb)	7·7	7·5	5·7

Verme suggests that most of the strain of nutritional stress falls in the last third of pregnancy and that a response to lowered nutrition is lengthened gestation, which allows small, undernourished fawns to grow longer and therefore bigger to face the rigours of parturition.

The last paper to be mentioned in this section on white-tailed deer actually differentiated between the level of protein given in the experimental diet. Murphy and Coates (1966) kept three groups of four does each on three levels of protein for 3 years. Despite the small numbers of animals used, which were unfortunately reduced by deaths during the study, the length of this experiment produced the following interesting results:

	1963			1964			1965		
Level of protein (%)	7	10	13	7	10	13	7	10	13
Producing does	4	4	3	4	3	3	3	3	2
Fawns produced	7	6	5	6	5	6	6	4	4
Fawn mortality	5	4	0	0	0	0	3	0	0

The mortality was of fawns, which were born dead or which died either at birth or a very few days afterwards. These data suggest that ovulation rate in these deer is less affected by protein levels than is immediate parturition survival. It would also suggest that the differences in embryo rates reported in wild populations of white-tails are thus due to differences in some other nutrient than protein.

The effect of range conditions in pregnancy in mule deer (*Odocoileus hemionus*) has been investigated in California and in Utah. Robinette and others (Robinette, Gashwiler, Jones, and Crane, 1955; Robinette, 1956; Julander, Robinette, and Jones, 1961) reported that on depleted range with poor nutrition the fertility is much lower than on good range. The ovulation rate (1·95) and foetal rate per doe (1·85) was higher on good range in southern Idaho on the Sublett unit than in the Antimony mountain unit of central Utah (ovulation rate 1·31; foetal rate 1·19). Comparison of two areas in Utah in different years indicated that the fertility was also affected by the nutritional condition of the previous winter. This is shown in the following foetal rates:

	Salt Lake area (*poor winter range, good summer range*)		Weber Canyon area (*poor winter range, very poor summer range*)	
	Yearling does	*Adult does*	*Yearling does*	*Adult does*
1949 (severe winter)	0·33 (9)	1·81 (27)	0·53 (17)	1·47 (96)
1952 (mild winter)	1·14 (70)	1·84 (107)	0·83 (46)	1·62 (85)

The sample sizes are given in brackets. It is perhaps significant that the greatest differences are noted in foetal rates of yearlings, suggesting that nutrition has a relatively more severe effect on this age class as far as pregnancy is concerned.

In California, Taber (1953) and Taber and Dasmann (1957) concentrated more on the actual type of vegetational succession and its effect on mule-deer reproduction. They presented the following data:

	Mean number of fawns per doe	Mean number of corpora lutea per doe
Chaparral. Very poor range with lowest available nutrition	0·77	0·82
Wildfire burn after one growing season. Fresh sprouting green feed	1·32	1·40
Wildfire burn after three growing seasons. Reduced nutrition	0·71	0·75
Shrubland, managed to produce herbaceous species, excellent feed	1·65	1·75

The two measures of fertility parallel each other closely, suggesting that here, as for white-tailed deer, the primary effect is on the ovulation rate.

This brief review of the effect of nutrition on pregnancy on deer of the genus *Odocoileus* would seem to indicate that this group is particularly worthy of greater investigation, in that their reproductive response to nutritional regimes shows how crucial is the level of nutriment during pregnancy. The genus appears in a number of areas of North America to be living in a rather delicate environmental balance with its food supply, and thus detailed studies of nutriment variations (protein content, mineral content, etc., of browse) and their tie-in with fertility, and thus recruitment, should prove more than valuable.

(vi) *Artiodactyls in tropical areas.* Unlike mammals living in more temperate climates, species living in the tropics are subjected to considerable nutritional

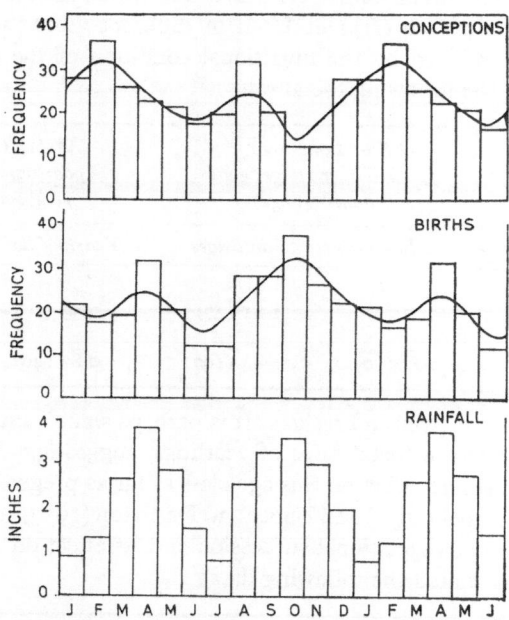

Fig. 41. The relationship of conceptions and births in hippopotamus (*Hippopotamus amphibius*) to mean monthly rainfall. From Laws and Clough (1965) *Symp. zool. Soc. Lond.* 15: 117–40.

variation as a result of rainfall incidence. Rain in these areas is followed by flushes of green and nutritious feed, so that it is not unusual to find that this is followed by an effect on reproduction. Previous chapters have described how rainfall incidence affected the onset and cessation of breeding. De Vos and Dowsett (1966) report that in three Zambian species (*Kobus vardoni*, Puku; *K. leche*, Lechwe; *K. ellipsiprymnus*, Waterbuck) breeding took place in all months of the year. However, during the rainy season from November to February there was a considerable peak in oestrus and mating activity.

Similarly, an effect of rainfall was found in the detailed study of reproduction of hippopotamus (*Hippopotamus amphibius*) carried out by Laws and Clough (1966) in Uganda. On the basis of intra-uterine embryos aged from growth curves, they showed that although breeding was continuous in this species, the incidence of conceptions varied with season, so that the pattern of births was strongly correlated with the rainfall incidence. This is shown in figure 41 from their paper. In the sense of Baker (1938a) this arrangement means that the rainfall incidence must be an ultimate factor governing breeding, as most births occur when feed conditions are best, due to the increased rainfall. The proximate causes of this pattern are unknown, but it would be very interesting to discover what the selective advantage is to breed at those times when young are born during the periods of low rainfall. This must exist, otherwise the hippopotamus would have evolved into a truly seasonal breeder. It may be that this species is at present changing the pattern of its breeding season.

(c) Other mammals

(i) *Rabbits*. The review of the effect of nutrition on reproduction in the laboratory rat by Russell, F. C. (1948) referred to above (page 174) also includes some information about the rabbit (*Oryctolagus cuniculus*) in the laboratory. Other than some comments about the importance of Vitamin K and a reference to the effect of manganese on ovarian function, the only relevant points in this review relate to the effect of protein. Rabbits placed on protein-free diets did not mate, and if the diet was imposed after conception abortion or foetal death resulted. However, rabbits kept by another worker on a protein-free diet (or on 2 per cent protein) did undergo oestrus and ovulation. A more recent paper by Lamming, Salisbury, Hays, and Kendall (1954) showed that Vitamin-A-deficient rabbits showed the following reproductive effects: (*a*) only 67 per cent of blastocysts 4 days after coitum were normal, compared with 97 per cent in controls; (*b*) there were 14 per cent less normal matings; and (*c*) 18 per cent less conceptions; (*d*) although the number of corpora lutea were not reduced, there was a significant loss of ova, as indicated by a reduced litter size (vitamin deficient $\bar{x} = 6\cdot33$, cf. control $\bar{x} = 9\cdot60$, significantly different at 1 per cent); (*e*) incipient Vitamin A deficiency resulted in a syndrome of resorption and abortion during late gestation, so that at 28 days after mating only $26\cdot4$ per cent of ovulations in deficient rabbits were represented by normal foetuses, compared with $74\cdot6$ per cent of ovulations in the controls.

In wild-rabbit populations the effect of nutrition on reproduction has already been referred to under the section on the breeding season. The excellent papers of Poole (1960) and Myers and Poole (1962) have already been discussed, and the interested reader is referred to page 108 and 133. The incidence of pregnancy as well as the onset and cessation of breeding in wild and penned rabbits are markedly affected by the amount of green grass available, but as yet the exact nutritional factors in green grass have not been elucidated. Watson (1957) found differences in fertility in two groups of rabbits on different pastures. On an area of excellent

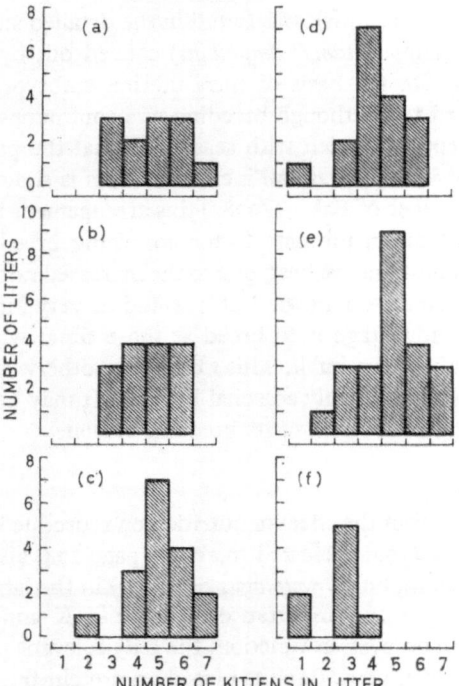

Fig. 42. Distributions of litter size in populations of rabbits (*Oryctolagus cuniculus*) fed on different foods (see text table). From Stodart and Myers (1966) *C.S.I.R.O. Wildl. Res.* **11**: 111–24.

pasture, 93 per cent adults and 52 per cent yearling females, were pregnant, compared with 62 per cent and 0 per cent similar females on a river-bed area of very poor grass.

An experimental investigation of the effect of different foods on breeding in penned rabbits has been reported by Stodart and Myers (1966), with the results below and in figure 42:

	Number of adult females	Number of litters per female	Mean litter size	Number of kittens per female
1962/63				
(a) Pasture only	4	3·0	4·8	14·3
(b) Pasture and green feed supplement	4	3·8	4·6	17·3
(c) Pasture and excess oat grain	4	4·5	4·8	21·8
1963/64				
(d) Pasture only		2·8	4·4	12·3
(e) Pasture and excess green feed	5	4·6	4·9	22·6
(f) No pasture, only excess hay and oat grain, no green feed	5	1·8	2·6	4·6

With the exception of the last treatment, the main effect was to alter the number of litters per female, which is related to the alteration of the length of the breeding season (see page 145). Palpation of females on the last treatment revealed a high incidence of resorption of foetuses and whole litters, whereas this was never recorded in females on other feeds. This prenatal mortality again demonstrates the necessity for green feed as a prerequisite to successful rabbit reproduction.

The effect of differences in soils on fertility in cottontails (*Sylvilagus floridanus*) was investigated by Williams and Caskey (1965) in Missouri. Using tracts only from females carrying the first litter of the season, they found that the mean litter size (3·8) of females from a population on a fertile soil was significantly higher than that for females from relatively poor soil areas (litter size, 3·2). Analysis of the samples showed that this was a true effect of soils, and not due to age or latitude factors. Analysis of manganese, calcium, and phosphorus in the carcasses did not, however, show any significant differences.

(ii) *Others.* There is a small amount of information in the literature on the effects of nutrition on pregnancy in some other species. Stones and Weibers (1965) reported an increased consumption of food in little brown bats (*Myotis lucifugus*) when they were pregnant, and Kallen and Wimsatt (1962) found that in the same species the circulating blood volume during pregnancy increased only by 17 per cent (cf. 40 per cent increase for pregnant humans) and commented on the surprisingly low metabolism of this bat while pregnant. Tamsitt and Mejia (1962) suggested that pregnancy rates were correlated with food abundance in a tropical bat (*Artibeus jamaicensis*). An indication of a metabolic shift, possibly related to nutrition, during pregnancy was mentioned by Sealander (1964). He found that the blood values of pregnant shrews (*Cryptotis parva*) and possums (*Didelphis marsupialis*) were lower than in non-gravid females. Finally, studies of Vitamin A deficiency on the cat (*Felis cattus*) have shown effects on pregnancy. On an intermediate deficiency level, where pregnancy very occasionally followed mating at the irregular heats observed, the female failed to gain weight after the 35th day of pregnancy and foetuses were aborted either about the 50th day or after term but with the dimensions of a 45-day foetus (Scott and Scott, 1964). As the normal placenta takes over the function of the regressed corpus luteum at about this time, it was suggested that there was a failure of placental function in these cats on Vitamin-A-deficient diets.

Keys, Brozek, Henschel, Mickelsen, and Taylor (1950) have reviewed the available evidence relating severe undernutrition to human pregnancy. Food restrictions associated with confinement in concentration camps result commonly in amenorrhea, but this is often associated with psychic factors (see page 257). There are a number of documented cases of drastic declines in birth-rates. During the Madras famine of 1877 there were only 39 births in a relief camp dealing with over 100,000 persons, and the normal birth-rate in the area of 29 births per year per 1,000 dropped to 4 births per year per 1,000. However, in this case, as in many others, there was a strong tendency for separation of the sexes, as the males moved out of the famine area to find food and work. Male separation cannot be invoked, however, to explain the drastic decline in the birth-rate found in Rotterdam, Holland, as a result of severe food restrictions in late 1944 and early 1945. It is not possible in most of these reports to determine exactly at what stage of the reproductive cycle lowered nutrition had its major effect.

CHAPTER 15

The Variation in Litter Size

A. LITTER SIZE AND LATITUDE

The breeding season at high latitudes is shorter than at low, so that the possible number of parturitions is reduced. In many field studies it has been noted that samples taken from higher latitudes inside a species range had larger litter sizes. It has been assumed, therefore, that this phenomenon is present to counteract the otherwise reduced productivity. Unfortunately, there are very few species in which the total productivity at various latitudes has been documented. This section will consider the most overall recent review and then present more recent information.

Lord (1960) collected all the available literature relating litter size to latitude in North American mammals, but, as his method of statistical analysis was somewhat unconventional, it is not possible to determine whether or not some of the correlations he presents are valid. Much of the range in litter size in single samples was concealed by the use of mean values with no indication of variation, and this was further accentuated by assigning a latitude to some species which merely happened to be the mean of the two latitudinal extremes of its north–south range. The correlation coefficients used are apparently based on scatter plots, where each point was not weighted for sample size, and where some points represent various litter sizes at various latitudes of a *single* species, while other points represent the average litter size and the mean latitude *for the entire species*. These methods mean that Lord's statements with regard to the significance of correlations in most non-hibernating prey species but not in hibernating or fossorial prey species or predators, while interesting, cannot be verified from the data he presents. In the discussion, Lord refutes the idea that the increased litter size with latitude is due to increased daylength, and claims instead that a more probable explanation is that there is a higher mortality at higher latitudes. There appears to be no concrete evidence to support either suggestion.

Since Lord's paper was published, a number of other papers have also been produced on this subject. They will be considered entirely on a species basis.

(a) Snowshoe hare (*Lepus americanus*)

Rowan and Keith (1956), in a paper mentioned by Lord, presented data indicating that litter size increased with latitude, ranging from a mean of 2·92 in Maine (44°) to 4·90 in Alaska (65°). Later Bookhout (1965) reported a lower mean litter size of 2·68 in Michigan (45°). Fortunately, the regional differences in reproduction of this species have been investigated in detail by Keith, Rongstad, and Meslow (1966). They emphasized that the first litter of the season in this synchronously breeding species is consistently lower than later litters and found no correlation between the size of the first litter and latitude, although the mean size of later litters did increase significantly with latitude. Comparison of the litters produced by hares from Wisconsin and from central Alberta, held in pens in Wisconsin, indicated that the variation seen was caused by genetic differences. However, the data below indicate that the total production did not vary in the same way:

Location	Latitude	Years	Mean litter size (later litters only)	Litters per adult female per year	Average No. young per adult female per year
Michigan	46°	1959–61	2·83 ± ?	2·42	6·5
Minnesota	46°	1932–38	3·17 ± 0·15	2·35	6·8
Ontario	46°	1959–61	3·48 ± 0·04	—	6·3
Montana	48°	1953–54	—	2·94	8·2
Newfoundland	49°	1954–62	4·23 ± ?	3·20	12·2
Alberta	54°	1961–65	4·49 ± 0·15	3·15	12·8
Alberta	56°	1949–57	4·06 ± 0·14	2·75	10·5
Alaska	65°	1955–56	4·33 ± 0·13	1·79	7·2
Alaska	65°	1958–61	4·63 ± 0·50	1·68	7·8

These workers comment that the pattern of change in total seasonal production of young (largest in centre of range and decreasing outward, see also Dodds, 1965) does not agree with Lord's (1960) suggestion that the increased litter size with latitude is due to increased mortality at high latitudes. They do not consider that year-to-year variations in litter size with population phase are important. Although significant differences have been reported (see page 249), they were unable to find consistent changes. Unfortunately, the statement that 'unpublished data from Rochester, spanning the 5-year period from a population peak to a low, likewise show significant annual differences in litter size which are uncorrelated with cyclic stage or population density' (page 955), is made without supporting data being presented, so that it is not possible to evaluate the importance of this variable. The reader is reminded that no consistent change in litter size with population phase has been reported in cycling microtine rodent populations (see page 248).

(b) Cottontail rabbit (Sylvilagus floridanus)

Lord (1961) compared the litter size from various previous studies and showed a general trend for litter size to increase with latitude, as did Barkalow (1962). However, much of the comparisons of annual mean to annual mean are invalid, because both these authors demonstrated considerable seasonal changes in litter size. As the monthly distribution inside the samples on which the annual means were based was in most cases unknown, this distribution would have produced a biased comparison (see Section B, below). Fortunately, Barkalow compared the monthly variation between his and previously published data as follows:

	Weighted mean embyro and litter counts							Weighted mean	Crude mean
	Feb.	Mar.	Apr.	May	June	July	Aug.		
Alabama and Georgia	2·6	3·3	3·8	4·6	2·7	2·4	2·3	3·2	3·1
North Carolina and Virginia	3·0	4·2	4·8	4·9	5·0	3·0	—	4·6	4·2
Connecticut	6·4	6·0	5·8	5·7	—	5·5	—	6·2	6·9

In each month there is a clear increase in litter size with latitude.

A detailed comparison of litter size in Missouri and Mississippi by Evans, Sadler, Conaway, and Baskett (1965) showed a similar state of affairs. Comparison of the size of the second litter showed: South-west Mississippi, 3·5; South-east Missouri, 4·8; Northern Missouri, 5·6; but these workers warned against attributing such differences necessarily to latitude, as differences of the same magnitude were found between areas close to each other. Russell (1966) reported that when live cottontails from two areas in Ohio (where litter sizes were different) were allowed to breed for a year in pens near one of these areas both groups showed the same number of corpora lutea per doe. This indicated that the breeding differences naturally found were not genetically based.

(c) Black-tailed jackrabbit (*Lepus californicus*)

French, McBride, and Detmer (1965) presented the following data from their own and previous studies:

State	Mean number of embryos per litter
Arizona	1·79, 2·24
California	2·3–2·5
Kansas	2·6
Idaho	3·35

(d) Northern vole (*Microtus oeconomus*)

A study of this species in Arctic Norway by Hoyte (1955) presented the following information not mentioned by Lord (1960):

Rosta   Lat. 68°   7·50 young per litter
Oyer   Lat. 62°   5·80 young per litter
Eidfjord   Lat. 61°   5·44 young per litter

(e) Golden mouse (*Ochrotomys nuttalli*)

In the northern part of its range (Illinois, Kentucky, North Carolina, Tennessee), ranging from 35° to 39°, this species has a litter size of 3·11 ± 0·19, which is significantly larger than in the southern range (Georgia, Florida, Texas) between 32° and 28°, where it is 2·47 ± 0·11 (Blus, 1966).

(f) Roof rat (*Rattus rattus*) and Polynesian rat (*R. exulans*)

The former species was found to have a mean litter size which significantly correlated with latitude, but in the latter the correlation was positive but not significant (Jackson, 1965).

(g) Bank vole (*Clethrionomys glareolus*)

The litter size of this species has been subjected to a detailed study by Zejda (1966). He compares previously published information on this species with his own results as follows:

Location	Mean litter size
Great Britain	3·95 ± 0·08 (158)
Brittany	3·84 ± 0·16 (43)
Czechoslovakia	4·90 ± 0·07 (388)
North Germany	4·98 ± 0·18 (57)
Kola peninsula, U.S.S.R.	5·27 ± 0·13 (102)
Moscow, U.S.S.R.	5·53 ± 0·15 (72)
Komi, A.S.S.R.	5·47 ± 0·22 (43)
Tatar, A.S.S.R.	6·10 ± ? (343)

In a very interesting discussion of this variation, Zejda emphasizes that although the litter size tends to increase with latitude, it also definitely increases from the west to east of the species range. He suggests that the actual climate and its effect on the length of the breeding season is more important than latitude, and uses the decrease in litter size found at higher altitudes in this species as confirmatory evidence.

(h) Sheep (*Ovis aries*)

In a study of Romney Marsh ewes in New Zealand, Averill (1964a) showed that the ovulation rate varied with season, but that between months there were consistent increases with latitude as shown below:

	Number of corpora lutea per ewe		
Month	North Auckland (36°)	Poverty Bay (39°)	Southland (46°)
March	1·20	1·20	1·46
April	1·35	1·51	1·78
May	1·21	1·39	1·62

The Southland rate was significantly higher in all months than that from North Auckland.

(i) Thirteen-lined ground Squirrel (*Citellus tridecemlineatus*)

The following data presented by Rongstad (1965) suggest that in this species the litter size may decrease with latitude, a reverse of the usual trend:

State	Mean litter size
Wisconsin	$8 \cdot 7 \pm 0 \cdot 3$ (34)
Manitoba	$8 \cdot 1 \pm 0 \cdot 1$ (269)
Minnesota	$8 \cdot 5 \pm 0 \cdot 1$ (129)
Kansas	$9 \cdot 0 \pm 0 \cdot 4$ (21)

However, the variation is so small that further data are needed for corroboration.

B. LITTER SIZE AND SEASON

In two of the examples above the change in litter size with latitude was constant, even though the litter size changed with season. In a number of seasonal and non-seasonal breeders there is a consistent alteration in the litter size as the breeding season progresses, and these will be described below.

(a) European rabbit (*Oryctolagus cuniculus*)

The most detailed information about change in litter size with season was presented by Watson (1957) for this species in New Zealand. He found no yearly differences, so that figure 43 gives data combined from the two seasons of his study. The shape of the curve shows that loss of ova is highest when there are

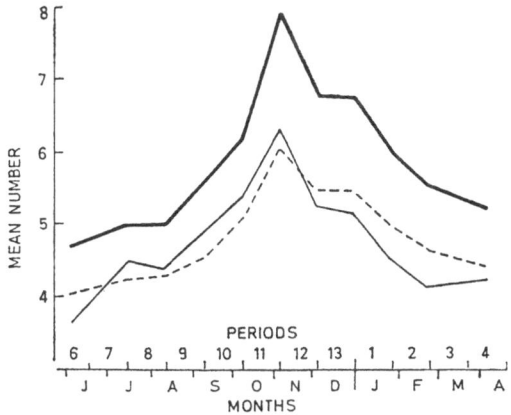

Fig. 43. Seasonal variation in litter size in rabbits (*Oryctolagus cuniculus*) *Thick line:* mean number of corpora lutea; *thin line:* observed mean number of embryos; *broken line:* expected mean number of embryos (see text). From Watson (1957) *N.Z. Jl Sci. Technol.* **38**: 451–82.

O

the greatest number of ovulations (and just afterwards), but this can be explained by the presence of an approximate correlation in Watson's data between the proportion of ova lost and the number ovulated in all seasons. The expected number of embryos is based on the annual total mean number of embryos per ovulation. Watson points out that the litter size is smaller in younger does, so that the increased proportion of yearlings which are pregnant later in the season may have resulted in a decrease in population litter size because of this age bias. However, old females also show a reduced litter size later in the season. Similar variations have been described for this species in Wales by Brambell (1944) and Stevens, M. N. (1952).

(b) Cottontail rabbits (*Sylivilagus floridanus*)

This species is interesting, in that during the breeding season it breeds synchronously, litters being produced in discrete separable groups which, at least in the early parts of the season, are attributable to synchronized matings. Thus a change in reproduction can be attributed to factors common to all females at any specific time. Lord (1961) showed that the litter size changed with season, but Conaway, Wight, and Sadler (1963) and Evans, Sadler, Conaway, and Baskett (1965) documented in greater detail the litter size and mortality changes in Missouri, with litter sequence as shown below:

	1961 season (Conaway et al., 1964)		1962 season (Evans et al., 1965)
Litter no.	Ovulations per female	Number of viable embryos per female	Embryo counts per female
I	3·64 ± 0·23 (12)	—	3·6 (38)
2	6·59 ± 0·06 (17)	6·20 ± 0·28 (65)	4·8 (17)
3	6·47 ± 0·19 (17)	6·24 ± 0·21 (14)	4·3 (15)
4	6·21 ± 0·34 (14)	5·50 ± 0·39 (14)	4·6 (15)
5	6·42 ± 0·39 (7)	5·71 ± 0·43 (7)	4·5 (16)
6	5·17 ± 0·57 (6)	4·00 ± 1·49 (3)	4·0 (5)
7	5·88 ± 0·42 (9)	5·20 ± 0·39 (5)	4·0 (2)
8	5·66 ± 0·72 (3)	—	—

The practical difficulties of obtaining large enough samples during the latter parts of the breeding season mean greater variation due to sample size, but a trend for reduced litter size after the initial peak is apparent. The incidence of prenatal mortality follows roughly the same trend.

(c) Black-tailed Jackrabbit (*Lepus californicus*)

In California Lechleitner (1959) described the number of visible implantations as varying seasonally from one to two in January and February to over five in the peak of the season in April. A more recent study by French, McBride, and Detmer (1965) (although again based on small sample sizes during the later parts of the breeding season) is of considerable relevance, in that it shows comparative data from five separate years collected from the same area in Idaho:

			Number of embyros per pregnant female			
Year	Feb.	Mar.	Apr.	May	June	July
1956 Mean	2·33	3·86	3·75	5·00	4·50	*
S.E.	0·43	0·67	0·75	0·00	0·64	—
N.	6	7	4	2	4	—
1957 Mean	*	2·43	4·00	6·25	5·67	3·50
S.E.	—	0·36	0·37	1·11	0·21	0·50
N.	—	14	18	4	6	2
1958 Mean	1·42	3·56	5·00	5·11	5·00	—
S.E.	0·16	0·30	0·49	0·35	—	—
N.	24	32	13	9	1	—
1959 Mean	1·20	2·08	3·43	3·66	3·60	*
S.E.	0·20	0·21	0·39	0·42	0·24	—
N.	5	24	14	12	5	—
1960 Mean	1·00	2·83	3·50	5·50	3·50	*
S.E.	—	0·30	0·37	0·25	0·28	—
N.	1	12	18	4	4	—

* No pregnant females in sample.

It will be seen that although the litter size increases and then decreases as the season progresses, the peak does not apparently occur at the same month in different years. Unfortunately, no information was presented in this paper on the change in litter size with age, and it is possible that the proportion of younger females in the monthly samples may have biased the picture of seasonal change.

(d) Snowshoe hare (L. americanus)
Details from the paper on this species by Keith, Ronstad, and Meslow (1966) have already been reported above, but the reader is reminded that the first litter is always smaller in size than later litters. This phenomenon has been reported in nine different studies on snowshoe hares listed by the above workers.

(e) European hare (L. europaeus)
Reynolds and Stinson (1959) document the following seasonal variation in Southern Ontario for mean litter size per pregnant female: January, 1·2; February, 1·7; March, 2·5; April, 4·0; May, 3·5; June, 3·9; July, 3·5; August, 3·5; September, 4·0.

(f) Bank vole (Clethrionomys glareolus)
The detailed investigation of Zejda (1966), referred to above, also analyses the effects of season, etc., on litter size. In Czechoslovakia the number of embryos per litter varies as follows:

April	5·09 ± 0·15 (55)
May	5·24 ± 0·10 (147)
June	4·82 ± 0·16 (67)
July	4·67 ± 0·17 (42)
August	4·73 ± 0·20 (30)
September	4·40 ± 0·21 (20)

Zejda further showed that litter size increased significantly with body size and weight, but that in this species the variation with age was small. The period of the year was of decisive influence on the litter size and not the actual age *per se*. Only in very old females was a significant drop in litter size seen. The succession of litters in a season also had no discernible effect, as no significant difference was found in the litter size of primiparous and multiparous females. All these comparisons suggest strongly that seasonal effects are most important in altering the litter size in the bank vole.

(g) Meadow vole (*Microtus pennsylvanicus*)

Kott and Robinson (1963) reported that the litter size of voles born in traps was 5·9 in May–June, 4·8 in July, and 4·0 in September in Ontario and that these differences were significant. In snaptrapped females the mean number of embryos in May–June (6·2) was significantly higher than in July–August (5·0).

(h) Levant vole (*M. guentheri*)

The variation in litter size with season can be seen on page 177 where data from Bodenheimer (1949) are presented. This worker also reported that laboratory colonies of the same species fluctuated between litter sizes of 3·8 and 6·3, but with no consistent pattern.

(i) Old-field mouse (*Peromyscus polionotus*)

The litter size of this species was documented by Caldwell and Gentry (1965) in South Carolina:

December–15 February	3·29 ± 0·42
16 February–May	3·24 ± 0·08
June–15 August	2·76 ± 0·19
16 August–November	3·48 ± 0·20

The litter size in June to August was significantly lower. Williams, Carmon, and Golley (1965) noted the following variation in this species with sequence of litters, which in most cases would parallel the seasonal changes, although it is unlikely that many females would survive in the field as long as these laboratory animals:

Order of pregnancy	1	2	3	4	5	6	7
Mean litter size	3·33	3·54	3·94	4·10	4·12	4·30	4·00

The drop between the last two litters was not significant.

(j) Deermouse (*P. maniculatus*)

A study by Brown, L. N. (1966) of the reproduction of deermice in Wyoming showed the following variation in fertility during 1964:

	Average number of corpora lutea	Average number of viable embryos
April	4·00 ± 0·50 (4)	3·75 ± 0·50 (4)
May	4·93 ± 0·35 (13)	4·50 ± 0·41 (8)
June	6·86 ± 0·65 (7)	6·80 ± 0·53 (5)
July	6·50 ± 0·57 (8)	6·40 ± 0·52 (5)
August	5·00 ± 0·00 (3)	5·00 ± 0·00 (3)

Prenatal mortality in the early stages of pregnancy did not vary with season, but the peak in litter size came in the middle of the breeding period.

(k) House mouse (Mus musculus)

The litter size in this species is normally fairly constant, and in commensal populations does not seem to vary with season. A study by Batten and Berry (1967) on the island of Skokholm off Wales showed that, unlike the usual situation, the litter size there did not correlate with maternal size, but it did vary with season:

Mar.	Apr.	May	June	July	Aug.	Sept.	Oct.
5·0	5·8	7·6	8·4	10·7	—	6·7	6·0

This population is also interesting in that post-implantation mortality is much lower than has been reported for other populations.

(l) Multimammate rat (Rattus (Mastomys) natalensis)

This species would seem to have the greatest reported variation in litter size with season. A study by Chapman, Chapman, and Robertson (1959) in what is now Tanzania showed that this species at Latitude 8° S had a number of embryos varying from three to sixteen. There was a slight tendency for the litter size to decrease after the peak of the breeding was reached in May, but the total sample was of only ten pregnant females.

(m) Muskrat (Ondatra zibethicus)

The number of young per litter, based on very large samples, was shown by Errington (1963) to vary seasonally in Iowa as follows: April, 5·8; 1–15 May, 7·4; 16–31 May, 7·3; 1–15 June, 6·5; 16–30 June, 6·5; 1–15 July, 6·6; 16–31 July 6·5; August, 7·2. The last figure is based on much smaller samples than the others, which show a clear trend to an early peak and then steady fall.

(n) Sheep (Ovis aries)

Examination of considerable numbers of Suffolk ewes at Cambridge, England, showed the following variation in ovulations (Averill, 1955):

Mating period	Ovulations per ewe
20–30 October	1·90
1– 17 November	2·30
18–30 November	2·37
15–31 December	2·00
20–31 January	1·88
1– 28 February	1·69
1– March	1·69

This work was followed up by Averill (1959) in later studies of Romney Marsh ewes in New Zealand, where the ovulation rates in 5–7-year-old ewes were shown to vary as in figure 44.

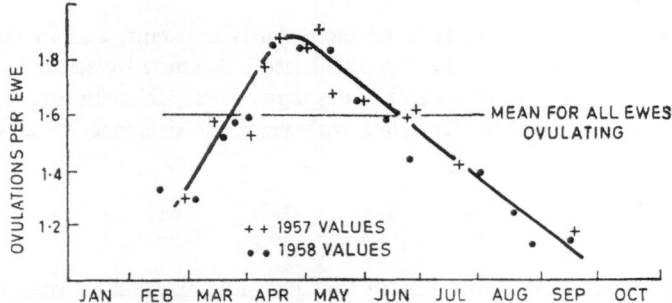

Fig. 44. The relationship between the time of year and mean number of ovulations per ewe. From Averill (1959) *N.Z. Jl agric. Res.* **8**: 575–83.

(*o*) Bandicoot (*Perameles gunni*)

The litter size in this Tasmanian marsupial varied significantly with sequence (Heinshohn, 1966) as follows:

(1) 2·14 ± 0·16, (2) 2·19 ± 0·18, (3) 2·80 ± 0·19, (4) 2·25 ± 0·49.

C. DISCUSSION

Although so much information is available as to the pattern and degree of fluctuation in mammalian litter size with latitude and with season, there is almost nothing known of the causes of such fluctuation. The physiological variation of litter size with age, and in some species with size or weight, is explainable in the first case in terms of a change in ovarian function with age (see Jones and Krohn, 1961) and in the second in terms of availability of uterine space. With the exception of studies such as those of Zejda above, the importance of these physiological variables in seasonal and latitudinal changes are as yet unassessed. For example, body size in mammals is known to increase with latitude (Bergmanns' rule), and it is possible that there is a relationship here with latitudinal changes in litter size. Although there can be no doubt that the general trend for litter size to increase with latitude is present (in at least lagomorphs and rodents) where the litter size is sufficiently big for such variation to be expressed, the evidence presented is far from conclusive that this is an even or consistent trend. For obvious logistic reasons, no single worker has assiduously collected information from one species over a wide latitudinal range, so that problems of comparisons of different sampling methods and methods of determining fertility render comparisons of the results obtained by separate workers somewhat difficult. Work such as that by Keith, Rongstad, and Meslow mentioned above points the way to obvious lines of investigation into the role of genetic variation in litter size. The observations of Pinter and Negus (1965) have indicated (see page 177) that the litter size of one rodent species can be significantly increased by placing on a greater daylength. This would suggest further experiment as to the comparative role of genetic and latitudinal origin effects and the actual effects of photoperiod.

In a recent paper Spencer and Steinhoff (1968) have attempted to explain altitudinal and latitudinal variation in litter size in terms of the variation in length of breeding season. They assume that a portion of the parental mortality will vary directly with the size of the litter produced, and suggest that a short breeding season will limit the maximum number of times a female can reproduce. This therefore presents an advantage to those females with a phenotype which produces larger litters. Long breeding seasons, on the other hand, favour those females which produce smaller litters. The contribution to the total rate of increase by those litters which were produced in the additional time in the longer seasons is supposed to be greater than that contributed by producers of smaller litters, because a larger proportion of parents producing such litters will survive to reproduce throughout the entire period of the breeding season. There appears to be no available evidence to prove or disprove the initial assumption behind this explanation of the geographic variation in litter size. However, it seems a fairly reasonable proposition and at the moment stands as a plausible explanation for the variation which has been described.

It seems likely from the available evidence that the seasonal alteration in litter size presents a more uniform pattern. In all species investigated the first and the last litters are the smallest, but there are apparently two main patterns in between. In some species the maximum litter size is greatest immediately after the beginning of the breeding season and then drops away gradually (i.e., curve skewed to the left), whereas in other species the peak litter size is in the middle of the season (i.e., curve symmetrical). The distribution and reality of these two postulated patterns await further investigation. The relative survival of the young to breeding of litters of various sizes would be interesting to investigate, to determine the selective advantages of maintaining the considerable variation in litter size. There is much controversy raging between ornithologists of a population bent as to the significance of clutch size and its variability. The interested reader is referred to the excellent paper by Mountford (1968) on the significance of litter size in mammals, in which he points out that the total number of weaned offspring is a combination of two factors, namely, the distribution of the different sizes of litter and the number of weaned offspring from each litter size. Obviously natural selection will favour genotypes that produce the greatest number of offspring from both these combinations of factors. The argument that the litter size which is most productive should also be the most frequent is considered fallacious by Mountford. In mammalian population studies, despite the greater practical difficulties involved, there is need for investigations of survival from litters of different sizes under different environmental conditions.

CHAPTER 16

Delayed Implantation

In 1935 Hamlett attempted to review all that was known at that time of the patterns of delayed implantation, and he assessed the suggestions previously published as to its ecological importance. He noted no common ecological factors and suggested that delayed implantation was a relatively useless characteristic, in which the hormonal situation governing the delay may have become closely associated with some other unknown characteristic which was of evolutionary importance. There is now available considerably more information on the patterns and taxonomic distribution of this phenomenon, but basically Hamlett's observation of a lack of *common* ecological relationships still holds good.

I. Taxonomic distribution of delayed implantation

This phenomenon occurs in a number of mammal species distributed over a wide range of taxons. It must first be strongly emphasized that delayed implantation is by no means closely related to the taxonomic relationships of a species. The representative (but incomplete) list below will demonstrate the peculiar distribution of species in which delayed implantation is definitely known to occur.

Species with delayed implantation		*Species with no delayed implantation*	
Marsupialia			
Setonix brachyurus	(1)	*Trichosurus vulpecula*	(1)
Bettongia lesueuri	(1)		
Protemnodon eugenii	(1)	*Antechinus stuartii*	(4)
Protemnodon rufogrisea	(1)	*Dasyurus viverrinus*	(4)
Potorous tridactylus	(1)		
Protemnodon bicolor	(1)		
Megaleia rufa	(1)		
Macropus canguru	(1, 3)		
Macropus robustus	(2)		
Chiroptera			
Eidolon helvum	(5)	No other species known	(6)
Edentata			
Dasypus novemcinctus	(7)	No other species known	
Rodents			
Mus musculus	(8)	Little information available	
Rattus norvegicus	(8)		
Phenacomys longicaudus	(9)		

Species with delayed implantation		*Species with no delayed implantation*	
Carnivora			
Martes americana	(10)		
Martes pennanti	(10)		
Gulo gulo	(10)		
Lutra canadensis	(10)	*Lutra lutra*	(14)
Enhydra lutris	(15)		
Taxidea taxus	(11)		
Meles meles	(12)		
Mustela erminea	(10)	*Mustela putorius*	(13)
Mustela frenata	(10)		
Mustela vison	(13)		
Ursus americanus	(16)	Little information	
Pinnipeda			
Phoca vitulina	(17)	*Odobenus sp?*	(17)
Halichoerus grypus	(17)		
Mirounga leonina	(18)		
Artiodactyla			
Capreolus capreolus	(19)	No other species known	

References in the above list and in figure 45 are as follows: (1) Sharman, Calaby, and Poole, 1966; (2) Sadleir and Shield, 1960; (3) Pilton, 1961; (4) Woolley, 1966; (5) Mutere, 1967; (6) Wimsatt, 1960; (7) Enders,'A. C., 1966; (8) Baevsky, 1963; (9) Hamilton, 1962; (10) Wright, 1963; (11) Wright, 1966; (12) Canivenc, 1966; (13) Pearson and Enders, 1944; (14) Stevens, M. R., 1957; (15) Sinha, Conaway, and Kenyon, 1966; (16) Wimsatt, 1963; (17) Harrison, R. J., 1963; (18) Laws, 1956; (19) Short and Hay, 1966; (20) Jonkel and Weckwirth, 1963; (21) Wright and Coulter, 1967; (22) Hamilton and Eadie, 1964.

Although there is little information in some groups, the absence of reports of delayed implantation can usually be assumed to mean that it does not occur. However, there has been at least one case where a delay in implantation was not detected by the original worker on the species but was subsequently found to be present. Comparison of species in general, such as *Lutra* and *Mustela*, indicates that delayed implantation is species specific, and species with very similar habitats and niches either may or may not extend their pregnancy.

A somewhat special case is reported by Bradshaw (1962) for the California leaf-nosed bat (*Macrotus californicus*) in Arizona. Females are fertilized in autumn, and soon afterwards the trophoderm of the blastocyst invades the uterine glands to form a type of placental association. However, embryonic growth is very slow indeed during the winter, and speeds up remarkably in March. Bradshaw has termed this pattern 'delayed development'.

II. The timing of delayed implantation

To elucidate the patterns of delayed implantation it is best to consider the timing of mating, implantation, and birth during the year. In the first group exhibiting *seasonal delayed implantation* each of the three stages occurs at fixed times of the year, and so presumably can be related in some way to environmental influences occurring at that time. This does not necessarily mean that there will be a common pattern of influences, however. In the second group which exhibit *Aseasonal delayed implantation* (whether or not mating is seasonal) the time of implantation can vary widely. In these cases physical factors in the environment will probably have little direct effect on the time of implantation (although indirect effects cannot be discarded), but biotic and particularly social factors may still be related to implantation.

(*a*) Seasonal delayed implantation

(i) *Synchronous delayed implantation.* Figure 45 indicates some of the patterns of pregnancy found in various mammals. In temperate terrestrial species there can be no doubt that an effect of the delay, which may or may not be its function,

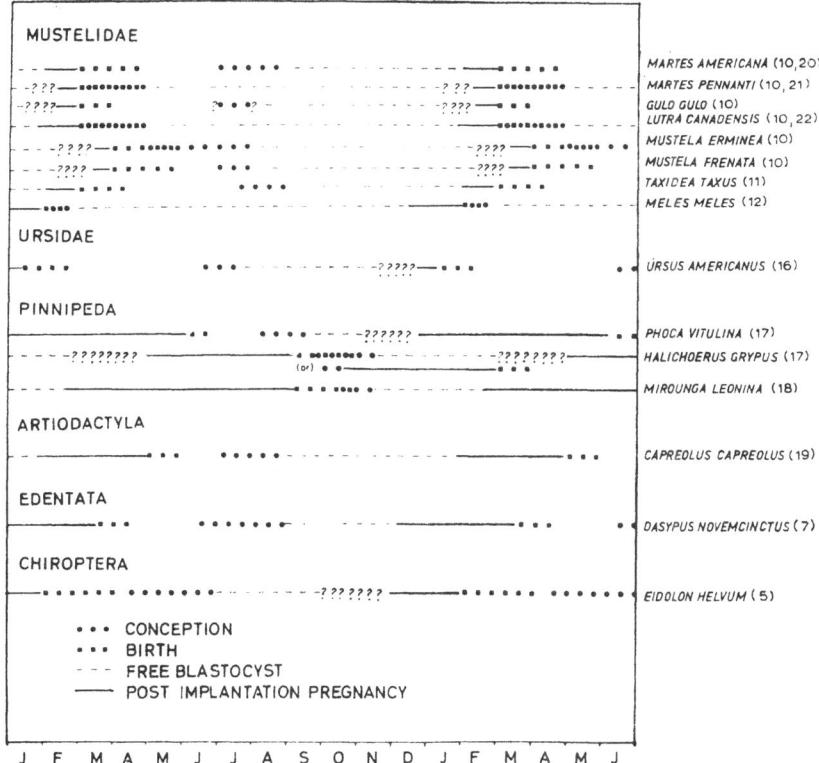

Fig. 45. Patterns of pregnancy with delayed implantation in certain mammals.

is that during the winter females are impregnated and pregnant but not subjected to the nutritional requirements of carrying a developing embryo. Implantation commonly occurs in late winter, so that birth follows in spring. The problem remains, however, as to why some species (whose length of gestation after implantation is sufficiently short) do not mate, implant immediately, and then give birth all inside the spring and summer. Information on the comparative growth rates of juveniles, and on their survival the following winter, is needed in groups such as the mustelids to determine if the early birth season has been selected because of a slow speed of juvenile development. It may be that the timing is critical and that juveniles cannot attain a sufficient size for survival by the next autumn and winter unless their birth dates are very early in the spring.

For most of the seals, delayed implantation results in birth and the next mating occurring very close together, as in *Mirounga leonina* (the southern elephant seal), thus enabling a reduced period of time for aquatic species to be out of their normal environment. However, in some of the more asocial species, such as *Phoca vitulina* (the harbour seal), this period is more extended, as animals will be hauled out on sandbanks and the like for more extended periods. In the former type of species it is often assumed that the period of time out of the water is as short as possible, because of the effects of terrestrial predation and reduced ability to feed. In actual fact, with the exception of man, a very recent predator, in evolutionary terms there seems to be little evidence of predation on present seal colonies which are located in generally inaccessible localities. However, the inability to feed probably acts to minimize the time out of the water, thus giving a selective advantage to the possession of delayed implantation.

There seems to be no explanation at the present time for the presence of synchronous seasonal delayed implantation in tropical species. The tropical bat (*Eidolon helvum*) shown in figure 45 was studied at lat. 0° 20′ N, and gave birth just prior to the major peak of rainfall, but there was no apparent reason why mating could not have occurred at the time of implantation and pregnancy occupy a more normal length of time.

The mink (*Mustela vison*) exhibits synchronous delayed implantation, but as this species has been almost entirely studied in captivity and much is known about the effects of light on its implantation, it will be considered separately below.

(ii) *Asynchronous delayed implantation.* In at least two and possibly more species the period of mating (and/or the period of births) is seasonal in the sense that it occurs over fixed, though long, periods of the year and yet the time of implantation does not occur at a fixed time in the annual cycle. The red tree mouse (*Phenacomys longicaudus*) breeds from spring to autumn (Hamilton, 1962), but its length of pregnancy can vary from 28 to 42 days if lactation stress results in a delay in implantation. This lengthy delay, for a rodent, is almost certainly the result of the peculiar diet of this species, which feeds almost entirely on fir needles. The finding of lactation-controlled delayed implantation in laboratory rats and mice means that other species of rodents, which are seasonal breeders, could possibly also exhibit short delays in implantation during breeding in the wild.

In the marsupials which exhibit delayed implantation the presence of a free blastocyst is correlated with the presence of a suckling joey in the pouch. If suckling ceases, the blastocyst implants and another pregnancy goes to completion (Sharman, Frith, and Calaby, 1966). Most of the larger marsupial species breed continuously and will be dealt with below, but in some, breeding is seasonal. In the quokka (*Setonix brachyurus*) mating occurs from January to March on Rottnest island, Western Australia, and births occur a month later, so that over 90 per cent of females have pouch young in July (Main, Shield, and Waring, 1959). However, during pouch life, if the joey is lost or removed, implantation can take place. Shield and Woolley (1963) have shown that if the joeys are removed in February 60 per cent are replaced by new joeys as a result of implantation of delayed blastocysts, but that this proportion decreases as the year proceeds, so that there are no viable blastocysts by August. These authors suggest that seasonal nutritional stress is the cause of a lack of survival of viable blastocysts in late pouch life and during the anoestrous season, as domesticated quokkas, quokkas in another population with no seasonal food short- age, and quokkas in the vicinity of garbage dumps all showed much higher rates of survival of blastocysts. Asynchronous implantation may also occur in *Pro- temnodon eugenii* (Berger, 1966) and *Potorous tridactylus* (Hughes, 1962), but the evidence for the seasonal nature of breeding in these two small wallabies is less substantial, and it now appears that *Potorous tridactylus* may actually breed continually, though at varying intensities (Sharman, Frith, and Calaby, 1966). However, *Protemnodon eugenii* certainly does have a period of anoestrus and can maintain viable blastocysts even through this period (Berger, 1966).

(*b*) Aseasonal delayed implantation

The presence of lactation-controlled delayed implantation in laboratory mice and rats infers that in these species, which breed continually in the wild, there will probably also be periods of delay. Sinha, Conaway, and Kenyon (1966) have reported that the sea otter (*Enhydra lutris*) has pups born during every month of the year, and that although the length of gestation is not known, the high proportion of delayed blastocysts found in tracts suggests strongly that implanta- tion is delayed and aseasonal in this species.

Most of the macropod marsupials breed continuously throughout the year, and yet can potentially implant a delayed blastocyst at any time that suckling ceases in the pouch. In those species such as the red kangaroo (*Megaliea rufa*) and the euro (*Macropus robustus*) which live in arid climates where the rain-induced period of good nutrition is erratic there is considerable pouch mortality during extended periods of drought (Ealey, 1963; Frith and Sharman, 1965; Sadleir, 1965a; Newsome, 1965, and see page 224). It has been suggested by Sadleir and Newsome that drought-induced pouch loss results in the implantation of de- layed blastocysts, so that the period of births can be effectively extended into the otherwise anoestrous season. This would be of considerable selective advan- tage, as females, though emaciated, would still have small pouch young ready in the pouch when the drought eventually breaks. Thus in these species nutri- tional stress may act secondarily to be the stimulus for implantation. In those species where nutritional stress is not as common, and where pouch mortality is not so marked or as synchronous, it has been suggested that ecological use of delayed implantation is to speed up the replacement of pouch joeys lost by accidental mortality (or, according to Marshall (1967), lost to predators). However, this explanation is difficult to accept, as the time difference involved is only the length of just under one oestrous cycle, i.e. usually a little less than a month.

III. The role of light in delayed implantation

The fixed timing of implantation of the delayed blastocyst in those species showing synchronized seasonal delayed implantation strongly suggests that photoperiodic influences are acting. Unfortunately, with the exception of some work on marten (*Martes americana*, Enders and Pearson, 1943; Pearson and Enders, 1944), all the available information on experimental and other observations of photoperiodic effects has been given on a single species under domestication, the mink (*Mustela vison*).

The length of gestation in this species depends on the date of mating. Pearson and Enders (1944) indicated that if mating on fur farms was carried out in early March in the United States, the length of gestation was 55–58 days, but this decreased steadily until matings around 22 March resulted in 45–47-day gestations. Hanson (1947) reported 60-day mean gestations in Swedish mink mated 2 March, and decreased to 45 days when the mating took place on 30 March. The most recent data come from Dukelow (1966), who reported the following relationship in mated mink from farms near the Great Lakes, U.S.A.:

Date of breeding	Mean length of gestation in days
10–14 March	56·8 ± 2·3 (12)
15–19 March	53·0 ± 2·7 (46)
20–24 March	50·3 ± 4·5 (25)
25–29 March	46·4 ± 4·4 (7)

In each case this degree of variation can be attributed to variation in the length of the delay period, as the length of time from implantation to parturition is known to be 30 ± 1 days (Enders, R. K., 1952).

Experimental alteration of light by Pearson and Enders (1944) was shown to change the period of gestation. Female mink placed on a fixed increase of $1\frac{1}{2}$ hours from 2 February and mated between 25 February and 5 March, resulted in a gestation of 49·4 ± 0·29 days in 10 females. Females unlit prior to the same mating date, but given the same photoperiodic regime after mating, had gestation lengths of 50·5 ± 0·49 days ($N = 19$). Unlighted controls had gestation lengths of 54·7 ± 0·80 days ($N = 20$), so that both the experimental treatments had reduced the delay period. Holcomb, Schaible, and Ringer (1962) compared the effects of extra lighting before and after mating and noted that females on extra light, prior to mating, but then under normal photoperiods, had longer gestations than females lighted before *and* after mating. The effect of photoperiod on mink has also been investigated by Beljaev, Klockov, and Zelezova (1963); Beljaev and Klockov (1965), and Aulerich, Holcomb, Ringer, and Schaible (1963). The former workers and Dukelow (1966) reported that the shorter gestations resulted in larger litters and postulated that this was due to reduced opportunities for prenatal mortality.

Thus increased light regimes prior to mating can advance the onset of mating, but extra lighting after mating is necessary to advance the time of implantation. Pearson and Enders (1944) have suggested that the effect of lighting prior to mating is to induce the pituitary to produce FSH. Copulation, early in the season, results in LH production; but because breeding has been advanced, either there is insufficient FSH to stimulate enough oestrogen to prepare the uterus for implantation or, alternatively, a high oestrogen level has not been maintained for sufficient time to allow completion of the uterine changes wrought by progesterone. When the photoperiod is increasing after mating, either artificially or naturally, the changes can then take place and implantation results. Similarly, Holcomb, Schaible, and Ringer (1962) suggested that 'the increase in light per day to which the mink were subjected, caused an increase of FSH production raising the estrogen level, thus hastening estrus. Later the suppression of secretion of estrogen from the ovary through altered production of gonadotrophins from the anterior pituitary may have allowed the full effect of progesterone from the corpus luteum to ready the uterus for the earlier implantation of the fertilized ovum and thus decrease the possibility of resorption of fertilized ova' (page 677).

IV. Discussion

There are thus two major environmental factors which need consideration relative to delayed implantation in mammals. Photoperiod can be strongly inferred to be the factor which times the implantation of the blastocyst in those species showing synchronous seasonal delayed implantation. Nutritional factors also seem to be closely related to delayed implantation in most of the species where it has been determined. In the case of those affected by photoperiod, it can be postulated that this is only acting as a timing mechanism in species who are subjected to regular seasonal changes in nutrition. Where the delay is directly related to lactation, in that the blastocyst does not implant during lactation, as in the marsupials, or where the blastocyst implants later due to a large litter and intense lactation, as in the rodents, it is perhaps not too wide off the mark to speculate that nutritional effects are also of importance. It would be of great interest to attempt to alter the delay period under lactation inhibition by manipulating dietary factors. Thus it is possible that the phenomenon of delayed implantation always acts to enable the female mammal to undergo those periods of nutritional stress, late pregnancy, and the whole of lactation at periods when environmental nutrition is maximal. The seal situation is slightly different, in that the lactating female is reducing her own body reserves and not feeding during a very intense lactation, but this can be explained as being due either to the time of late pregnancy being a period of good nutrition and/or that lactation is so short and intense that time would not be sufficient for the female to feed and produce the highly concentrated milk simultaneously.

The problem that is still unsolved is the reason in a number of species as to why mating does not take place a few days before the onset of implantation. The phenomenon could be more realistically called in many mammals 'advanced mating'. In some species the males are apparently infertile at this time, but in others, although information is scanty, potentially mating could occur. Indeed, in the grey seal (*Halichoerus grypus*) mating does actually occur at this time (Harrison, R. J., 1963). This species, which apparently can have a normal or delayed implantation, is worthy of a detailed investigation as to the ecological advantages or otherwise of the phenomenon. Comparative studies of population dynamics, and thus selective forces between species living under the same environmental conditions which do and do not exhibit delayed implantation, should result in a greater understanding of its ecological role, especially with regard to male fertility. Two examples of such pairs living in comparable environments are red deer (*Cervus elaphus*) and roe deer (*Capreolus capreolus*), and river otter (*Lutra lutra*) and sea otter (*Enhydra lutris*).

PART FOUR

Parturition, Lactation, and Immediate Post-natal Survival of Young

CHAPTER 17

The Effects of Weather from Parturition to Weaning

The stage of development of the neonate mammal governs its degree of independence of the physical environment at birth and immediately afterwards. In some mammals the precocious young are born blind with little hair and with little ability to regulate their own temperature. In other species the young are born in a much more advanced form and are almost completely equipped to cope immediately with a detrimental environment. An example of the range of variation in development is the extremely well-advanced leveret of the hare and the poorly developed rabbit kit.

Unlike previous parameters, environmental effects on young after birth tend to be absolute in action, so there is little graded response between mortality and survival. Even if the species concerned produces litters of more than one, detrimental effects act on the litter as a unit in most cases, so that there is little grading in the response, at least at the individual level. In the population sense a physical factor may have a graded effect on the survival picture, but it should be remembered that although this is measured in terms of survival of young, the factors are always acting directly on the mother, so that it is *her* response to the environment which controls juvenile survival at this stage, and not the direct response of the young themselves.

For this reason most of the evidence considered in this chapter has a very important behavioural element, relating the effect of physical factors to maternal behaviour and thus survival of the young. In many cases the environmental factors simply act as the executioner, the fate of the young being determined by the mother's behaviour. A particular example of this is the effect of weather on rabbit litters described below. In precocial young, maternal behaviour (nest construction, keeping the young warm) is of vital importance to neonatal survival, but this factor is less important in species whose young are born in a more advanced state.

(a) Rabbit (*Oryctolagus cuniculus*)

During work on penned rabbits in Australia, Myers and Poole (1962) noted considerable mortality in nestling kits following heavy rains. Bull (1961) also noted these deaths in New Zealand rabbits – the mortality being caused by flooding of the nesting burrows and being especially prevalent if associated with cold weather. This cause of death was also recorded by Mykytowycz and Gambale (1965) in large 45-acre paddocks where the effects of confinement would be minimal. Behavioural observations under penned conditions (Mykytowycz, 1959, 1960) indicated that females who were low in the dominance hierarchy were forced to dig their nesting burrows in sites where flooding was more likely to occur. Therefore, although rain and cold caused the death, the selection of those young which succumbed was a function of the social situation.

(b) Sheep (*Ovis aries* and *O. nivicola*)

Mortality factors affecting the new-born lamb have been reviewed in a detailed article by Moule (1954) and in other articles by Moule (1950a) and Smith (1964d), in which a variety of climatic factors have been noted to contribute to lamb deaths. Adult sheep are resistant to the effects of high temperatures, but new-born lambs cannot survive for longer than 2 hours at 100° F dry-bulb, 87° F wet-bulb. However, lambs rapidly gain the ability to control their temperature, so that at 1 week of age they can survive 6 hours at 99° F (d.b.) and 70° F (w.b.). Temperatures during lambing frequently exceed 100° F (d.b.) during midday in Queensland, and Moule has noted rectal temperatures of over 105° F in lambs under 3 days old. As a result of the inability to regulate temperature, many new-born lambs die during periods of heat. On one occasion considerable lamb mortality was associated with over 6 hours per day of temperatures over 100° F (d.b.) and 75° F (w.b.) being maintained for up to 7 successive days. Behavioural factors are also important here, as the ewe usually leaves her new-born lamb in the shade. If she chooses a position near to water so that the lamb is not left alone for long periods while she is going to drink, survival is higher than if there are prolonged absences from the lamb whose heat budget is further thrown off balance by water deficiency and who then leaves the shade to seek its mother.

At the opposite end of the temperature scale cold and wind exposure were noted by Moule and Smith as causes of mortality. If lambs are born in high winds the amniotic coat tends to dry on the fleece, thus decreasing its insulation value. Moule described one occasion during lambing where the new-born lambs were exposed to 35-m.p.h. winds with an ambient temperature of 42° F (d.b.), 35° F (w.b.). Many lambs showed rectal temperatures of 85–88° F, and a number succumbed. Alexander and Williams (1966) showed that under cold conditions when the fleece was wet the lamb's ability to find the ewe's teat was considerably reduced, so that the time to first suckling would be increased by this type of weather, with concomitant mortality due to delayed suckling. Cherniavski (1962) has reported considerable mortality of lambs of the snow sheep (*O. nivicola*) in Koryak highlands, U.S.S.R., when there is continuous rain and cold winds from the north.

A form of accidental mortality related to weather conditions was also noted by Moule (1954). Lambing in March after a period of prolonged rainfall (5·6 inches in 2 weeks preceding lambing and 4·4 inches during 11 days of lambing) resulted in the deaths of 39 of 349 lambs due to their being bogged in muddy soil from the moment of birth, and not being able to rise or move to the mother. Ludbrook (1963) has reported that a 1–2-day-old buffalo calf (*Syncercus caffer*) was also bogged in mud and then deserted by the mother.

(*c*) Caribou (*Rangifer tarandus*)

This species gives birth after migrating, under particularly difficult conditions which have been described on page 156. Although the young are born in a very advanced state, there is considerable evidence that weather factors result in a degree of mortality. Kelsall (1957) and Kelsall and Loughrey (1957–58) reported that at least 35 per cent of dead calves examined by them had died immediately after birth in blizzards, and probably a considerable proportion of the 40 per cent which were classified as dying due to material desertion also died due to extreme cold and high winds. Both Pruitt (1961) and de Vos (1960) report deaths of caribou calves during blizzards, and Pruitt quotes Russian reports that the entire calf drop is occasionally lost if birth coincides with bad weather.

Experimental work by Hart, Heroux, Cottle, and Mills (1961) on caribou calves showed that temperature regulation was well established at birth, but the calves altered their metabolic rate when weather conditions became detrimental. A lowering of the temperature alone to $0°$ C caused the metabolic rate to double, but when this was combined with wind and precipitation the rate increased fivefold and the calves became hypothermic and usually died. These workers commented that if the calves are sheltered these effects are considerably reduced, but on the barren grounds of Northern Canada (where the field observations of the previous workers were made) shelter is almost non-existent.

(*d*) Musk-ox (*Ovibos moschatus*)

The very severe winter of 1959–60 in Greenland resulted in a high incidence of neonatal mortality due to deep snow and cold (Hall, 1964). In the next winter there was only a small depth of snow, and calf mortality decreased. Tener (1965) has noted that the young of this species in Canada can be born in really extreme cold, and quotes one birth at a temperature of minus $27°$ F, but he emphasizes that there must be strong selective factors acting against the birth of musk-ox young at times when deaths will occur due to cold.

(*e*) Seals

Morton-Boyd, Lockie, and Hewer (1962) reported that the position on the breeding ground where grey seals (*Halichoerus grypus*) reared their pups often governed their chances of survival. During strong winds and heavy seas, off north Rona, Shetland, pups on exposed rocks were often washed into the sea and drowned. The proportion which did so would have depended on age, but no estimate was possible. Heavy seas also forced moving seals to use restricted passages to and from the breeding site (see page 242), with the result that pup mortality increased in these passages due to aggression between adults. Lugg

(1966) noted only a very low mortality (2 pups per 81 births) in a colony of Weddell seals (*Leptonychotes weddelli*) in the Vestfold hills, Antarctica. However, this breeding assemblage was extremely well protected from the sea, and much higher rates of pup mortality have been reported in other colonies of this species where conditions are more exposed.

(*f*) Other species

Buckner (1966) noted that the numbers of juvenile *Sorex cinereus* found were inversely correlated with the rainfall of the previous month in S.E. Manitoba. During heavy rain many nestlings were apparently drowned in the nest. Similarly, McCarley (1966) found that a number of litters of *Citellus tridecemlineatus* in northern Texas were drowned due to exceptionally heavy rainfall in May 1965. Survival of a few litters occurred when they were located on higher ridges. It would be interesting to know, in species such as this, whether social position governed littering site, and thus susceptibility to natural hazard, as has been mentioned above for the rabbit.

Extremely arid conditions resulted in the death of young pronghorns (*Antilocapra americana*) in a particularly dry year in Oregon (Buechner, 1961). It is possible that the actual cause of death was insufficient nutrition in this case. Seasonally arid conditions probably affect lactating animals more than non-reproducing individuals. Cox (1965) found that lactating pipistrelle bats (*Pipistrellus hesperus*) were continually drinking at waterholes all night during midsummer, whereas this activity was reduced at other times of the year.

(*g*) Discussion

Perhaps the most drastic of observable selective pressures are those occurring immediately after birth. Therefore it is rather remarkable that even the small number of observations above have been made showing the detrimental effects of weather at this time in some mammals. It might be postulated that documentation of this type of mortality is an indication of a species being unfitted to its environment, or as being in a state of altering the time of parturition so as to avoid the weather effects. Most mammals have well-developed behavioural mechanisms to minimize the weather effects on the new-born young by building nests, maintaining lairs, or simply keeping the young under temporary shelter. These behavioural devices act to separate the new-born young from the physical environment, thus rendering weather effects minimal in the majority of cases.

CHAPTER 18

The Effects of Nutrition from Parturition to Weaning

From parturition onwards the female mammal produces increasing quantities of milk, the nutritive content of which requires a correspondingly increasing level of nutritive intake. At this time the food and water supply is perhaps more crucial to the survival of mother and young than at any other time in the entire reproductive process. The change-over in the nutritive intake of the young from its mother's milk to solid food also constitutes a period of great dependence on the nutrition available, and here especially water supply can be crucial in drier environments. This chapter will review the available information on the effects of variation in nutrition on lactation and, through lactation effects, on the survival of the suckling young.

(a) Rodents

In laboratory rats there is a very considerable increase in the food consumption as lactation progresses, and it has been estimated that the energy required by a female rat for the 22 days of lactation is 279 Calories, compared with requirements for 79 Calories for the 22 days of gestation (Russell, 1948). High levels of protein are necessary for survival of young rats to weaning. Experimental reduction of protein in the diet (or giving purified protein, such as casein or soya-bean) results in considerable lactational upsets. A particular effect is not a reduction in the level of the protein in the milk but in the volume of the milk produced. The most important mineral is calcium, and females on low-calcium diets tend to drain their own calcium reserves while maintaining a reasonable calcium content in the milk. The evidence regarding the importance of fatty acids in the diet and their effects on lactation is equivocal.

In one wild rodent species, the bank vole (*Clethrionomys glareolus*) the bio-energetics of reproduction in the laboratory have been studied by Kaczmarski (1966). Figure 46 shows an increase in the energy value of food taken during the latter stage of lactation. Kaczmarski estimates that raising a litter of four to weaning requires an extra 289 kcal compared with an extra 75 kcal to support a litter of five through gestation. Kaczmarski comments that the 'ecological efficiency' of this species (i.e. the increase in the weight of the litter or the 'production' in terms of the additional energy assimilated by the female) is relatively high. For new-born young this ratio indicates 11·0 per cent efficiency and for weaned young 14·6 per cent efficiency.

The effect of nutrition on lactation in the wild equivalent of the laboratory rat (*Rattus norvegicus*) has been reported by Davis (1951b) in his comparison of city populations, where the level of nutrition available from refuse cans, etc., was much higher than for rats living on farms, where the main diets were horse-feed and manure. The data below show the difference in lactation between these

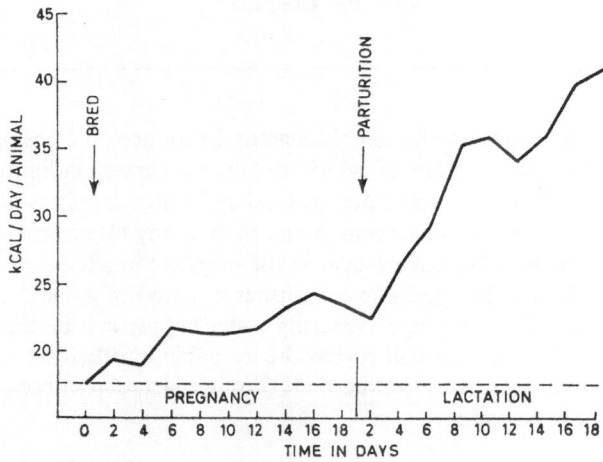

Fig. 46. Daily food assimilation of female bank voles (*Clethrionomys glareolus*) during pregnancy and lactation. The curve represents a mean of ten females. The average level of energy assimilation in non-reproducing females of the same size (17·5 kcal/day) is marked with a broken line. From Kaczmarski (1966) *Acta theriol.* 11: 409–17.

populations – the incidence of lactation being calculated by multiplying the standardized mean prevalence of lactation by 365 and dividing the total by the number of days that lactation can be determined (36):

	Farm rats	City rats
Percentage of adult females pregnant	19·2 (312)	23·7 (915)
Percentage of adult females lactating	22·2 (307)	43·2 (918)
Incidence of lactation (mean)	2·2	4·3
(range of two S.E.'s)	1·9–2·5	4·2–4·8
Percentage loss of litters	49	10
Estimated number of young weaned per year	13·6	35·0

The differences in lactation incidence are thus further reflected in the survival of the suckling young in wild *R. norvegicus*.

Greenwald (1957) has suggested that nutritional conditions can govern the size of voles (*Microtus californicus*) at weaning. He noted that in 1952 in California (a year when the rains were erratic and grass growth was poor) the mean weight of the ten smallest females trapped was 17·85 g and of males 17·98 g. In 1953 there were good rains, so that vegetation was much more abundant. In that year the equivalent mean weights were: females, 13·58 g; males, 15·19 g. The assumption here is that the smallest animals are trapped immediately after weaning.

Many of the heteromyid rodents live in desert areas where nutrition and water supplies are of major importance in reproduction. In their comparative review of this group, Eisenberg and Isaac (1963) suggest that the smaller litters found in the desert-living genus *Dipodomys* when compared with other heteromyids may be a result of lactational strain on the mother. *Dipodomys merriami*

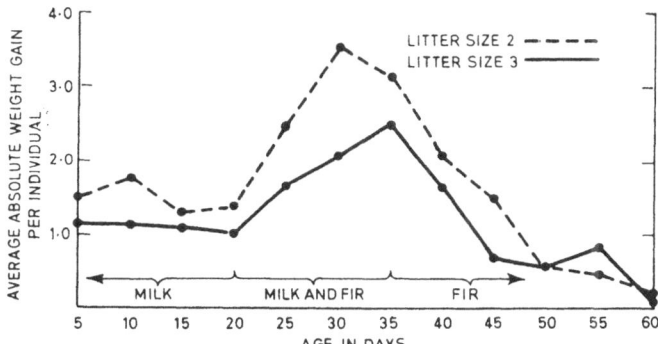

Fig. 47. Weight gains in young red tree mice (*Phenacomys longicaudus*) in litters of two and three. From Hamilton (1962) *J. Mammal.* **43**: 486–504.

(a typical desert form) has a small mean litter size for a rodent (2·2) and a relatively short length of gestation (19–22 days). There appears to be no information in these species as to the variation in lactation lengths possible under various conditions of drought or as to the post-natal mortality resulting from any such variation. Vaughan (1962) noted that the timing of breeding in the plains pocket gopher (*Geomys bursarius*) in Colorado was such that post-natal care and lactation occurred under optimal nutritional conditions.

A unique species, as far as its source of nutriment is concerned, is the red tree mouse (*Phenacomys longicaudus*), which feeds entirely on Douglas Fir needles. Hamilton (1962) has described how litters of two in this species always outweigh at birth and outgrow litters of three (see figure 47). The female produces a constant biomass per litter, suggesting that she is at the maximum level of lactation possible on this peculiar diet. Weaning takes place from 30 to 40 days after birth in this species compared with 17–19 days in a close relative, *P. ungava* (the tundra vole). The shape of the growth-gain curves in the figure strongly suggests that this species is living on a diet of very little metabolic value.

(b) Sheep

An immediate effect of lambing and the onset of lactation is a considerable increase in the feed intake of the ewes. Hadjipieris and Holmes (1966) showed

the following voluntary intakes on various diets. The hay used was of a particularly nutritious variety, and thus reduced the necessary intake, but nevertheless the increase is marked. The amount of milk produced by the ewe and its

	Digestible organic matter intake per day (g)
Dry ewes on hay	780
Dry ewes on grass cubes	1,140
Lactating ewes (twins) on grass cubes	1,850
Lactating ewes (twins) on grass cubes and hay	1,487
Lactating ewes on pasture (estimate)	2,535

onset after parturition can also be modified by nutritional means. McCance and
Alexander (1959) kept two groups of ewes on different levels of nutrition during
pregnancy, so that the well-fed group gained 10 kg and the poorly-fed group lost
the same amount. The effect on lactation was as follows:

	Milk secretion rates (ml/hr)	
	Days 1 and 2 after lambing	Days 4 and 5 after lambing
Well-fed ewes	82·3 ± 16·2	47·9 ± 4·9
Poorly-fed ewes	13·0 ± 3·2	14·3 ± 3·4

These results are referred to by Schinkel (1963), who also emphasizes the importance of the level of nutrition in the last part of pregnancy with regard to the
amount of colostrum in the mammary gland at birth. Thus, under experimental
conditions, the activity of this gland is very closely controlled by the present and
past nutritional status of the ewe.

Nutrition and water supply is of major importance to the production of milk
in ewes in tropical or arid areas. In a series of papers studying the effects of
environment on the reproduction of Merinos in Queensland, Smith (1964b,
1964c, 1964d) has emphasized the role of high temperatures (see page 216) on
lactation and lamb survival. However, the desertion of young and a large proportion of lamb losses at certain seasons are also related in part to the poor
nutrition or more possibly the poor water supply available to the ewes. As evidence for a nutritional effect and that lactation puts a considerable strain on poorly
fed ewes, the following data are presented on the duration of anoestrus after
lambing at different seasons of the year:

	Mean duration of anoestrus after lambing (days)	
Period of lambing	Lactating ewes	Ewes which lost lambs
November 1961	97	81
March 1962 (best feed conditions)	41	39
July–August 1962 (poorest feed conditions)	195	176

If ewes suckling lambs were given extra nutrition in the form of lucerne the period of time to first oestrus after parturition was significantly reduced, indicating that the strain of lactation on ovarian activity could be alleviated by a supplemented diet.

(c) Deer

There is little experimental evidence relating nutrition to lactation in deer. In the experiments of Murphy and Coates (1966) already referred to (page 186), the authors record that on two occasions, at the lowest protein level (7 per cent), female deer (*Odocoileus virginianus*) died during lactation, suggesting a drain on their metabolism, Youatt, Verme, and Ullrey (1965) found that even acute feed shortages during pregnancy did not reduce the nutritive value of the milk produced by this species, so that the new-born fawns were not affected. However, in these latter experiments the does were only on reduced intakes during pregnancy and were placed on a normal level during lactation.

On the natural range the degree of physical deterioration of does during pregnancy (which is related to the level of nutrition available) governs almost entirely the survival of fawns. As an indirect measure of this, Verme (1962) reported that if does lost fawns near to parturition their physical condition immediately started to improve, and like the sheep mentioned above, they returned to oestrus earlier. This meant that does on poor range bred earlier than does on good, but because this usually resulted in greater fawn losses, it did not lead to any permanent shift in the breeding season. In another study of the variation in fawn production in white-tails, Marburger and Thomas (1965) found that during years of normal rainfall but overstocking (and hence poorer relative nutrition) the numbers of fawns seen per doe steadily decreased. A year of drought resulted in a considerable mortality of both adults and suckling fawns, relieving the overstocked situation so that nutrition availability increased, and so did the fawn crop. Unfortunately, it is not possible to determine from this paper as to how much of the variability of the fawn crop was due to nutritional stress during lactation, it being more than likely that the litter size was also affected. However, during the drought a failure of lactation would certainly have been a cause of fawn mortality.

An isolated observation completes the available information on the effect of nutrition on deer lactation. Cowan (1956) noted that female *Odocoileus hemionus* fed more intensely for a short period just prior to the birth of the fawn. They were observed at this time feeding earlier and earlier in the afternoon and even midday, presumably to increase their total nutritional intake just prior to the onset of lactation.

(d) Rabbits

In studies of penned populations of wild rabbits (*Oryctolagus cuniculus*) the effect of green feed on onset and cessation of breeding has been well documented. However (independent of the density involved), when the green feed is drying out at the end of the breeding season a common feature is a decrease (culminating in cessation) of lactation, with subsequent death of the kittens being nursed (Myers, 1958; Mykytowycz, 1960; Myers and Poole, 1962). In experiments giving similar populations different types of supplementary feed (see page 190), Stodart and Myers (1966) reported that in one control group which received no supplement, when the pastures dried off, all the 2-week old litters contained undersized young, and at least one kitten had died in each litter as a result of milk starvation.

(e) Macropod marsupials

The prolonged period of pouch life and lactation compared with the length of gestation in marsupials makes them particularly interesting with regard to the effect of environmental variables on this period of the reproductive cycle. In addition, the permanent attachment of the young joey to the mother's nipple for the first part of pouch life and its very close association with the mother for the rest, means that the determination of post-natal mortality and its association with lactation failure may be determined very accurately. As the various parameters of growth are now known for a number of species of macropod (*Setonix brachyurus*, Shield and Woolley, 1961; *Macropus robustus*, Sadleir, 1961; *Megaleia rufa*, Sharman, Frith, and Calaby, 1964), it is possible to determine the date of birth of pouch young very accurately, and by comparison of samples of females taken at known time intervals to estimate the seasonal incidence of pouch loss.

Such loss has now been demonstrated to occur in a number of species of macropod during periods of lowered rainfall and very poor nutrition. Ealey (1963) and Sadleir (1965a) described losses of the larger pouch young of the euro (*Macropus robustus*) as the dry season lengthened in the arid north-west of Western Australia. A drought period in Southern Queensland resulted in the death of all joeys of the grey kangaroo (*Macropus giganteus*) born between September 1964 and November 1965. (Kirkpatrick and McEvoy, 1966.) Frith and Sharman (1965) compared the reproduction in three populations of red kangaroos (*Megaleia rufa*) living in areas of New South Wales and Queensland where the nutritional and water conditions were very different. The greatest level of pouch mortality was related to the nutrition and water available, being greatest on an area where there is a long drought period.

In an excellent study of the same species in central Australia, Newsome (1965) showed that the death rates of pouch joeys was related to a measure of the degree of drought intensity called the index of aridity which has already been described (see page 138). The proportion of joeys older than 3–5 months old decreased progressively as the period of drought lengthened. By using probit analysis, Newsome obtained a significant correlation between the survival of pouch young and the index of aridity, and this is shown in figure 48.

In a thesis submitted in 1961 the writer proposed that these desert kangaroos demonstrated a previously undescribed type of breeding pattern in response to a peculiar environmental situation. The reader is referred back to page 50, where the different sorts of breeding season were categorized. It seems that a large mammal, with a long period of time between conception and weaning, living in an environment when the optimal season for breeding is not only short but completely unpredictable in time (as occurs in deserts), is faced with a considerable problem to produce the maximum number of viable young. This is overcome by maintaining a continuous series of conceptions throughout the year, and thus producing a continual supply of small pouch young which during many periods may suffer a heavy mortality. When rains fall, however, this system means that the maximum number of young are available in a sufficient stage of development whereby they can profit from the continuing lactation of the female

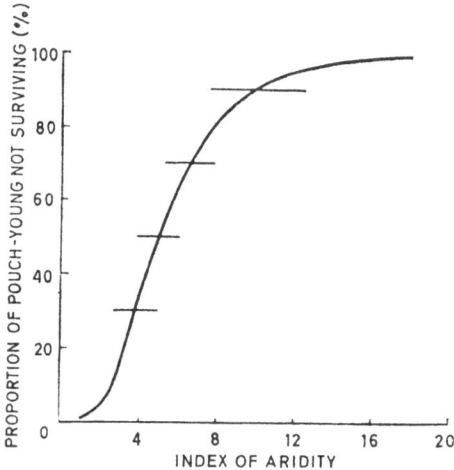

Fig. 48. Theoretical regression of the proportion of pouch young of the red kangaroo (*Megaliea rufa*) which do not survive to leave the pouch on the increase in the index of aridity (see text). The 5 per cent confidence limits of the increases in the indicies of aridity required to induce 30, 50, 70, and 90 per cent responses are shown as horizontal lines. From Newsome (1965) *Aust. J. Zool.* **13**: 735–59.

and the excellent feed conditions outside. The picture is thus one of continual staggered breeding and high post-natal mortality but at the same time maximum production of viable young. This idea was published later (Sadleir, 1965a) and has been also discussed for other areas by Newsome (1965) and Frith and Sharman (1964).

(*f*) Other mammals

A number of desert mammals described by Bodenheimer (1957) give indications of the strain of milk production. Reduced supplies of feed in semi-arid areas such as South-west Africa results in reduced lactation. Bodenheimer reports that in that region elephants (*Loxodonta africana*), zebras (*Equus zebra*), and several species of antelope produce very little milk, and in the case of the elephants are thus not able to adopt strange calves, a practice found in other areas of higher nutrition. In little brown bats (*Myotis lucifugus*) Kallen and Wimsatt (1962) found that lactating females were considerably lighter than after lactation has ceased, and had a low red blood cell-count, indicating nutritional strain. Females of this species were found by Stones and Wiebers (1965) to consume a great deal more food while lactating. There is very little information relating nutrition to lactation in wild carnivores. Jordan, Shelton, and Allen (1967) showed that in years that the moose (*Alces alces*) of Isle Royale, Ontario, had a good calf crop, the litters of wolves (*Canis lupus*) were larger, and pup mortality was generally closely related to the availability of calves as food during parturition. Indirect but suggestive evidence of the importance of nutrition during lactation for black bears (*Euarctos americanus*) comes from data presented by Miller, R. S. (1963) for litter sizes found near the Saskatchewan River:

		Body weight of cubs (lb)	
Size of litter	Sample size	Mean	Range
1	1	17·0	—
2	4	12·9	12·0–4·0
3	6	9·0	8·5–9·5
4	4	5·4	4·5–6·0

The role of nutrition in human lactation and its effect on fertility has been the subject of a considerable amount of discussion by anthropologists and human demographers. Lactation has been suggested as causing a reduced fertility while the baby is being suckled, although some workers disagree with this suggestion. Solien de Gonzalez (1964) has shown clearly that interpretation of the data available has been confused due to a lack of understanding of the effect of nutrition on the mother. She points out that women in advanced economic areas have a short lactation period, and the babies are fed supplementary food early in life. In these cases there is little reduction in fertility. In poorer and more primitive communities lactation occurs for much longer periods and, even if the mothers are undernourished, the lactation performance is good from the point of view of quality and quantity of milk. Solid foods are given to the baby at a much older age, so that the general lactation strain on the mother is more intense and prolonged. It would appear that the studies that have been done on lactation amenorrhea have been carried out in high-income Western societies where the intensity of lactation is less than in primitive communities. Solien de Gonzalez therefore suggests that in primitive communities on poor nutrition intensive lactation does result in reduced fertility, and she presents evidence from New Mexico and Italy to support this idea. Lactation can also affect the timing of seasonally breeding primates (see page 163).

(g) Discussion

Reduced food intake or deficiencies of any vital section of the diet of a lactating female result in detrimental effects on the ability to rear young. However, the evidence from laboratory and domestic mammals indicates that, up to a certain point the female will drain her own body reserves of the various components needed to produce milk to a remarkable degree, before any lowering in milk quantity or quality can be determined. An interesting problem is apparent when comparisons are made over a wide range of mammalian reproductive systems. At one end of a range of systems it is apparent that the lactating female is producing a very high quality or quantity milk over a very short period of time, and thus cannot possibly derive the nutriment in this milk from her diet at that time. Indeed, in certain species, such as the pinnipeds, the female does not usually feed during lactation, but whether or not she does feed, the main part of nutriment in the milk in this type of lactational situation is derived from body stores, and thus indirectly from intake some time before lactation commenced. At the other end of the range females may be seen to be actively increasing their food intake during lactation, and thus deriving a large part (possibly the majority) of the nutriment in the milk from direct feeding. These two extremes are thus of a direct and indirect effect of the nutritional regime on lactation effectiveness. It would be of considerable interest to compare the energy efficiency of these different types of mechanism.

In some mammal species (red tree mouse, black bear) there is evidence from the growth rate of the young of different litter sizes that the output of the female is strained to its utmost in rearing larger litters. It would be of interest to determine the relative total survival and thus contribution to the population situation of smaller well-fed litters compared with larger litters of smaller weaning size. This leads into the largely unexplored regions of the role of energetic efficiency in terms of lactation and nutrition. There is a need for investigations and comparisons between species with different roles in the community (predators, prey, etc., herbivores, cf. carrion feeders, etc.) as to the efficiency with which they convert their nutritive intake into a biomass of young. For example, it would be interesting to determine at what point in different species the female stopped draining her own reserves and the suckling young died. It might be suggested that perhaps this point is related to the probability of that female producing a subsequent litter, which would survive to breeding, or to the delicate balance of probabilities of her own as opposed to her present offspring's survival. The efficiency of different litter sizes in terms of the conversion to biomass has been already mentioned. Another variable of interest would be the relative efficiency of an early weaning at a low body weight as opposed to a late weaning at a higher body weight. In each case the ultimate efficiency could be determined only in terms of the chances of the young contributing genetically to later generations.

CHAPTER 19

The Role of Predation in Survival to Weaning

Depending on its stage of development, and ability to move and defend itself, the new-born mammal is particularly susceptible to predation. However, the degree of detection of this predation by biologists is a function of the size of the female and of her behaviour at parturition. Most of the known examples of documented predation on new-born young given below are of a very limited number of species, because in practically all other cases the female is so cryptic at birth and during early lactation that predation may be suspected but cannot be easily verified or studied in detail.

(a) Artiodactyls

In Africa there have been many reports of females giving birth in concealed positions in long grass or away from the rest of the herd. Ansell (1960) mentioned that reedbuck (*Redunca arundinium*), puku (*Kobus vardoni*) and hartebeest (*Alcephalus lichensteini*) all conceal their young until quite well grown. Similar behaviour is almost universal in the smaller African deer. However, in the larger deer, such as the wildebeest (*Connochaetes taurinus*), Ansell describes how the young are born in a very precocious state and run with the herd from birth. In one instance a new-born wildebeest was seen running off with the herd less than five minutes after it has first attempted to stand.

Despite this precocity, it is apparent from the work of Estes (1966) that young wildebeest are still subjected to a considerable amount of mortality due to predation by the spotted hyaena (*Crocuta crocuta*). This species is highly selective. Estes mentions that:

'Observations on hyaena hunting behaviour and kills indicate that it selects new calves and mainly ignores those more than a few days old, which are far harder to catch, unless they are separated from a herd. When pursued, a female with calf invariably runs into the nearest herd. Survival chances for new calves are best when there are numerous slightly older ones that make it hard for a hyaena to single out and keep a particular quarry in sight. By selecting against calves born outside the season, particularly ahead of time, hyaena predation thus acts to maintain a sharp peak; by the same token, mortality is far more severe in small herds than in large aggregations. Indeed the constantly much higher percentages of calves found in concentrations throughout the year, is one of the most significant observed differences between aggregates and small herds' (page 1000).

African hunting dogs (*Lycaon pictus*) also act as severe predators on new-born wildebeest (Estes and Goddard, 1967).

However, as mentioned before, most African artiodactyls conceal their young at birth to avoid predation. Estes (1967) mentions that in *Gazella grantii* and *G. thomsonii* there is a degree of elaborate behaviour to reduce the chances of detection. The mothers eat all faeces of the young and clear up placental remnants and birth fluids after parturition – actions presumed to reduce the odour around the fawning site. Females of these species sometimes co-operate to help drive off smaller hunting predators such as jackals (*Canis mesomelas*). The impala (*Aepyceros melampus*) also suffers considerable mortality of new-born young which can be attributed to predation (Dasmann and Mossman, 1962a). This species normally gives birth in very thick bush, and the young are only ever found in very thick cover. Despite this there is a degree of lion (*Panthera leo*) predation (Schenkel, 1966a). An avian predator, the African hawk eagle (*Heira-äetus spilogaster*), is reported as feeding on new-born lechwe (*Kobus leche*) (Allen, 1963).

In North America the majority of artiodactyls also are cryptic at parturition and hide their new-born young, although young caribou (*Rangifer tarandus*) are born after migration in exposed areas and are subject to predation by wolves (*Canis lupus*) (Kelsall, 1958). Wolves also predate young moose calves (*Alces alces*), and during summer Pimlott (1967) reported that over 80 per cent of wolf scats contained moose hair and 70 per cent deer hair. Atwell (1964) observed a wolf actually killing a 2–3-day-old moose calf despite its defence by the mother, and comments that the calf was apparently normal and healthy. Mule deer (*Odocoileus hemionus*) hide during birth and immediately afterwards in dense brush (Dasmann and Taber, 1956) to avoid predation. Johnson, D. E. (1951) noted that most calves of the elk (*Cervus canadensis*) in Montana hide up in sagebrush and show a very marked preference for the edge of one vegetation type or other. He noted two instances of black bears (*Euarctos americanus*) killing calves, and found other evidence of this species preying repeatedly on young elk. Einarson (1956) records the fascinating observation that just prior to the fawning season every island in the Athabasca River of Alberta had one doe preparing to give birth. He attributes this behaviour to avoidance of potential wolf predation. Apparently the degree of vegetational growth can also affect the incidence of predation in mule deer, as if feed is scarce the doe must forage farther away and for longer periods from the hidden fawn, thus leaving it more exposed to predation.

Predation thus is a major mortality factor in juvenile artiodactyls, independently of the degree of precocity of the young. Unfortunately there have been very

Q

few studies which have attempted to measure the degree of predation and its variation with the availability of other food for the predators or with the number of predators. It would seem from the work of Estes and others that young artiodactyls are a preferred food of canine predators.

(b) Rabbits

The numerous studies that have been carried out in Australia on wild and penned rabbit (*Oryctolagus cuniculus*) populations have already been referred to at some length (pages 108, 132). In penned populations predation was particularly high at the end of the breeding season, as the kittens were apparently in very poor condition due to food shortages. They were seen eating on the surface right through the day and were repeatedly killed by ravens (*Corvus coronoides*) (Mykytowycz, 1960). In large open pens a number of litters were predated by foxes, *Vulpes vulpes* (Mykytowycz and Gambale, 1965), a predator also reported in New Zealand (Bull, 1961). Myers and Parker (1965) noted that the degree of fox predation in wild populations depended largely on the type of habitat in which the warrens were located. If burrows were dug in sandy soil predation was high, as the foxes could easily dig down to the nestlings.

A detailed study of rabbit predation was reported by Calaby (1959) in Victoria, who was most concerned with the role of the little eagle (*Hieraäetus morphnoides*). Between 6 May and 4 June 1958, 6 eagles over 450 acres killed an average of 5 to 6 rabbits per day, except when the weather was cold and windy, inhibiting rabbit and eagle movement. Although 8 eagles were hunting over the same area between 2 June and 7 June, cooler weather with mists reduced the average kill per day to 2 kittens. From May to July an average of 6 eagles killed an estimated total of 260 rabbits (about 4 per cent of the kit population). Aerial predators would seem to be less important for the European rabbit than terrestial ones. Myers and Schneider (1964) observed that 67 per cent of kittens died in two unpenned warrens over a single breeding season, by far the majority of deaths being due to feral cats (*Felis cattus*), which continually hunted over the area. Other predators were goshawks (*Accipiter fasciatus*) and owls (*Tyto alba*), but these were not as important.

There is little information on the role of predation in neo-natal mortality in other lagomorphs, although it is undoubtedly of some importance. Fitch (1947) describes one instance of a rattlesnake being discovered in the throes of eating a young cottontail (*Sylvilagus auduboni*) inside a nest, having already consumed three others.

(c) Other mammals

It is of interest that the same sort of cryptic behaviour and hiding of young which has been presumed to be due to predation avoidance is also found in at least three of the large predators. Schenkel (1966b) has noted that lionesses (*Panthera leo*) keep their litter hidden in dense thickets or rock outcrops until about 7 weeks after parturition. Robinette, Gashwiler, and Morris (1961) noted similar secretive behaviour in the female cougar (*Felis concolor*). The female lynx (*Felis lynx*) makes her breeding dens on 'islands' of mixed forest in the U.S.S.R. which are completely surrounded by large mossy bogs and are therefore very difficult of access (Iurgensen, 1955). It is possible that, at least in the lion, this sort of behaviour is due to intra-specific factors as colloquial reports suggesting that male felids are occasionally cannabalistic to cubs.

Other miscellaneous observations suggest predation on young. Chittleborough (1958) presented some evidence of the immediately post-parturition loss of humpback whales (*Megaptera nodesa*) and attributed this either to sharks or killer whales (*Orcinus* sp.). The fact that many mammals give birth only at night has been commented on by Rossdale and Short (1967). They noted that 86 per cent of a very large number of foal (*Equus caballus*) births occurred between 1900 and 0700 hours and suggested that this may be due to concealment from predators. A more likely explanation would be to prevent the drying out of the embryonic membranes during the lower daytime humidity.

Finally, predation on the young has been suggested as being the cause of a decline in a population of *Mus musculus* when the area was invaded by *Microtus californicus* (Lidicker, 1966). Experimental fieldwork by DeLong (1966) showed that when *Microtus* was present there was only an average of 1 nestling per lactating female *Mus* compared with 2–3 per lactating female when *Microtus* were absent. Laboratory observations showed that adult *Microtus* did actually seek out, kill, and eat *Mus* nestlings.

(*d*) Sheep (*Ovis aries*)

A number of studies by Smith (1964b, 1964d, 1965a) have noted the role of predators in mortality of lambs in Queensland. It is impossible to determine the degree of primary predation, and so that although the following predators were recorded as killing lambs, maternal behaviour or debilitation due to combinations of water shortage and hot weather were almost certainly the primary causes of death: pig (*Sus scrofa*), fox (*Vulpes vulpes*), dingo (*Canis familiaris*), wedgetailed eagle (*Aquila audax*), raven (*Corvus coronoides*), and even ants.

A recent study by Alexander, Mann, Mulhearn, Rowley, Williams, and Winn (1967) on the activities of predators in lambing flocks in South Australia showed that foxes avidly scavenged on foetal membranes and dead lambs, but were easily frightened away from lambs by aggressive ewes. Only one attack on a healthy lamb was seen. Crows also eat many foetal membranes, but will frequently attack any lamb which is separated from its mother. However, if these attacks were directed at healthy lambs no injuries were seen, and only comatose lambs succumbed. Thus the lamb condition and maternal behaviour govern predation effectiveness.

(*e*) Discussion

Only very scattered and rather irregular observations have been made of the importance of predation on mortality of new-born mammals. With a few exceptions, general statements such as 'new-born mortality due to predators occurs' or 'females exhibit young-concealing behaviour which is presumed to be due to the activities of predators' are all that can be made for most mammal species. The role of predation in population regulation has been debated at very great length by theoretical biologists, sportsmen, and pastoralists, to the level of nausea, but the data on which this debate is based, at least in the case of the juveniles, are astonishingly small. There is evidence in certain bird studies (Jenkins, Watson, and Miller, 1963, for example) that predation is acting as an executioner only for a proportion of the population which is going to die for reasons of behavioural exclusion. Such mechanisms may well occur in mammals. Although it has been colloquially observed many times that predators tend to kill mainly sickly adults, there is little information as to the effect of the physiological condition of the new-born on its susceptibility to predation.

CHAPTER 20

The Effects of Miscellaneous Factors from Parturition to Weaning

It seems possible that in a number of different species of mammals there is some juvenile mortality resulting from purely accidental factors. However, as has been shown when the role of behavioural selection of nest site governed mortality due to weather factors, it is more than possible that while deaths of neonates appear accidental, behavioural events prior to death were the ultimate cause. For example, weather (page 217) and density (page 242) factors have been shown to result in considerable mortality of juvenile pinnipeds on the breeding grounds. In addition, Mathisen, Baade, and Lopp (1962) have noted that in colonies of Stellar's sea lion (*Eumetopias jubata*) the bulls occasionally roll on to pups, crushing them to death, and there is some mortality from pups getting themselves firmly wedged into rock crevices. Rand (1955) reported similar deaths in pups of the Cape fur seal (*Arctocephalus pusillus*), although much of the mortality here was caused by fights between mothers who had trouble in recognizing their own young. Wirtz (1968) has described considerable mortality of pups in populations of Hawaiian monk seals (*Monachus schauinslandi*) and suggests that these deaths are the result of fighting among adults. Accidental factors would also seem to be the cause of the death by drowning at very high tides of nestlings of the shrew (*Sorex vagrans*) in salt-marshes in California (Johnston and Rudd, 1957).

In view of the fact that maternal death at parturition is not an unknown phenomenon in humans, and indeed was a major cause of mortality prior to the advent of modern medicine, it is strange that there is so little information as to this type of death in mammals. In domestic stock, maternal death is well known, and is especially associated with breeds bred for size, such as Hereford cattle. Only a single reference has been discovered to this phenomenon in wild mammals. Knowlton and Michael (1965) found a very recently dead white-tailed deer (*Odocoileus virginianus*) which showed signs of recent birth, and although it had eaten the placenta, showed indications of dying immediately after giving birth.

The death rate of human neonates is found to increase with altitude. Grahn and Kratchman (1963) found a significant correlation between the neonatal death rates in the United States and the altitude. If $Y =$ the neonatal death rate per

1,000 live births and $A =$ the altitude in feet, then $Y = 17 \cdot 51 + (1 \cdot 04 \times 10^{-3} \pm 0 \cdot 15 \times 10^{-3})A$. Attempts failed to find correlations with other factors, such as radiation-induced mutation or injury; and these authors suggest that the lowered oxygen concentration at higher altitudes causes reduced foetal growth and increased deaths immediately after birth. Mazess (1965) reported similar correlations over a greater range of altitudes in Peru.

PART FIVE

Density and Social Factors in Breeding, Pregnancy, and Survival to Weaning

CHAPTER 21

Social Factors in Adult Reproduction: Introduction

The social environment surrounding a mammal can have major effects on its reproductive ability. The elements which constitute this environment have a more intense action when individuals are living at high densities, although there are many dangers inherent in using density as a measure of the action of these social elements in a group of mammals (see page 24). The reproductive response to a detrimental social situation can range from a mere reduction in the live weight at birth, with little effect on survival of the neonate, to a cessation of oestrous cycles so that breeding completely stops. This range of responses overlaps the classification of breeding phenomena, which are used in this book to document responses to other environmental variables. However, if for reasons of consistency the onset or cessation of breeding was the only parameter considered here, and the effect of social factors on other parameters such as the proportion of females pregnant, the degree of prenatal mortality and litter size (to name a few), were relegated to another section on social factors and pregnancy the result would be an arbitrary and artificial division. The majority of papers considering the effects of social factors on reproduction document the various levels of reproductive response within a continuum, so that they would need under this system to be repeatedly referred to under separate sections based on subdivisions of the possible range of responses. For this reason the following chapters will consider the effects of social factors on reproduction from oestrus to weaning. It is realized that much the same line of reasoning could be logically applied to the rest of the book, but experience in organizing the material available has shown that actually the papers describing the action of factors such as climate, nutrition, etc., do fall more naturally into the separate units which make up the various stages of the reproductive processes.

The type of behaviour exhibited by mammals in the social environment is affected by the physical environment in which the social behaviour takes place. This part of the book will thus consider separately the effects of social factors on breeding in natural and in experimental populations. It will be seen from what follows that in many cases the experimental situation is far removed from that which the animal could experience naturally.

CHAPTER 22

Social Factors in Adult Reproduction in Natural Populations

It is possible roughly to categorize animals into cyclic and non-cyclic species. In certain mammals repeated observations have shown that the numbers present in any one locality fluctuate from year to year in a more or less regular manner. Examples include the 3- or 4-year cycle in numbers of certain microtine rodents (*Lemmus, Dicrostonyx, Microtus*) and the 10- or 11-year cycle of some lagomorphs (*Lepus*). Although there are changes in numbers present inside each year in these species, there is a fairly regular cycle of high-density years and years of lower density. These marked changes in numbers have been the subject of lengthy field studies and are now beginning to be well documented. This does *not* apply to many other species of mammals whose changes in numbers from year to year are not so obvious. Therefore, in by far the majority of mammal species there are no field studies involving estimates of density of more than three years duration. This lack of data means that although it is fairly easy to classify a species from the literature as being truly cyclic, the categorization of species as non-cyclic is usually fraught with presumption. Therefore the reader is warned that the categorization of species used below will probably prove inaccurate with further field research. The distinction is maintained, however, in the belief that the regularity of density changes in cyclic species should mean a biological system at work which may be of a different nature than mere density changes would suggest – whereas in 'non-cyclic' species other mechanisms, probably more closely related to the density alone, may operate on the reproductive processes.

A. 'NON-CYCLIC' SPECIES

(a) Lagomorphs

The wild European rabbit (*Oryctolagus cuniculus*) normally starts breeding in North Wales about February (see figure 16, page 71). In his classic study of the reproduction of this species, Brambell (1944) noted that the breeding season lasted for 15 weeks in 1941 and for 17 weeks in 1942. In a later study from the same area, Lloyd (1963, 1964, 1967) found that the season in 1957 lasted for 22 weeks and suggested that this extension was related to the decline in rabbit numbers caused by biological control programmes involving myxomatosis. Figure 49 from Lloyd's thesis shows that comparisons from different areas and years indicate a short breeding season when numbers are dense and a long season when they are low. The estimates of density are unfortunately subjective in most cases. In years of high density (1959, 1962) the incidence of pregnancy and the length of time during which the females were pregnant were also reduced. Under conditions of high density the ovulation rate was lowered, and this was

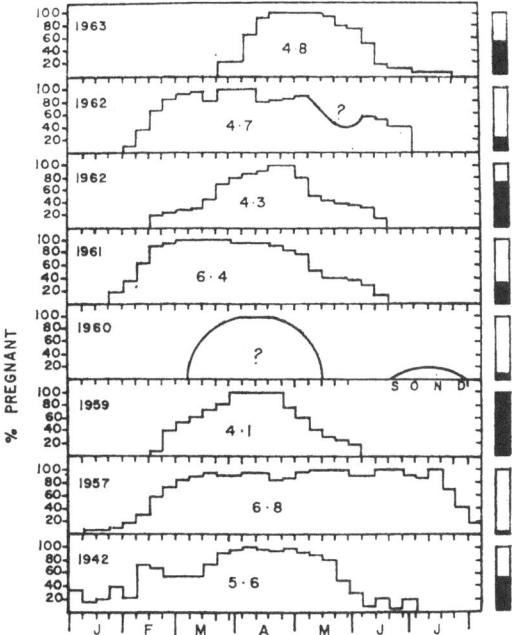

Fig. 49. The pattern of breeding in the wild rabbit (*Oryctolagus cuniculus*) in various years in North Wales. The figures in the centre of each histogram indicate the annual mean litter size, and the height of the black column to the right of the histograms indicates the relative mean annual densities in each year. The two histograms for 1962 were from separate populations at different densities. Modified with permission from Lloyd (1964).

reflected in reduced litter sizes as Lloyd found little significant difference in the degree of prenatal mortality in his samples. Prenatal mortality was lower in Lloyd's samples of low-density populations than in Brambell's samples of higher densities. He mentions that the ovulation rate did vary with the age of the doe, but the nature of the variation depended on the population density, suggesting the importance of social factors which have been investigated by various biologists in Australia (see page 263).

Probably the most detailed studies which have ever been performed on the reproduction of a wild mammal were those carried out in Wales by Brambell and his associates on the rabbit. In papers which are models of thoroughness of biological investigation and the theoretical problems inherent in the analysis of the data produced (Brambell, 1948; Brambell and Mills, 1949; and Allen, Brambell, and Mills, 1947; to mention some), they document the reproduction of this species with special reference to its high prenatal mortality. Unfortunately, in the context of this book, they present very little data on the environmental variables present when their various samples were collected, so that the influence of the environment on the fluctuations they describe cannot be determined. Nevertheless, the papers do document the type of variation in reproduction seen in a mammal, and as such should be read by all workers trying to present their reproductive data in the most accurate way.

Evidence of social factors affecting breeding was found for the North American swamp rabbit (*Sylvilagus aquaticus*) by Conaway, Baskett, and Toll (1960). In Mingo swamp, Missouri, they found that rainfall caused seasonal flooding of the habitat and consequent reduction in living area. This meant that the rabbits were sometimes suddenly crowded together by a natural phenomenon, and this effected their breeding as shown below:

		Rain-fall (inches)	Maximum water acreage	Corpora lutea per female	Maximum pre-im-plantation loss * (%)	Minimum partial litter resorption * (%)	Minimum total litter resorption * (%)	Minimum total embryo loss (%)
1956	March	1·80	3,100					
	April	4·05	3,200	4·0 (9)	4·4 (6)	5·6 (5)	38·8 (9)	46·0
	May	4·05	3,500					
1957	March	3·10	1,500	3·7 (37)	3·1 (18)	14·9 (12)	0·0 (17)	17·9
	April	11·20	12,000	—	—	14·3 (6)	61·9 (20)	70·0
	May	12·00	14,000					
1958	March	6·80	10,500					
	April	4·85	10,500	3·7 (16)	2·2 (12)	8·6 (6)	0·0 (16)	10·8
	May	5·27	9,000					

(The sample size is given in brackets.)   * As a proportion of ovulated ova.

In 1957 there was a very sudden flooding between 1 April and 8 April which crowded the rabbits into open fields. This was followed by a sudden increase in the resorption of whole litters in certain females, suggesting that under crowding the rabbits were being subject to some intense social stress, which had a marked effect on their reproduction.

(b) Rodents

The work of Southwick (1958) on house mice (*Mus musculus*) living in different densities in English corn ricks has already been referred to (see page 25), where the numbers of mice in various ricks were given. Considerable differences were found in the fertility, litter size, and prenatal mortality, as shown below:

	Low density	Medium density	High density	Very high density
Proportion of females with visible pregnancies at autopsy (%)				
12·6–17·5 g	40	38	35	24
17·6 + g	73	58	63	61
Mean litter size	6·23	5·68	5·62	5·11
standard error	±0·20	±0·15	±0·10	±0·20
Proportion of pregnancies with resorbing embryos (%)	14	16	15	27

Although the last class is based on a single rick when compared with from seven to twenty-one ricks for the other classes, the average proportion of females pregnant at very high densities was significantly lower than in the others. When separated into weight classes the different density categories were not significantly non-homogeneous. There was, however, a very significant drop in mean litter size with increasing density, and the proportion of pregnancies with resorbing embryos was significantly higher in the very high density rick when compared with the other density classes.

Another relevant paper on house-mice densities in corn ricks is that of Rowe, Taylor, and Chudley (1964) They only dealt with four individual ricks, but at much higher general densities than the previous paper describes. The females in their lowest density rick (less than 4·0 weaned mice per cubic metre) had a mean litter size of 5·29, which was significantly higher than in the other three ricks, where the density was over 8·0 mice per cubic metre and where the mean litter sizes were 4·51, 4·83, and 4·75. These data were similar to those presented by Southwick, but the proportions of females found with pregnancies showing resorbing embryos (38, 22, 34, and 29 per cent) were higher than he reported. There was no indication of a continued increase of this parameter with density as Southwick had reported at lower densities.

Observations on two other rodents under non-commensal conditions are relevant. Dunmire (1960) trapped *Peromyscus maniculatus* at different altitudes and found that on the basis of trap success the deermice were living at different population densities. This was reflected in the incidence of prenatal loss as shown below, although the litter size remained much the same (slightly increased with altitude):

Altitude (ft)	'Population density' (captures per 100 trap nights)	Percentage resorption
12,400	5·4	2·6
9,800	8·1	17·4
7,100	12·3	20·0

As the proportional resorption decreased with altitude, it is unlikely that any factors connected with altitude were the cause of prenatal mortality. In a study of kangaroo rats (*Dipodomys ordi*), McCulloch and Inglis (1961) found that during one exceptionally good winter in the Southern Great Plains of Texas and Oklahoma the population of this normally dispersed species increased rapidly during the breeding season. They suggested that increased density caused a decrease in the litter size from 3·37 early in the season to 2·76 later on.

(c) Other mammals

Buckner (1966) reported that the breeding season of the shrew (*Sorex cinereus*) was longer in a year of high density than in normal years, the opposite of what might have been expected. In a study of elephants (*Loxodonta africana*) under conditions of much higher density than normal, Laws (1966) suggested that the finding of the mean interval between calves of about 7 years as opposed to about 4 years for less-dense areas may have been due to density affecting reproduction in this species. A similar suggestion, for the same species, was made by Buss and Savidge (1966), who noted that the annual increment based on large samples detected by aerial observation varied inversely with density:

Locality	Density of elephants (per square mile)	Annual increment (%)
Queen Elizabeth park	2·2	8–9
Tangi–Karuma area	2·8	7–7·5
Murchison Falls park	4·5	6–6·5

In the latter area the density was so high that other aspects of behaviour were affected and migration was restricted.

The effect of density itself on the types of behaviour exhibited by wild mammals may affect reproduction. Davies (1953) described a very small breeding group of grey seals (*Halichoerus grypus*) in Cardigan Bay, Wales, which produced only 5 pups in 1950 and 1 in 1951 from a group of 50 adults (including 20 adult females). He suggested that this colony of a socially breeding species is so small in numbers that its low density on the breeding beach is detrimentally effecting conception.

In dense colonies of grey seals pup mortality is related in an interesting manner to density (Morton-Boyd, Lockie, and Hewer, 1962). On North Rona in the Shetland Islands the densely populated areas were those routes over which adults travelled to and from the sea. Each movement of a transient animal through the harems would result in considerable aggression, and in the meleé many pups were killed accidentally. There was a rough relationship between the density of pups with their mortality:

Breeding area	Proportion of pups dead of those born (%)	Average density (pups/acre)
Fianius North	19·5	140
Fianius Central	19·2	120
Fianius South	13·0	58
Sceapull	12·4	25

In certain of the main gullies which were direct routes of access to the sea up to 29 per cent of the pups died due to adults fighting. There is also an effect of weather, as during high seas the seals were forced to use only a few of the normal means of access to the breeding grounds, and again mortality of pups was common (see page 217). Observations by the writer on grey seals on Gasker, Outer Hebrides confirm that the actual density was less important than the effect of the access routes to the sea. Coulson and Hickling (1964) also noted that the mortality of pups was closely related to their density relative to the length of shore line available. They give the following data for islands in the Farne group, Northumberland:

Breeding area	Proportion of pups dead of those born (%)	Mean number of pups per 100 m. of shore line
Staple	18·6	77
North Wamses	12·5	42
Brownsman	11·9	14
South Wamses	9·8	10

They also noted that this sort of mortality in grey seal pups increased during the second half of the breeding season, when the density on the breeding area was higher.

Leuthold (1966) found that the males of Uganda kob (*Adenota kob*) in dense populations formed a lek system and defended territories up to 30 metres in diameter on communal territory grounds. In low-density populations the male behaviour changed, and they defended much larger (100–200-metre diameter) territories with much less interaction between males, and therefore presumably less interference to mating. Leuthold suggested that this would be more efficient in breeding. It would be of interest to compare the incidence of pregnancy from these different density populations and see whether the change in male behaviour does have an effect. A most interesting phenomenon in this respect was reported by Sugiyama, Yoshiba, and Pathasarathy (1965) in troops of Hanuman langurs (*Presbytis entellus*). Births occurred during the whole year in this species in India, but different troops showed different incidences of births. A change in the leader male of a troop led to an immediate period of aroused sexual excitement, with the result that there was a significant rise in the frequency of conceptions after such a leadership change took place.

(d) Discussion

The information relating social factors to breeding in non-cyclic natural populations does not present anything like a uniform picture. Most of it is based on the house mouse, a commensal living at greater densities than described for feral populations (see Anderson, P. K. (1961); and see figure 56 and page 272) and in a medium which consists of its food supply; and on the rabbit, a species whose density is again largely affected by human agricultural practice allowing considerable increases in numbers. In these two species these 'higher densities than normal' resulted in very detrimental effects on reproduction. With the exception of the swamp rabbit study, where an unusual, though natural phenomenon resulted in suddenly increased density (which was almost certainly the cause of an increased resorption rate), the other evidence from the remaining species is somewhat equivocal and the effect of density on the breeding changes observed suggestive rather than conclusive. The section which follows shows much clearer and better documented relationships between density and social factors and reproduction.

B. 'CYCLIC' SPECIES

(a) Microtine rodents

A number of species of rodents are subject to approximately 4-year cycles, during which in 1 or 2 years their numbers are so great as to reach observable peaks. The best known of these occurs in various species of lemming. Field studies of these species have shown that there appear to be two fairly constant patterns of alteration in the duration and timing of breeding which are always found at the same phase of the population cycle. The first and best documented of these is a very early curtailing of the length of the breeding season during the peak year of high density.

During his observations of a sudden fall in numbers of *Lemmus trimucronatus* in Alaska, Rausch (1950) collected a sample of dead animals very early in the year which showed no signs of any reproductive activity. Kalela (1961) found that only 54 per cent of mature female *Lemmus lemmus* were pregnant in the August of the peak year of 1959 in Finnish Lapland, and none were pregnant in September. He had earlier reported that during the peak and decline years of a cycle in *Clethrionomys rufocanus* in the same locality the breeding season was much shorter than in the year preceding. Similar results were reported by Zejda (1961) for *Clethrionomys glareolus* Czechoslovakia, by Bergstedt (1965) for the same species in Sweden and by Chitty (1952, 1955) for *Microtus agrestis* in Wales. Zejda (1967) has also pointed out that this curtailing of breeding affected the age of the rodents present in autumn, as during the peak years the short breeding season is confined to spring, so that the population is still composed of older animals in the autumn; whereas in non-peak years the longer breeding season results in the population in autumn consisting of animals largely born in that year.

In the report of verbal discussion following Chitty's 1955 paper on *M. agrestis* evidence was presented showing that in two small areas where the voles were at different densities but at the same stage of the population cycle (end of the breeding season of the peak year), there was a difference in the fertility of females and males, suggesting that breeding had been more detrimentally affected in the denser population. Kalela (1961) suggested that the difference in actual cessation of breeding of *L. lemmus*, which he found in two different fields at the same stage of the cycle, indicated that the actual density achieved at the end of the season was the cause of the early cessation of breeding. Contrary evidence to this will be presented below. Indeed, Chitty (1955, 1962) has pointed out that any general explanation of the causes of cycles in microtine rodents must account for a number of phenomena during the year of peak abundance which are not, however, directly related to density. These are (besides the curtailing of breeding referred to) the presence of higher body weights during the peak year and a differential mortality between the two sexes.

The second pattern of alteration of the breeding season related to the phase of a population cycle, is the presence of breeding during the winter preceding the peak year. This usually means that in the autumn there is little reduction in breeding, the voles continuing to reproduce through the winter, so that the population is at a relatively high density when the spring of the peak year is reached. Chitty and Chitty (1962) deduced this sort of over-winter reproduction in Welsh populations of *M. agrestis* from an investigation of body weights during different years. In the United States workers on populations of *Microtus pennsylvanicus* (Golley, 1961; Beer and Macleod, 1961) and of *Microtus montanus* (Murray, 1965) have described reproductive phenomena indicating that these species also breed through the winter preceding a peak. Zejda (1964) described a population of *Clethrionomys glareolus* which was already reproducing in February of a peak year.

Perhaps the most complete documentation of the breeding season is that of Krebs (1963, 1964). He followed the breeding patterns of the brown (*Lemmus trimucronatus*) and varying lemming (*Dicrostonyx groenlandicus*) for 4 years at Baker Lake in the Canadian Arctic. The relative densities based on a number of independent estimates for two species are shown in figure 50. This indicates the year of very high density, in this case 1960, and the years of lower density before and after, and demonstrates that the density changed inside the years of the cycle, especially in the year of decline. These alterations in numbers should be borne in mind and compared with figure 51, which shows the variation in the length of the breeding season at different phases of the cycle. Summer breeding in both species always starts around the middle of June, when the snow melts, and both of the aforementioned patterns of alteration in the breeding season are well demonstrated by this figure. The abrupt cessation of breeding in the peak year of 1960 is marked, but breeding also ceased very abruptly in the following year

R

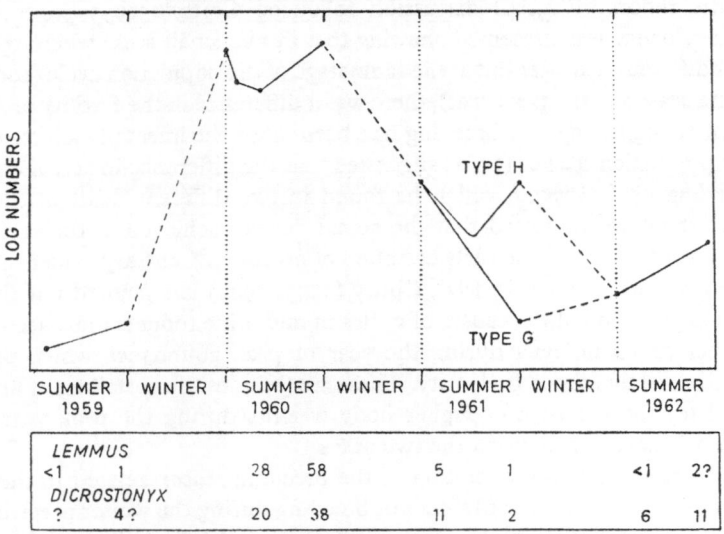

Fig. 50. The generalized density changes in populations of the brown (*Lemmus tri-mucronatus*) and varying lemming (*Dicrostonyx groenlandicus*) at Baker Lake, N.W.T., Canada, over 4 years of a cycle. The numbers are based on exhaustive trapping of the species concerned. From Krebs (1964) *Tech. Pap. Arct. Inst. N. Am.* **15**: 104 pp.

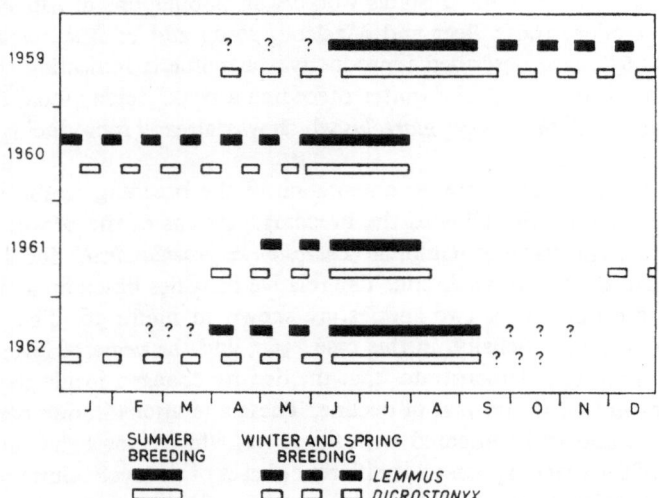

Fig. 51. The variation in the length of the breeding season in two species of lemming (*Lemmus trimucronatus, Dicrostonyx groenlandicus*) over different years of a population cycle. Based on data presented by Krebs (1964) *Tech. Pap. Arct. Inst. N. Am.* **15**: 104 pp.

of decline. This is good evidence that curtailment of the breeding season in lemmings is *not* due to density *per se*, as suggested above, since numbers were much lower in the autumn of 1961 than 1960. The second pattern of alteration in the presence of winter breeding is also well demonstrated. Although spring breeding under the snow seemed to occur in all years, extensive winter breeding was found only in the years of increase, and was completely absent over the winter of 1960–1. *Dicrostonyx* bred over the following winter too, but the decrease in numbers of *Lemmus* was so dramatic that none could be caught at this time. Krebs reviews the literature relating to the alterations of breeding season in cyclic populations and found that the two patterns of shortened summer breeding in the peak year and lengthened winter breeding in the years of increase are relatively commonly reported. There are a few papers (for example, that of Hamilton, 1937, 1941) in which exactly opposite effects are described. Nevertheless, it would seem that the majority of workers report a common pattern of reproductive alterations in cyclic rodents involving changes in the time to attainment of puberty (see page 27) and changes in the length of the breeding season. It is important to consider at this point that these changes may not be related to density but to intrinsic behavioural changes in the population as suggested by Chitty (1962) and which are proposed as the *causes* of the fairly regular alterations in yearly observed density.

In another investigation, Krebs (1966) reported the changes in the length of the breeding season of *Microtus californicus*. He suggested that his observations that breeding can occur with good recruitment during the dry season of some years but can be absent in other years, and that the onset of breeding may be delayed in some years by as much as 10 weeks after the sprouting of fresh green grass, can both be explained in terms of intrinsic changes in the populations concerned, in a similar fashion to lemmings. He emphasized that 'these changes in reproductive parameters are not necessarily of great importance in determining density changes but they are important symptoms of what is going on in the population' (page 266). As yet the nature of these changes is unknown, but the meagre evidence available would suggest some alteration in the level of aggressive interaction between individuals.

There has been some controversy in the literature as to whether the proportion of females which are pregnant at different stages of the population cycles vary significantly. Despite the assertion of Christian, Lloyd, and Davis (1965) that the proportion of females pregnant declines with density, all the evidence presented to date indicates that changes in the proportion pregnant are an artifact due to samples being taken which included immature animals. As the time to reach maturity is known to vary with population phases (see page 27), this phenomenon alters the proportion apparently pregnant. Indeed, a close reading of the papers quoted by Christian *et al.* (Rausch, 1950; Chitty, 1952; Kalela, 1957, 1961; Zejda, 1961, 1964; Krebs, 1963) gives no evidence that these original authors were confused on this point. Krebs (1964) gives clear data showing no alteration in the proportion of females pregnant which cannot be explained by changes in the length of the breeding season or by changes the in ages to puberty. A similar state of affairs was found by the detailed investigation of Koshkina (1965) for the red vole (*Clethrionomys rutilus*) in the U.S.S.R. This worker also concluded that the variation in the intensity of reproduction was insignificant between different years.

The evidence relating the litter size to the stage of the population cycle likewise gives no indication of any constant relationship. Kalela (1957), Krebs (1963, 1964), Zejda (1964), and Koshkina (1965) found no significant alteration in litter size with the phase of cycle. There were the usual seasonal and age alterations in litter size (see page 197), as also reported by Murray (1965) for *Microtus montanus*, but when comparable groups (same ages, same stage of breeding season) were compared the litter size was not seen to alter significantly. Patric (1962) found a significant difference in litter size in an undesignated species of *Clethrionomys* in years of different densities, but his data lump embryo counts of females collected throughout the whole year, with no evidence as to the proportions during different seasons or of different ages. Yearly comparisons are invalid.

The degree of prenatal mortality occurring in cyclic rodents at different stages of the population cycle does not seem to have been investigated by any other worker than Krebs. Fortunately he has presented excellent data for two species of lemmings of the various types of prenatal loss. Total prenatal loss, where the entire litter is noted to be resorbing, was noted only in summer young of *Lemmus trimicronatus* during the peak summer, and the summer of decline. Partial prenatal loss, where at least one of the litter remained alive, was estimated (see Brambell, 1948) by comparing the corpora lutea count with the number of implantation sites and the numbers of living embryos. The data are given below, with the sample size in brackets. The reader is referred back to figure 50 for the stage of the population cycle.

	1959	1960	1961	1962
Lemmus trimicronatus				
Pre-implantation loss				
Proportion of litters showing loss (%)	17 (18)	31 (67)	15 (34)	15 (41)
Proportion of ova lost (%)	4 (102)	5 (476)	2 (216)	2 (257)
Post-implantation loss				
Proportion of litters showing loss (%)	6 (18)	18 (67)	9 (34)	10 (41)
Proportion of embryos resorbing (%)	2 (98)	4 (451)	2 (211)	2 (251)
Dicrostonyx groenlandicus				
Pre-implantation loss				
Proportion of litters showing loss (%)	64 (11)	44 (25)	56 (34)	45 (44)
Proportion of ova lost (%)	17 (77)	20 (178)	18 (242)	11 (287)
Post-implantation loss				
Proportion of litters showing loss (%)	18 (11)	20 (25)	32 (34)	18 (44)
Proportion of embryos resorbing (%)	8 (64)	6 (143)	9 (198)	7 (255)

Although *Dicrostonyx* showed significantly higher losses than *Lemmus* in all measures, there were no significant differences between years. However, there was a significant increase in both pre- and post-implantation losses for the latter species during the peak year of 1960. It is to be hoped that later workers on microtine cycles will continue to investigate this aspect of reproductive changes to determine whether or not an increase in prenatal mortality in the peak year is a common feature in different species.

In microtine rodents there are thus some changes in the nature of reproduction at different stages of the population cycle which appear to be truly related to the cycle itself in that they are repeatedly reported by different workers on different species in different areas. As yet the causes of the cycles themselves are still the subject of considerable debate among ecologists, the amount of which is certainly not proportional to the weight of concrete field data available. The observations of crashes in numbers in the absence of predators, disease, gross food shortage, and the like leave as a possible alternative Chitty's hypothesis (1962) of an intrinsic change in quality (reflected in reproductive and other changes), although the almost untestable (and therefore almost unburyable!!) hypothesis of changes in food quality remains to cloud the issue further. Current evidence indicates that the most valid explanation is of an alteration in behaviour being the common cause of cycles, so that in the light of other behavioural alterations of reproduction, as outlined above and below, the changes seen in microtine cycles can be related to these mechanisms.

(*b*) Muskrats

During his lifelong studies of the muskrat (*Ondatra zibethicus*) Errington (1951, 1963) noted that this species underwent an approximately 10-year cycle of numbers. During the various stages of the cycle the litter size of muskrats varied significantly, as is shown in the following data collected over more than 20 years:

Year	Population phase	Litter size	Year	Population phase	Litter size
1935	Low	6·64 ± 0·44	1946	Low	6·40 ± 0·41
1936	Low	6·42 ± 0·57	1947		7·73 ± 0·47
1937		7·29 ± 1·48	1948		7·30 ± 0·25
1938		6·53 ± 0·55	1949		8·09 ± 0·30
1939		7·00 ± 0·35	1950		7·95 ± 0·24
1940		7·38 ± 0·27	1951	High	8·17 ± 0·15
1941	High	8·19 ± 0·46	1952	High	8·01 ± 0·22
1942	High	8·41 ± 0·28	1953		7·45 ± 0·17
1943	High	7·91 ± 0·16	1954		7·20 ± 0·30
1944		6·95 ± 0·12	1955	Low	7·12 ± 0·30
1945	Low	6·91 ± 0·22	1956	Low	6·35 ± 0·22
			1957		7·58 ± 0·34

This large amount of data is particularly revealing about the fluctuations of litter size, because, although there is some seasonal variation in this parameter,

over 70 per cent of all litters are born inside a 2-month period (May, June) and
there is only slight variation in the age of maturity, very few females reproducing
in the year of their birth (see page 12). Therefore, unless there is some considerable variation in the litter size with age over the first year and there are repeated
differences in the proportion of ages in the various yearly samples (which seems
very unlikely), the change in litter size reported by Errington indicates a significant alteration in the breeding potential of the muskrat. It is interesting also that
it is the reverse of what may have been expected, in that the litter size was highest
in the years of greatest density. Indeed, Donohoe (1966) reported that if the
density of muskrat populations was artificially increased as a result of a raising
of the water level due to dam building there was no discernible effect on the litter
size or on male fertility. Once again there appears to be little information
available on any changes in behaviour which may be related to the cycle in numbers in the muskrat.

(c) Lagomorphs

The snowshoe hare (*Lepus americanus*) also undergoes a long cycle of density
changes of about 10 years in duration. Unfortunately, this species has not been
the subject of any long-term studies which have reported on the reproductive
patterns, but there is scattered information about reproduction at certain phases
of the cycle. Rowan and Keith (1956) reported the following data from hares in
Alberta:

1949 increasing numbers	4·11 embryos/pregnancy
1950 peak numbers	3·93 embryos/pregnancy
1951 'crash' decline	3·33 embryos/pregnancy
1952 numbers still lower	2·20 embryos/pregnancy

In a more detailed study of reproduction, but only lasting over three seasons
(1959–61), Newson, J. (1964) compared the reproduction of two populations on
Manitoulin Island, Ontario. Each population underwent a slightly different set
of density changes, so that comparison can be made between years and between
populations of different densities in the same year. The parameter showing the
greatest variation was mean number of ova ovulated as shown below:

		Month of ovulation			
		April	May	June and July	Total
1959	East – peak	2·29 ± 0·08	3·95 ± 0·15	3·00 ± 0·44	9·24
	West – increasing	2·76 ± 0·21	4·33 ± 0·21	4·02 ± 0·13	11·10
1960	East – declining	2·43 ± 0·12	3·06 ± 0·13	3·25 ± 0·18	8·74
	West – peak	2·45 ± 0·22	3·78 ± 0·14	3·06 ± 0·22	9·29
1961	East – declining	2·41 ± 0·09	3·17 ± 0·14	2·84 ± 0·15	8·42
	West – declining	2·28 ± 0·13	3·45 ± 0·10	3·22 ± 0·16	8·95

The east population had a lower ovulation rate, started breeding later, and had a higher rate of prenatal mortality than the west population. As these results were consistent throughout the study these changes could not have been due to direct density difference but were probably related to population phase. Only the ovulation rate showed any significant difference which could be related to population phase in both populations.

There is indirect evidence that another species of hare, the blacktailed jackrabbit (*Lepus californicus*), may also exhibit something like a 10-year cycle in numbers. French (1964) and French, McBride, and Detmer (1965) studied this species in Idaho. These authors unfortunately give little concrete data to support their statements on density changes, but the reproductive data is given below:

Year	Population phase	Length of breeding season (days)	Proportion of mature females pregnant (%)	Average number of embryos per pregnant female
1956	Increasing	135	76·7 (38)	3·65
1957	Increasing	130	59·5 (108)	3·91
1958	Increasing	124	52·7 (150)	3·34
1959	Peak	120	43·5 (171)	2·77
1960	Rapid decline	132	49·4 (80)	3·44

These data are somewhat difficult to interpret, because there was seasonal variation in the prevalence of pregnancy and in the mean litter size. Although the authors obtained statistical significance between litter size in the different years, this was based on data which included mature females with no litters as having a litter size of zero. This practice has therefore made it impossible to determine whether or not the true litter size actually varied significantly or not. The breeding-season length was based on the period of time from the first to last pregnancy, and does seem to have decreased as the population increased in a similar manner to the vole populations, but no information is given of the degree of individual variability in this parameter.

A potential effect of density during the snowshoe hare cycle was discussed in an interesting paper by Bider (1961). He comments that the density of hares will affect the period of time necessary to result in a single male/female contact, and on the basis of random movement, calculates that for one encounter in a 2-week period the density would need to be 32 hares per square mile, in a 3-week period, 22 hares per square mile, and in a 4-week period, 14 hares per square mile. Although the absolute nature of these figures may be questioned, as hares will not move randomly, Bider has drawn attention to a possible effect that lowered density may have in reducing the probability of mating contacts and thus potentially affecting the process of reproduction following mating. This sort of problem deserves further investigation, particularly with regard to the altered behaviour patterns of female hares during oestrus and the general movement patterns of individual hares at the beginning and end of the breeding season.

CHAPTER 23

Social Factors in Adult Reproduction in Experimental Populations

---

The effects of density and thus of social factors on reproduction in mammals have been investigated experimentally by three types of technique. The first is to release into a very large cage or population pen a small number of animals and to document the resultant changes in numbers and the effect of these changes on reproduction. The second is to combine groups of animals together into very small cages and to document the resultant changes in reproduction. The third is to attempt to remove a certain age or sex from a wild population and to see the effects of this treatment. Each of these techniques has its own advantages and limitations, but in combination they should allow for considerable information as to the role of social factors and density on mammals. Unfortunately, as will become apparent, practically all the research that has been done is concentrated on only two species, both dependent on human activities for a major part of their environmental requirements.

This section will also include a discussion of the effects of animal contact on the physiology of breeding. This work carried out by physiologists interested in the effects of external stimuli on reproductive physiology is nevertheless very relevant to the role of social contacts in reproduction.

A. COMBINED POPULATIONS

(a) Mice

The assembling of male or female mice (*Mus musculus*) in unisexual groups in small population cages results in a marked inhibition of reproductive function. Christian, Lloyd, and Davis (1965) have reported that the weights of testes and accessory glands of mice decrease linearly as the logarithm of the density increases, although Bailey (1966), who combined males in groups of from one to thirty-two animals per cage while keeping the density constant, found that the testis weight decreased with numbers and not with density. Assembled populations of wholly female mice also demonstrate reproductive inhibition. Parkes (1963) has called the induced pseudopregnancy the Lee-Boot effect, after its discoverers. Females housed in groups of four or five in small cages show an increased incidence of pseudopregnancy, so that their next oestrus may be prolonged for a week or more. In larger groups of up to thirty the cycle of most females becomes very irregular, and many females go into anoestrus. Whitten (1966) has reported an average of 2·9 oestrous cycles per female in individually caged animals, compared with 0·6 cycle per female when caged as a group. He pointed out that in each experiment certain females continued to exhibit fairly normal oestrous cycles, suggesting a possible role of social status. It occurred, however, when tactile, visual, and auditory stimuli were removed, and it was not due to the increased pseudopregnancy effect found in smaller groups of mice. Similar decreases in the incidence of oestrus were earlier reported by Whitten (1959) and also Lamond (1959).

There are numerous papers reporting on the effect of male mice on female reproduction, and the reader is referred to several excellent reviews of the physiological aspects of the subject (Parkes, 1963; Parkes and Bruce, 1961; Bruce, 1967; Whitten, 1966). What follows only partially covers the literature, but is necessary for comparison with other mammals.

If males are introduced into dense populations of caged females the length of the mean oestrous cycle is shortened and the incidence of pseudopregnancy reduced. This effect persists if all tactile stimuli are removed and if the females are blind or deaf (Whitten, 1966). The stimulus from the male in this case would seem to act during the period of follicular growth of the female's cycle, as the length of time the contact is made governs its efficacy. If male mice are paired with females previously kept by themselves the incidence of mating does not fall evenly over the following nights. Called the Whitten effect by Parkes (1964), this was first reported by Whitten (1956). He described the timing of the finding of vaginal plugs after male introduction as follows:

	Vaginal plugs found after					
Total plugs	Night 1	Night 2	Night 3	Night 4	Night 5	Residue
317	43	44	146	42	15	28

The oestrous cycle of female mice thus tends to be synchronized by the presence of the male. This effect has been further investigated by Whitten and reported in a series of papers (1958, 1966). The incidence of mating on the first night after joining with the male can be increased if the male is kept in a small basket inside the females' cages for two previous days or if the females are placed in cages recently contaminated by males. Investigations by means of oestrous smears show that the effect is not carried over into the subsequent cycle, and that generally the oestrous cycle is shorter in the male presence. The effect of synchrony is further heightened if the females are kept in large groups prior to male introduction. Figure 52 shows that if the females, as well as having been grouped, are exposed to a confined male or to his urine, then the incidence of vaginal plugs is highest 2 days after joining. It would seem from these and similar experiments by Marsden and Bronson (1965b) that the synchronization of the females' cycles is caused by the smell of the males' urine and that urine from males of other species has no effect. Marsden and Bronson presented experimental evidence, however, which suggests that the synchrony of oestrus found in these experiments is due more to the release of the oestrus-suppressing effect of crowding in unisexual female groups than it is to the stimulation from the odour of male urine.

The final category of socially facilitated effect on reproduction in mice has been called the Bruce effect (after its discoverer) by Parkes (1963). This category

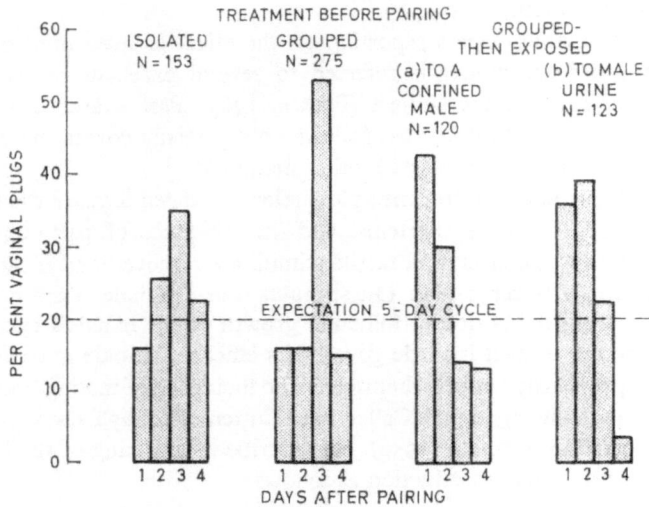

Fig. 52. Frequency distribution of vaginal plugs in days after pairing. The female mice (*Mus musculus*) had been previously isolated, grouped, or grouped but exposed to confined males or male urine. Values are expressed as percentages of the total number of vaginal plugs occurring within four days. From Whitten (1966) *Adv. Reprod. Physiol.* 1: 155–77.

of the effect of males on pregnancy is the subject of many papers, a few of which are: Bruce, 1960, 1961, 1963, 1967; Parkes and Bruce, 1961; Marsden and Bronson, 1965a; Chipman, Holt, and Fox, 1966; Chipman and Fox, 1966; Dominic, 1967. If a female mouse is mated to a stud male which is then removed and a strange male of the same or of a different strain is introduced into her cage, this results in a blocking of implantation so that pregnancy is interrupted and the female returns to oestrus. This reaction is caused by the odour of the male urine which is apparently very volatile, as the effect is only found when a female is in fairly frequent contact with fresh urine. The blockage will occur only if the strange urine is present during the first 4 days after mating with the original male.

Bruce (1967) has reviewed the mechanism behind both oestrous synchrony and pregnancy block. The effect of the odour of male mouse urine is to stimulate the female hypothalamus so that the output of FSH and LH increases and oestrus and ovulation follow. The secretion of pituitary prolactin is restricted by the same stimulus, and implantation is prevented. She suggests that the effect of other females on the female cycle is similar, in that there is also here an increase in prolactin output and thus reduced oestrus also brought about by the odour of urine. Apparently the male effect overrides the female effect (see above). Although the actual origin of the substance producing these effects is not known in mice, its production is under the control of the gonads.

In addition to the effect of the male in synchronizing the oestrous cycle of the female, a recent paper by Beilharz (1968) has shown that the male can cause an increase in the litter size which occurs at this synchronized oestrus. In experiments involving very large numbers of females the average litter size on days 19, 20, 21, 23, and 24 or later (after the introduction of the male) ranged from 9·41 to 9·76 young per litter. However, on day 22 the average number of live young per litter was 10·31, a significantly higher figure. Beilharz suggests that the effect on the number of ova shed may be much greater, as uterine conditions may impose a limit on the foetuses which could be carried to term.

In much of this work the effects seen do not occur in *all* the females, and it would be interesting to follow up this fascinating physiological story with a behavioural investigation to try to determine whether the social status of a female controlled the degree of reproductive response to a male's urine. The recent paper by Muller-Velten (1966) has given as indication as to the reactions of the mouse to a 'fear scent' ('Angstgeruch') which is found in the urine and is set free only during emotional excitement of a frightened mouse. The length of time the fear scent lasts together with the differential response to it between sexes suggest that the same phenomenon is being dealt with here. It also seems possible that the crowding of female mice during pregnancy may affect the development of their offspring. Keeley (1962) has reported that the offspring of crowded female laboratory mice were slower to respond to unusual stimuli and showed a general decreased awareness. The effect persisted for up to 100 days of age, independent of the subsequent conditions the young mice were reared in.

(b) Other mammals

The effect of the male in synchronizing oestrus had been well recognized in sheep before the research above on mice was carried out. Schinkel (1954) and Sinclair (1950) reported that if the ram is introduced to a group of ewes early in the breeding season they will tend to come into oestrus together. Shelton (1960) mentioned a similar effect in goats (*Capra hircus*). Experimental work by Watson and Radford (1960) showed that the smell and sound of the rams are sufficient to cause synchronization and, like mice, the ewes tend to come into oestrus prior to the 17th day after introduction if they are allowed some degree of association with the rams before actual introduction. Control ewes kept separate from the ram until introduction exhibited a peak in mating between 18 and 26 days after introduction. The introduction of the ram will also advance the onset of observable oestrus (Radford and Watson, 1957). The degree of synchronization involved by a sudden introduction of rams will be seen in figure 53, which is taken from a paper by Thibault *et al.* (1966). The number of ewes mated rises steeply after 11 days in this large sample of observations. The presence of the ram will also shorten the duration of oestrus by up to 50 per cent (Parsons and Hunter, 1967). Skinner and Bonsma (1964) found that the mean time to the day of mating after introduction of a bull to a group of cows was 19·3 days in groups of cows alone and 9·5 days in groups run with vasectomized bulls. It would thus seem that bulls may have some synchronizing effect in cattle. Similarly, evidence presented by

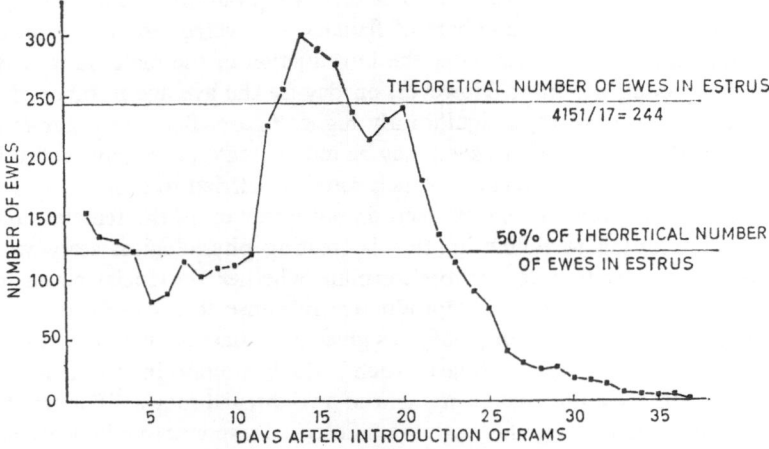

Fig. 53. The number of ewes mated during a 35-day period after introduction of rams. Data was collected on 4,151 Merino ewes during spring breeding over 6 years. From Thibault *et al.* (1966) *J. Anim. Sci. (Suppl.)* **25**: 119–142.

Polikarpova (1945) suggests that the presence of the boar may synchronize oestrus in sows. The presence of boars in adjoining pens or in the same paddocks resulted in over 93 per cent of sows coming into heat inside 16 days. Groups of female nutria (*Myocastor coypus*) kept in small pens had a very variable inter-oestrous period, but the introduction of a male caused the majority of females to come into heat in a few days (Ehrlich, 1966).

Cooper and Haynes (1967) found that the presence of the male rat causes an earlier incidence of oestrus in the female whose oestrous cycles have beome irregular due to a low nutrition diet. Bronson and Marsden (1964) reported that male deermice (*Peromyscus maniculatus*) can induce synchronous oestrus in grouped females.

The effect on reproduction of grouping females in high densities has been noted for a number of mammal species. Chitty and Austin (1957) found that female voles (*Microtus agrestis*) caged in breeding pairs had short oestrous cycles characterized by spontaneous ovulation, and the females were only receptive to the male at one stage of the cycle. Females kept in groups of from three to twelve exhibited a prolonged oestrus with induced ovulation and coitus at any time. An opposite result was reported by Breed (1967), who found that the vole was an induced ovulator regardless of the social environment. In his review of the literature, this worker suggested that there is no satisfactory evidence for any species of *Microtus* being a spontaneous ovulator. The length of time that oestrus was exhibited was found to vary in caged Arctic fox (*Alopex lagopus*) females, depending on whether they were housed as individuals or in pairs (Samkov, 1964).

Although it is complicated by the effects of lowered nutrition, the confinement of human females in large numbers in high densities is known to result in an increase in the incidence of amenorrhea, and presumably this is related to ovarian malfunction. In their review of human starvation, Keys, Brozek, Henschel, Mickelsen, and Taylor (1950) suggest that psychic factors are more important here than nutritional ones, because the amenorrhea developed immediately after internment before nutritional effects could have become apparent, and also because certain sectors of the female population, such as nuns, did not develop the amenorrhea.

Finally, the effect of strange males on blocking pregnancy has also been reported in the deermouse (*Peromyscus maniculatus*) (Eleftheriou, Bronson, and Zarrow, 1962; Bronson, Eleftheriou, and Garick, 1964). Another interesting effect on the reproduction of this species was a considerable increase in prenatal mortality when the mated pairs were crowded together (Helmreich, 1960). Gilbert and Bailey (1967) found a high conception rate and reduced prenatal mortality in female ranch mink (*Mustela vison*) when they were kept visually isolated from other females.

(c) Discussion

From the discoveries on laboratory mice of the effect of the odour of the male's urine on various phases of female reproduction has arisen what Parkes (1963) has called the discipline of 'exocrinology', which is the study of the physiological effects of externally produced substances termed by Whitten (1966) *pheromones*. These are defined as 'substances, or mixtures of substances, which are produced to the exterior by an animal and may be received by a second individual of the same species, in which they produce one or more specific reactions'. He suggests that the two effects outlined for unisexual female groups (pseudopregnancy in smaller groups, anoestrus in larger) may be caused by a single pheromone at two dose levels. The effects caused by male mice (shortened oestrus, inhibition of spontaneous pseudopregnancy, synchronization of oestrus, and pregnancy block) are all the result of earlier than normal development of follicles ahead of oestrus and ovulation, and these could also be explained by the action of a single pheromone. The role of pheromones in reproduction of wild mammals will be discussed later (see page 273).

It should be pointed out that in certain rodents which live in naturally dispersed populations there is evidence from laboratory studies that the male may have a considerable inhibiting effect on the exhibition of female oestrus. Continued proximity to males by females of the following species: *Liomys pictus*, *Perognathus californicus*, *P. flavus*, *Dipodomys panamintus*, *D. nitratoides*, and *D. merriami* results in a cessation of oestrous cycles and a period of prolonged anoestrus (Eisenberg and Isaac, 1963).

B. FREELY GROWING BUT CONFINED POPULATIONS
(*a*) Mice

Most studies of the effects of increasing density on reproduction have been carried out with house mice. The earlier experiments by Strecker and Emlen (1953) were complicated by a limited food supply. Breeding ceased as soon as the food limit was reached, and was characterized by an inhibition of female reproduction with involuted reproductive tracts in sacrificed females. A later experiment allowed egress from the limited food supply pen, and in this case there was no interruption to breeding (Strecker, 1954).

A more definitive experiment was carried out by Southwick (1955a), who studied three mice populations in which the unlimited food and shelter was placed evenly through the pen and three in which it was confined to limited areas. He found no differences in reproduction which could be attributed to spatial arrangements, but the reproductive patterns in each population varied as each increased to high densities. In four of the six populations the birth-rates slowly declined, but for different reasons in each of two pairs. In the first the rate of food consumption decreased as the density rose, due to a few dominant mice inhibiting the feeding of subordinates. The males produced fewer sperm, and fewer females were ovulating. In the second pair, however, no feeding hierarchies developed, and the food consumption increased with density. Southwick suggests that the birth-rate in these pens decreased either because of interference with copulation, due to the very large numbers of males present in a small area, or because of failures in early embryonic development.

Another effect of high density was that, when the level of aggression rose to an average of one fight per hour per mouse, the litter mortality increased enormously (Southwick, 1955b). The behaviour of adults also altered, so that too many attended to the litters, and the type of nest built was altered to the detriment of the survival of young. If the normal bowl-shaped or covered-over nests were built (as at lower densities) infant survival was between 60 and 90 per cent. If only one or two adult females attended the litter the nest was usually platform-shaped and survival averaged 59 per cent. The attendance on the litter of several females or a single male usually resulted in a poorly constructed platform nest, but the survival dropped to a mean of 18 per cent. If the density was so great that the female could not build a nest there were usually no survivors. Brown, R. Z. (1953) reported the following proportions of young weaned in a similar experiment: platform nest, 0 per cent; bowl-shaped nests, 31 per cent; covered nests, 82 per cent. These studies have demonstrated the importance of behavioural factors and their variation between populations in freely growing populations. They should be borne in mind in all the freely growing populations described below.

The birth-rate has been found to change with increasing density in a number of experiments on penned mice. Christian, Lloyd, and Davis (1965) review most

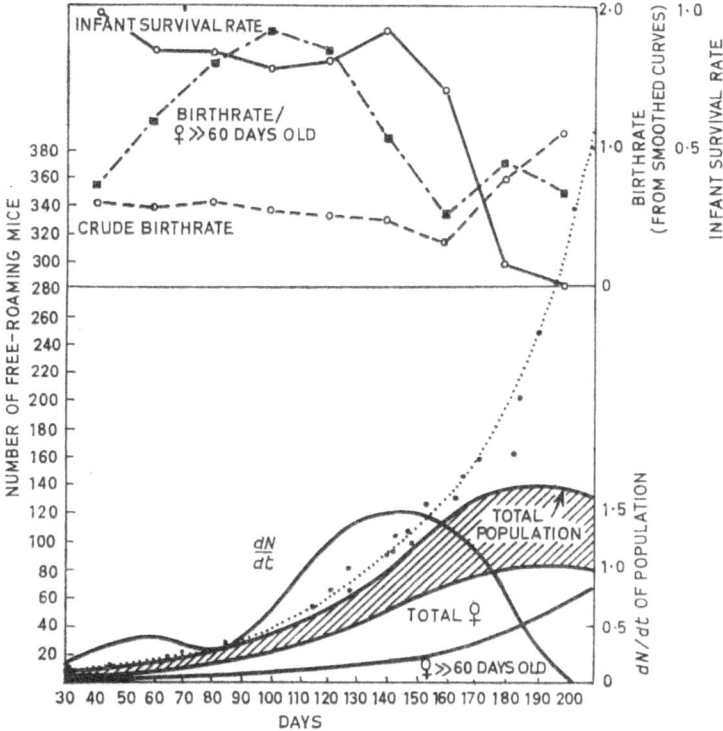

Fig. 54. Demographic data from a freely growing confined population of house mice (*Mus musculus*) to show the crude birth-rate, birth-rate per female over 60 days of age and nursling survival in 20-day intervals. The difference between the total population size (top line of darkest area) and cumulative births (broken line, bottom part of graph) is the cumulative mortality, mainly of nurslings. From Christian, Lloyd, and Davis (1965) *Rec. Prog. Horm. Res.* **21**: 507–78.

of the literature on this subject and give examples from the work of Christian and his co-workers. Figure 54 shows the manner of change in the birth-rate in a typical mouse population described in this paper. There is at first an increase and then a decline as the population increases. A drop in the birth-rate with increasing density has been found in many laboratory populations. Figure 55 combines data from six populations studied by Christian, and shows clearly the inverse relationship between birth-rate and density.

The birth-rate (the mean number of young produced per unit adult female over a fixed period of time) is a variable which is relatively easily measured under the conditions of a penned experiment. It results from a number of interacting reproductive processes which could individually, or in combination, be themselves affected by the increasing density. Three of these are probably the most important. Firstly, the rate will be affected by the number of females in the

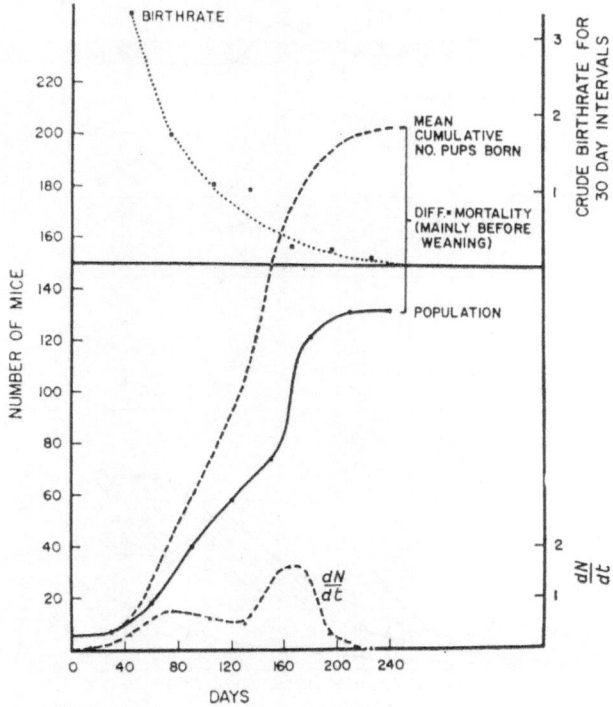

Fig. 55. Plots of combined demographic data from six freely growing confined populations of house mice (*Mus musculus*) with roughly equivalent upper asymptotes and equivalent developmental times to reach asymptotic levels. From Christian, Lloyd, and Davis (1965) *Rec. Prog. Horm. Res.* **21**: 507–78. Subsequent work (Christian, *pers. comm.*) has shown that with data from further populations the combined curve (solid line) is more sigmoid, with the lower limb developing more slowly and reaching a point of inflection above 60 per cent of the asymptote value. The birth-rate points are similar with further data, but the curve is smoother.

population which are reproductively active (i.e., ovulating and undergoing cycles of behavioural receptivity) during the time considered. Secondly, it will also be affected by the mean ovulation rate of the reproductive females, and thirdly, by the degree of prenatal mortality (and, incidentally, the mortality at birth or between birth and the time of assessment of the birth-rate). These second and third factors are often combined and expressed as the litter size which results from them. There is only little evidence of the *relative* importance of these three parameters in the variation of the birth-rate with density (but see figure 55), although the state of reproductive inhibition should be determined before the causal mechanisms can be elucidated.

One study in which the individual parameters contributing to the birth-rate decline were investigated was that of Crowcroft and Rowe (1957). Four population pens of 6 ft square area were each started with one male and two females. The birth-rates declined in pens as shown below:

	Average number of young born per female in period			
	0–5 months	6–10 months	11–15 months	16–20 months
Pen A	8·5	7·2	3·8	2·9
Pen B	12·5	4·2	0·0	0·0
Pen C	11·3	2·6	4·0	2·4
Pen D	12·2	6·4	1·7	1·9

In pens B, C, and D the birth-rates after 6 months were significantly lower than in the first 5 months. Each population reached maximum levels after 8–9 months, although excess food was continually available. The birth-rate decline was the result of all three parameters changing with density. When the females were killed at the end of the experiment Crowcroft and Rowe found that many females in pens A, B, and C had imperforate vaginae and no corpora lutea in their ovaries. There was also evidence of changes in prenatal mortality and litter size in the terminal samples:

	Percentage resorption	Apparent mean litter size
Pen A	13·8	4·2
Pen B	7·0	4·0
Pen D	4·8	4·4
Wild mice (Laurie, 1946)	2·7	5·6

It is also relevant that effective breeding completely ceased in populations B, and yet infant survival remained high in all populations and male fertility was unaffected by density.

Further evidence of the alteration in various parameters with density in mice comes from a later paper by Crowcroft and Rowe (1958). They confirmed that the post-mortem examination of females from a confined and dense population showed a higher proportion of non-reproductive females than found in wild populations and then carried out another experiment in which females from a dense population where reproduction had declined were allowed to spread into a much larger pen. Reproduction immediately increased, but so quickly that some litters must have been conceived before the females were allowed to disperse.

S

This suggests that prenatal mortality was a major cause of a decreased birth-rate in the dense colony. In another experiment a similar mouse population was divided into two equal-sized colonies. The non-fecund females were either transferred to another large pen with a few mice (i.e to reduced density), wherein they started breeding immediately, or to another new small but crowded pen (i.e., roughly same high density), where they still failed to reproduce. A final experiment involved the release into a large pen of a completely adult dense population which had not produced litters in 6 months. The majority of females conceived in the first week after dispersal. Some similar experiments were reported by Petrusewicz (1958).

Thus there can be little doubt that if mice are allowed under conditions of confinement to increase their numbers to very high densities the reproduction is inhibited considerably through detrimental alteration of various parameters. It is also apparent that there is considerable variation in the nature of the response seen, in that reproductive impairment occurs at variable densities and is effected through different parameters. Those studies which have investigated behavioural effects indicate strongly that this variation can be largely explained in terms of behavioural differences in response in mice populations. One behavioural factor which apparently increases to the detriment of reproduction is the level of aggression. Vessey (1967) was able to renew population growth in mouse populations which had stabilized by adding a tranquillizer (Chlorpromazine) to the feed. This substance has no effect on reproduction in mice, but its addition resulted in a considerable increase in infant survival, although the exact manner in which this was related to decreased aggression was not stated.

(b) Other rodents

Lidicker (1965) compared the reproduction in penned populations of four species, and found that as density increased, breeding ceased completely in one *Peromyscus maniculatus* population but was only reduced in another, while breeding continued with little effect in two *Peromyscus truei* populations and one *Oryzomys palustris* population. A single *Mus musculus* population showed reduced fertility with increasing density, but the main effect was to inhibit sexual maturation (see page 24). In the pens where breeding was not affected by increased numbers the populations appeared to be regulated by increased juvenile mortality. This paper indicates considerable interspecific and intraspecific variation in the effect of density on reproduction.

A particularly interesting paper by Clarke (1955) described the processes taking place in two experimental vole (*Microtus agrestis*) populations, one being started at four times the density of the other. As soon as breeding began, both populations increased in size, but the growth rate of the larger was less than that of the smaller, so that as time went on the two populations became more equal in density. The larger population was considerably greater in its second than in its first year, yet the maximum size in the second year was only slightly greater than in the first. The experiment started in April 1950, and the low-density population bred until

December, while the high-density population stopped breeding in August. Breeding in the low-density population began again in March 1951, while breeding in the high-density population started in April/May.

In addition to affecting the length of breeding the birth-rate was also impaired by density, as shown below:

Age group in days	Dense population		Low population	
	Litters/female	Young/female	Litters/female	Young/female
28–56	0·1	0·5	0·8	2·8
56+	0·0	0·0	0·9	4·2
84+	0·2	0·7	0·4	1·0
112+	0·0	0·0	0·7	2·2
Gross reproductive rate	0·3	1·2	2·8	10·2

The last line represents data from the first breeding season only. Aggressive fighting and considerable wounding were seen in both populations, but were more prevalent in the larger and more dense one.

(c) Rabbits

Myers and Poole (1962) have reported extremely detailed work on penned populations of wild rabbits (*Oryctolagus cuniculus*) in Australia. This paper is very important because it describes in considerable detail the social structure of the population studied and emphasizes the importance of behavioural documentation in an understanding of density effects. In 1957 these workers set up three pens of 2 acres with the following populations of adult rabbits (Myers and Poole, 1959):

Pen A   2 males and 4 females
Pen B   4 males and 6 females
Pen C   9 males and 10 females

The pens were lit at night and were continuously observed over more than 2 years. Animals were individually marked and their reproductive status determined by repeated handling and palpation. In 1957, after the initial litters, every litter but one was conceived at a post-partum oestrus, so that females were continuously reproducing. In 1958, when the pen densities had increased, this continuous breeding stopped and there were a number of intervals subsequent to the first known pregnancy for certain females during which no litters were produced. Myers and Poole call these 'interruptions to breeding', as they represent barren intervals from the point of view of population increase. These intervals were then analysed relative to density and social status as shown below:

	Year 1957			Year 1958		
	Pen A	Pen B	Pen C	Pen A	Pen B	Pen C
Number of females present	4	6	10	23	33	28
Females suffering interrupted breeding (%)	0	0	10	60	85	14
Number of intervals of interrupted breeding	0	0	1	18	43	4
Mean days out of reproduction per female suffering interrupted breeding						
Dominant animals	0	0	0	19	48	—
Other animals	0	0	0	19	39	—
Overall mean	0	0	42	19	41	14
Mean number of days out of reproduction per adult female	0	0	4·2	8·6	35·0	2·0

The breeding season in Pen C during 1958 was so short that comparisons with regard to social status could not be made.

As the population density increased in the second year there was a marked increase in the overall interruptions to breeding, which seemed roughly proportional to the final density. The increase in the numbers of days out of reproduction was not due to an increase in the *number* of intervals of interrupted breeding but due to an increase in the *length* of the individual intervals. In the highest density pen (B) the dominants were actually out of reproduction for a longer period of time than the subordinate animals. The mean interval of interrupted breeding, which was higher at the high densities, could be explained by one of two possible factors. Either there was an increased rate of absorption during the earlier stages of pregnancy (when palpations were inaccurate) or there was a tardiness in returning to behavioural oestrus following resorption or failure to conceive at the post-partum oestrus. Behavioural observations suggest that the latter explanation is the more likely one, as a number of females went for long periods of time without exhibiting any signs of sexual activity.

The density of the population of rabbits also had an effect on pregnancy, in that it increased the resorption rate, as shown by the data below from Myers' and Poole's paper:

	Year 1957			Year 1958		
	Pen A	Pen B	Pen C	Pen A	Pen B	Pen C
Number of females present	4	6	10	23	33	28
Number of known pregnancies	16	19	33	72	150	82
Proportion of known resorptions (%)	0	0	0	4	13	2
Number of presumed post-partum matings	16	19	34	87	187	56
Proportion of females failing to litter following presumed post-partum matings (%)	0	0	3	21	31	9

The data from pen C in 1958 is again somewhat different from the other pens, as this population was starving at that time. The degree of interference with reproduction was roughly proportional to the density in the second year. However, the rate of resorption and the rate of failure to litter following presumed post-partum mating was not related directly to the social status of females in the pens. Although the female rabbits set up territories in areas of different favourability depending on their social status, the rate of resorption was not apparently related to the nature of the territory. The very best area in pen B was that occupied by group B_1, but this group had the highest rate of resorption. Indeed, the authors comment that 'a pregnant female can dig her burrow, build and line her nest and drop her litter on the same night, and will break group territory to do so' (page 241). It is interesting that different results were reported by Mykytowycz (1959, 1960, 1961) for similar penned experiments but with smaller pens. He found similarly that the rate of resorption was higher at high densities but that socially inhibited females lost many of their litters through intra-uterine mortality. Myers and Poole (1962) have suggested that this may have been due to the subordinate does in this case being all younger animals, as they had also shown an age effect on resorption. Mykytowycz found that, although dominant females inhibited the reproduction of subordinates in the early part of the breeding season, during the later months of breeding the subordinates managed to breed successfully and litter in isolated breeding stops.

The result of these various factors acting as the density increased was that the fecundity of the does in dense populations was reduced. Although the mean number of litters per female and mean litter size varied little between the pens in any year, the mean number of kittens produced during 1958 was only slightly higher than in 1957. Compensation for the difference in the length of the breeding season between these 2 years means that the fecundity during the dense year of 1958 was reduced. There was good evidence that social status had a marked effect on fecundity, as the dominant does in pens A and B during 1958 produced significantly more kittens than subordinate does. Most of this difference was due to the older and more dominant females having more litters of a larger mean size than the subordinate females. A similar picture was also found by Mykytowycz (1961).

The paper by Myers and Poole has been considered in such detail because it shows clearly that under penned conditions the density of rabbits has very considerable effects on their pattern of reproduction. It also shows that behavioural factors related to the social position of individual animals inside the pen are reflected in differentials in reproductive patterns of dominant and subordinate animals. For example, Staples (1967) reported that although ovulation in laboratory rabbits will normally not occur unless intromission is effected, if females are left in pairs mounting by dominant does can cause ovulation in submissive does, but never in other dominant does. In a similar way to the experiments on mice by Southwick, it is apparent that the density alone is not the exclusive cause of inhibition in reproduction, but merely serves to accentuate social factors.

As well as the behavioural effects in these penned experiments, there is evidence that the living area can affect reproduction in rabbits. In an important paper which reviews his earlier work and the whole problem of the effect of density on sociality and health in mammals, Myers (1966) reports experiments to investigate the effect of the living area. The results of the first of these are shown below:

Pen	Population density (rabbits/acre)	Mean number of litters per female	Mean litter size	Mean number of kittens born
1A	200	1·0	5·0	5·0
1B	200	1·4	5·2	7·3
2A	50	1·8	5·3	9·6
2B	55	2·0	4·9	9·0
3A	35	1·6	5·0	8·0
3B	35	1·8	4·9	8·8

The production of kittens was significantly lower at the highest density than at the two lower densities. This is much the same sort of result as before, but as certain aspects of the data suggested that the living area itself had an effect, Myers carried out a second experiment, the results of which are shown below:

Treatment No.	1	2	3	4	5
Number of replicates	3	6	12	5	5
Population size					
Number of males	8	4	2	2	2
Number of females	12	6	3	3	3
Enclosure size (square yards)	450	225	123	225	450
Density per acre	200	200	200	100	50
Mean number of acts of aggression					
per hours	17·0	18·1	8·8	4·2	2·6
Loss of litters (*in utero*) (%)	34	42	35	37	35
Mean litter size	5·4	4·8	4·3	4·4	5·2
Mean number of young produced					
per female	11·8	8·8	9·4	10·2	12·9
Survival of young to 30 days (%)	17	28	31	33	28

These data are particularly valuable because of the replicates which were set up of the five types of treatments. Treatment 3 can be considered as a single breeding group living in the smallest amount of space. Treatments 1 and 2 represent increasing space with density held constant, and treatments 4 and 5 represent increasing space with the numbers of animals held constant. It will be seen that space affected litter size and therefore fecundity, as it increased with increasing space independently of density. Survival of the young was reduced at the highest density, even in a large space.

C. UNCONFINED FIELD POPULATIONS

The use of field experiments, despite inabilities to control all the necessary variables, adds extra information to our understanding of the role of the environment on reproduction in mammals. Unfortunately, the total number of reported experiments is very few indeed, which is a pity, because this is a singularly fertile field for research. Some observations have been already made with regard to deliberate or unintentional alteration of the social environment (see page 32) and its effect on the onset of puberty. This section will report on five papers where effects on pregnancy, etc., were documented.

A paper by Davis (1951a) on the relation between the level of population and pregnancy in the rat (*Rattus norvegicus*) of Baltimore is often quoted as indicating the effect of density on reproduction. In actual fact, this worker determined the level of populations of rats in various city blocks and then removed a high proportion of the population. These populations were then sampled, some 2–3 months later and termed as increasing populations. Decreasing populations were categorized by normal trapping. The point here is that the 'increasing' category was the result of experimental manipulation, but as no information was given as to the categories of animals removed from these populations and whether or not they were proportionately balanced to the same categories in the original population, it cannot be determined as to whether the manipulation duplicated a natural increase. The results of samples of large numbers of rats are given below:

	Stationary populations	'Increasing' populations	Decreasing populations
Proportion of adult females in maximal breeding season (spring) (%)	14·4 (268)	41·6 (69)	25·3 (96)
Proportion of adult females during rest of breeding season (%)	23·1 (174)	26·0 (194)	32·7 (66)
Total proportion of females pregnant (%)	18·6 (442)	29·8 (263)	30·9 (162)
Incidence of pregnancy * (mean)	3·8	6·1	6·3
(Range of 2 standard errors)	3·2–4·3	4·9–7·2	4·8–7·8
Proportion of females lactating (%)	43·0	42·4	39·9
Mean litter size	10·34	10·29	9·93

* For explanation see page 220.

The main effect on reproduction was thus seen in the prevalence of pregnancy, but as the prevalence of lactation was approximately equal in the three types of population phase, there must have been considerable mortality at birth or during early lactation in the second two groups.

Two papers refer to populations of woodchucks (*Marmota monax*) which were manipulated on the Letterkenny Army Ordnance depot in Pennsylvania (Snyder, R. L., 1962; Davis, Christian, and Bronson, 1964). The manipulations and data are given in tabular form below:

Manipulation	Area G Females removed 1957, 1958	Area C Adult both sexes removed 1957–60	Area D Very few adults removed (effectively a control)
Density per 215 acres in March 1959	74	157	73
Proportion adult (%)	55	39	72
Sex ratio	57% male	—	Marked female surplus
Litter size, 1959	2·81	3·64	3·62
Reproductive rate, 1959	1·20	2·21	2·97
Proportion of litters re- sorbing (%)	29	9	7

After 1959 the populations returned to approximately equal densities, and the birth-rate in population C declined, due to the removal of adults, leaving only yearlings to breed at a lower rate. However, survival of young in this area was high. In population G many females in 1959 failed to breed, and the incidence of pregnancy in yearling females was very low. As a general conclusion, Snyder suggested that 'the intensity of the stimuli from social pressures depends upon the structure of the population in terms of age composition and sex ratio as well as upon its density' (page 515).

In studies of populations of deermice (*Peromyscus maniculatus*) the writer noted (Sadleir, 1965b) that although many females were detected as pregnant in the field, the numbers of juveniles were very low during most of the breeding season, indicating considerable mortality between birth and first entering the traps after weaning. Experimental removal of adults showed that juvenile survival increased indicating that adult aggression was the cause of juvenile mortality. This work was later confirmed by much more detailed experiments carried out by Healey (1967). It would seem that in this species aggressive activity of adult deermice can result in considerable mortality in juveniles, although whether the majority of this mortality falls before or after weaning is not known.

CHAPTER 24

The Relevance of Observations from Experimental Populations to Natural Populations with Regard to the Role of Social Factors in Adult Reproduction

If data collected from the laboratory are to be used to decipher mechanisms of interaction between the environment and the individual which are then to be applied to an understanding of environment/animal interaction in the wild, the degree to which the laboratory environment can be compared with the field environment must first be established. This comparison may be made by considering the two situations in terms of their spatial construction ard biological composition. The latter is composed of the numbers and type of animals used. Spatially the laboratory arrangement is very different from the field situation on two main counts. The first is the obvious one of confinement, i.e. the inability of the experimental animal to emigrate from the area. Under field conditions increases in numbers are almost inevitably followed by movement from areas of high density. The second is the inability in the laboratory to duplicate the spatial complexity of the field environment. This can be avoided by the use of populations which are confined in field areas, such as has been done in Australia in rabbit population studies (Myers and Poole, 1959; Mykytowycz, 1959). It can be of considerable importance in its effects on the frequency of individual contact between the animals in the experimental area. If the environment is not complex the experimental animals will be unable to get away from other individuals, which will in turn affect the social arrangements in the pen or cage. The lack of environmental complexity will also affect the location of feeding and watering facilities, which will in turn have effects on the social development of the laboratory colony. There is also evidence from studies such as those of Myers (1966) and Bailey (1966) that the total amount of space available will have an effect independently of the density. This may mean that there are even more qualities of the spatial environment which cannot be duplicated in the laboratory. Laboratory and penned experiments are thus conducted in environmental conditions which can never be considered as absolutely comparable with field conditions.

The type and past history of the animals used in laboratory experiments are often very different to the general background of individuals in wild populations.

The problem of the origin of the individual used in laboratory or penned experiments can be easily overcome by using wild-caught animals, but this procedure itself presents considerable difficulties. Under natural conditions animals in the wild, which are surviving at times of high densities, have been previously subjected to a multitude of environmental effects, both physical and social. Not only have they survived as a result of their knowledge of their physical environment but also they have fitted into the current social system which the population around them has developed. The artificial selection of wild animals from the field and their combination at the start of penned or cage experiments constitutes a very real difference in the background of the laboratory and field population. The variability of the speed of population increase from small numbers in duplicated experiments can often be explained on the basis of behavioural incompatability or inability of the individuals selected to form a stable social structure under the conditions of the experiment. There appears to have been no attempt other than in certain primates to transplant into a pen or cage a wild social family group or to enclose a wild social unit in a penned experiment. This should prove a very interesting comparison with an artificially combined group of animals in which both are allowed to increase in numbers.

The types of behaviour shown by experimental animals can give a good indication as to the 'naturalness' of the experimental set-up when this behaviour is compared with the behaviour seen in wild populations. Following their work on penned populations of rabbits, Myers and Schnieder (1964) and Mykytowicz and Gambale (1965) studied the behaviour of unconfined natural warrens of wild rabbits. They found no evidence of differences in behaviour which could be attributed to the lack of a boundary fence around the natural warrens. On the other hand, Anderson, P. K. (1961) has compared the social structure in house-mouse populations over a very wide range of densities and finds considerable differences in the types of social structure seen. He suggests that there are three main types of behavioural organization. Over higher densities a social hierarchy develops where a single male is dominant. At intermediate densities a system Anderson calls 'togetherness' is found, where although the density was high, there was no evidence of aggressive behaviour or of hierarchal development. At lowest densities the mice become territorial in the sense that there is evidence of mutual exclusion with regard to living area. The range of densities involved and the changes in reproductive parameters seen are shown in figure 56. Anderson suggests that, although it is more than likely that the same patterns of individual behaviour may be the root of both territorial and hierarchal social organization, the development of the second system is due to the lack of opportunity to emigrate and the lack of escape cover. In other words, it would seem that this sort of behaviour in mice is partially or wholly induced by the spatial conditions. Unfortunately the picture is not clear, because hierarchal systems also occur in commensal populations of wild mice in ricks and human habitations. These are natural populations in the widest sense, but are somewhat special cases with regard to the superfluity of food and shelter available. There seems to be very little in the literature on species other than the rabbit and mouse with regard to the differences in behaviour exhibited in pens, captivity, and the wild. The evidence from these two species suggests that the behaviour observed is much the same in its gross pattern, but that probably more detailed investigations will find differences in degree.

A very common impression gained from reading papers describing cage and pen experiments is that the actual density involved is very much greater than

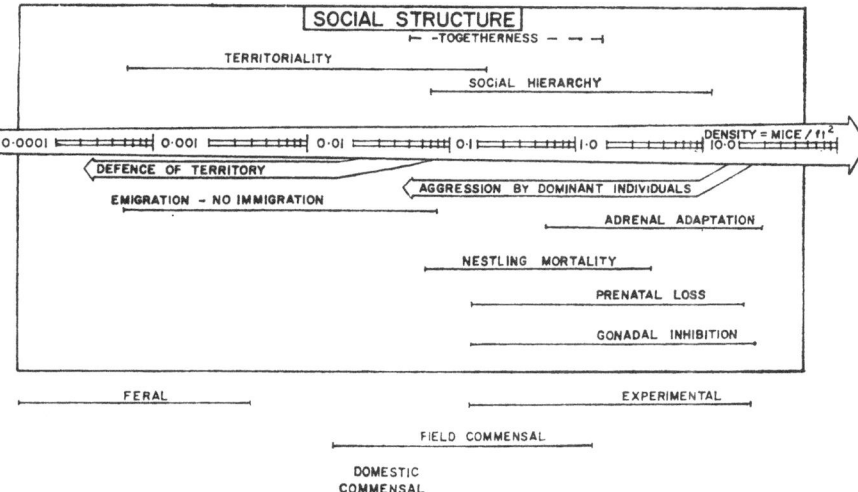

Fig. 56. Relationships between habitat, density, social structure, and regulating factors in *Mus musculus* populations. Density is shown in the logarithmic scale across the centre. The range of density records for feral, domestic, field commensal, and experimental populations reported in the literature are given at the bottom of the figure. Types of social structure are shown above and population responses are shown below the density scale. References may be found in the original article. Modified from Anderson, P. K. (1961) *Trans. N.Y. Acad. Sci.* **23**: 447–51.

densities found in nature. In contrast, the data from the Australian rabbit work show that the densities in the pens were roughly comparable with densities found in the wild. Similarly, as will be seen from figure 56, some of the experimental mice population work was done at densities of the same order as found in commensal populations. By the same token, however, many experimental mice colonies were at much higher densities than found naturally.

The relevance of a laboratory study to field situations can be further gauged by comparing the types of reproductive pattern found in penned populations at high densities with those exhibited in the field. Unfortunately, with the exception of those species which are cyclic and are not considered here, there seem to have been very few field studies where density was reported relevant to reproduction, and even fewer where the wild populations were reported at high densities. Indeed, it would seem that much of the natural development of mechanisms of population control or regulation has resulted in populations never reaching very high densities. This may be the reason why the only two mammals which have shown these effects in the wild, at least as far as can be judged from the literature, are those which are living in environments which have been greatly affected by human interference. Both the rabbit and the mouse are species which live in areas where from the point of food supply alone man has improved the available environment considerably. This may have resulted in these two species being able to increase in numbers to such high levels that density effects become obvious, whereas other species of mammals are limited in numbers before the reproduction is affected.

Bearing this point in mind, a comparison will now be made of the reproductive patterns shown by the rabbit. Myers and Poole (1962) found increased interruptions to breeding in penned populations as density increased, which they suggested was due either to a reduced incidence of oestrus or to an increased incidence of failure to conceive at oestrus. Interruptions to breeding were not mentioned as such by Lloyd (1963, 1964) in his study of density effects in wild-rabbit populations, but he did find that the breeding season was reduced in length during years of high density. Myers and Poole found no alteration in the length of breeding. Although these workers found that dominant females had larger litters than subdominants, the mean litter size did not change significantly when the density increased in the penned populations. In contrast, Lloyd found a considerable reduction in litter size in high-density years. Increased density in penned populations resulted in a marked increase in prenatal mortality. Although a high incidence of prenatal mortality has been known to be a feature of the reproduction of wild rabbits for a number of years (Brambell, 1944; Watson, 1957; Poole, 1960), there seems to be no definite evidence relating it to density in wild populations. Although the level of prenatal mortality was lower in Lloyd's study than the level reported in other field studies, he found no direct relationship between it and density. These patterns in the rabbit show such diversity between wild and penned studies that it is unwise to reach any conclusion about the relevance of the penned studies to wild populations in this species.

For the house mouse the comparison of penned to wild populations presents a similarly confused picture. In their laboratory experiments, Crowcroft and Rowe (1957) found that high densities resulted in some populations in the females ceasing to undergo oestrous cycles. On the other hand, Southwick (1958) found that the proportion of females which were fecund did not alter with density when they were adult in wild rick populations (see page 241), and similar results were presented by Rowe, Taylor, and Chudley (1964). The difference in the lower weight classes with density can be attributed to differences in the attainment of puberty. Both penned and wild populations showed considerable reduction in litter size at high densities. Resorption of litters was common in dense laboratory populations, and was noted by Southwick in his highest density rick, but Rowe *et al.* found no increased resorption in their two higher-density ricks. The evidence for house-mouse populations is therefore also equivocal.

It would therefore seem that comparison of the sorts of reproductive patterns exhibited by penned populations with those of dense field populations results in no valuable conclusions with regard to the relevance of hypotheses derived from observations on penned populations to the situation in the wild. It would appear that there have been very few conclusive studies on wild mammals relating density to reproduction or even relating social factors to reproductive success. Eisenberg (1967) has suggested that the reproductive and behavioural responses seen when rodents are confined together in a restricted space will depend largely on the type of social structure characteristic of the species involved. Naturally asocial species will thus exhibit reproductive breakdown at much lower densities than species living naturally in fairly dense societies.

This part closes with a brief mention of the new science of 'Exocrinology' and the significance of pheromones. There is no direct evidence that pheromones affect reproduction in wild mammals, but the wideness of the taxonomic distribution of effects on reproduction (i.e. synchronization of oestrus by males) which are directly comparable to effects definitely resulting from action of pheromones in the laboratory would suggest that many mammal species other than those investigated produce pheromones. Fraser (1968) has reviewed the role of pheromones in the reproductive behaviour of ungulates. Any standard text on general mammalogy (Bouliére, 1954; Corbet, 1966) will indicate the large number of odoriferous glands present in all types of mammals. These glands are suspected to have considerable importance as sources of odours for demarking territories and other social functions. Indeed, in one species, the rabbit (*Oryctolagus cuniculus*), some exceedingly excellent work by Mykytowicz (1964, 1965) has shown experimentally the importance of the submandibular or chin glands in territorial and social phenomena. Mykytowicz has demonstrated that these glands are most active during the breeding season and that their size reflects the social position of the individual rabbit. The evidence from this and other studies is that mammals produce a number of odours which are related in some way to their breeding. Although it is somewhat speculative, it would seem more than likely that future research will demonstrate that odours act as pheromones and can influence the reproduction of wild mammals at high densities in exactly the same way as occurs in the laboratory. Indeed, it is possible that these pheromones may have very important effects through their influence on reproduction on the whole mechanism of population regulation. This is a field of the greatest potential interest, and it is to be hoped that future research will show the degree of importance of exocrinology to the endocrinology of reproduction in wild mammals.

PART SIX

Human Interference

CHAPTER 25

The Role of Humans in the Ecology of Breeding in Mammals

Domestication of many mammals has involved the deliberate alteration of many factors known to affect breeding. These have been dealt with in previous chapters and will not be reconsidered here. This chapter will be concerned with the little that is known of the indirect and direct effects that man's alteration of the environment has on wild species of mammals.

Agricultural pursuits, such as irrigation and the planting of crops of high nutritive value, have already been seen (pages 149, 177) to alter the nutrition available to several species of small mammals and thus to increase their ability to reproduce. For example, Prasad (1956) mentioned that the breeding season of the Indian gerbille (*Tatera indica*) is related to the indirect effects of human agricultural practice, in that breeding responds to the planting of seed crops at the end of June. The stomach contents at this time show that the gerbils are feeding heavily on sown seed.

Detrimental effects on reproduction by agricultural practice have also been reported. Twigg (1965) noted that *Holochilus sciureus* in British Guiana nested in sugar-cane fields under the cane trash. When this trash is removed many nests are revealed, and this disturbs the nesting females. Burning of the cane prior to harvest results in a heavy incidence of mortality in the nestling mice. Kuhn, Wick, and Pedersen (1966) noted that dairy cattle commonly trampled on the mounds above the nest of Townsend's mole (*Scapanus townsendi*) in Oregon, and this was the highest cause of nestling mortality.

Another series of indirect effects on reproduction in mammals may possibly occur through the application of pesticides or herbicides in agricultural pest control. Although a great deal is known about this subject in birds and that certain pesticides can have detrimental effects on avian reproduction, the writer has been unable to discover very much in the literature relating pesticides to mammal reproduction. Flux (1965a) suggested that a very high incidence of ovarian tumours on hares (*Lepus europaeus*) in New Zealand may have been due to treatment with DDT or other insecticides, but gave no supporting evidence. Snyder, B. D. (1963) noted that the application of different amounts of endrin to populations of the vole (*Microtus pennsylvanicus*) reduced the densities at intense applications and particularly reduced the numbers of females. Both 0·6 lb endrin

T

per acre and 2·0 lb endrin per acre significantly reduced the number of litters produced up to 2 months after treatment. Bernard and Gaertner (1964) fed laboratory mice (*Mus musculus*) on varying concentrations of DDT and found that the litter size did not decrease when the mice were fed up to 300 ppm DDT, but that significantly more females produced no litters at all when fed on 200 or 300 ppm than when fed no DDT or 100 ppm DDT. In a recent paper Morris (1968) reported detailed experiments on the effects of different concentrations of endrin on deermice (*Peromyscus maniculatus*). Although adult survival was directly related to the level of endrin (up to 7 ppm) fed, litter production frequency and litter size were not affected. At higher endrin levels, however, there was some evidence of postnatal mortality before weaning, and Morris suggests that this may be due to endrin being concentrated in the milk. Any survey of journals such as the *Journal of Wildlife Management* will show that there is increasing documentation of the presence and quantities of pesticides in wild mammals. There is a need for study in the field, and under simulated field pesticide concentrations in the laboratory, of the effects that these substances are having on reproduction in wild mammals.

Human activities also sometimes have direct effects, whether they be deliberate or not, on mammal reproduction. Particular examples come from human interference in seal colonies. Personal experience and comments such as that of Davies (1953) indicate that grey seal (*Halichoerus grypus*) breeding colonies are very much disturbed by the presence of humans. Not only is pup mortality increased by the movements of disturbed adults but there was some evidence that the presence of a field party (which included this author) on Gasker, Outer Hebrides, Scotland, may have delayed parturition in a number of cows in the early breeding season of 1966. Harrison-Mathews, in a discussion following a paper by Hediger (1962), has noted that the grey seal copulates on land normally in colonies at North Rona, Shetland Islands. However, in smaller colonies off the coast of Wales and Cornwall, where the animals are subjected to much more human interference, copulation regularly occurs in the water, and over a much shorter period of time. Human interference has also been reported as affecting other seal species. For example, Rand (1955) suggested that sealing parties which stampede the rookeries of the Cape fur seal (*Arctocephalus pusillus*) in South Africa may be the cause of the very high level of pre-implantation intrauterine mortality found (34 per cent). Although most sealing occurs in the winter, there is some going on in early summer at the time of implanation. The deliberate killing of very young mammals is not common, although some young seals are killed for their pre-moult skin and Erickson, Nellor, and Petrides (1964) have reported killings of suckling black bear (*Euarctos americanus*) cubs in Michigan.

There is one report of disturbance by intensive trapping and by the movements and activities of the trappers of beaver (*Castor canadensis*) in Idaho. Leege and Williams (1967) compared the productivity of one colony subjected to such harassment with an undisturbed colony and noted that there was a reduced proportion of females whose uterine tracts showed scars of pregnancy in the former colony.

Finally, in at least two types of mammal there have been deliberate attempts to alter reproduction by administering anti-fertility agents as a method of population control. Balser (1964) reported that penned trials with diethylstilboestrol were successful in terminating pregnancy in coyotes (*Canis latrans*). In a field trial 5,000 baits were dropped in areas of New Mexico wherever coyote sign was noted. After a period to allow the anti-fertility agent to take effect, dead samples were taken from the baited area and a control area some miles away. All the bitches from the control area (13) were undergoing normal pregnancies, but 16 of the 20 from the baited area had pregnancies blocked either by a failure of implantation or resorption of embryos. It appears that the administration of anti-fertility agents is thus a possible method of predator population control. Some experimentation has also been carried out on female wapiti (*Cervus canadensis*) to determine the effectiveness of chemosterilants on the control of reproduction in this species (Greer, Hawkins, and Catlin, 1968). These workers injected diethyl-stilboestrol into 36 pregnant females, but were only able to cause the cessation of pregnancy by this technique in 30 per cent of the animals. Such an approach would require considerable modification before it could be applied to the regulation of herds of wild wapiti. Control of male fertility by substances such as cadmium chloride, which interferes with testicular function, is also possible, and it has been attempted by the author and E. A. Smith on grey seal (*Halichoerus grypus*) bulls, but with equivocal results.

Bibliography

ADAMCZEWSKA, K. A. (1961) 'Intensity of reproduction of the *Apodemus flavicollis* (Melchior 1834) during the period 1954–1959' *Acta theriol.* **5**: 1–21.

ADAMS, W. H. (1960) 'Population ecology of white-tailed deer in north eastern Alabama' *Ecology* **41**: 706–15.

ALDERMAN, G. (1963) 'Mineral nutrition and reproduction in cattle' *Vet. Rec.* **75**: 1015–18.

ALEXANDER, G. (1956) 'Influence of nutrition upon gestation length in sheep' *Nature, Lond.* **178**: 1058–9.

——, MANN, T., MULHEARN, C. J., ROWLEY, I. C. R., WILLIAMS, D. and WINN, D. (1967) 'Activities of foxes and crows in a flock of lambing ewes' *Aust. J. exp. Agric. Anim. Husb.* **7**: 429–36.

—— and WILLIAMS, D. (1966) 'Teat-seeking activity in new-born lambs: the effects of cold' *J. agric. Sci., Camb.* **67**: 181–9.

ALLANSON, M. (1958) 'Growth and reproduction in the males of two species of gerbil, *Tatera brantsii* (A. Smith) and *Tatera afra* (Gray)' *Proc. zool. Soc. Lond.* **130**: 373–95.

—— and DEANESLEY, R. (1934) 'The reaction of anoestrous hedgehogs to experimental conditions' *Proc. R. Soc. (Ser. B)* **116**: 170–85.

ALLEN, D. M. and LAMMING, G. B. (1961) 'Nutrition and reproduction in the ewe' *J. agric. Sci., Camb.* **56**: 69–79.

ALLEN, L. D. C. (1963) 'The lechwe (*Kobus leche smithemani*) of the Bangweulu swamp' *Puku* **1**: 1–8.

ALLEN, P., BRAMBELL, F. W. R., and MILLS, I. H. (1947) 'Studies on sterility and prenatal mortality in wild rabbits I. The reliability of estimates of prenatal mortality based on counts of corpora lutea, implantation sites and embryos' *J. exp. Biol.* **23**: 312–31.

ALTLAND, P. D. and HIGHMAN, B. (1968) 'Sex organ changes and breeding performance of male rats exposed to altitude: effect of exercise and physical training' *J. Reprod. Fertil.* **15**: 215–22.

ALTMANN, M. (1960) 'The role of juvenile elk and moose in the social dynamics of their species' *Zoologica, N.Y.* **45**: 35–9.

AMOROSO, E. C. and MARSHALL, F. H. A. (1960) 'External factors in sexual periodicity', chapter in *Marshall's Physiology of Reproduction*, vol. 1, part 2, A. S. Parkes, Ed. (Longmans: London), 877 pp.

ANDERSON, J. (1945) *The Semen of Animals and its Use for Artificial Insemination* (Imperial Bureau Animal Breeding and Genetics – Edinburgh), 151 pp.

—— (1964) 'Reproduction of imported British breeds of sheep on a tropical plateau' *Proc. Vth. Int. Congr. Anim. Reprod. (Trento)* **3**: 465–9.

ANDERSON, P. K. (1961) 'Density, social structure and non-social environment in house-mouse populations and the implications for regulation of numbers' *Trans. N.Y. Acad. Sci. (Ser. 11)* **23**: 447–57.

ANDERVONT, H. B. (1944) 'Influence of environment on mammary cancer in mice' *J. Natn. Cancer Inst.* **4**: 579–81.

ANSELL, W. F. H. (1960) 'The breeding of some larger mammals in northern Rhodesia' *Proc. zool. Soc. Lond.* **134**: 251–74.

ASDELL, S. A. (1964) *Patterns of mammalian reproduction* 2nd edn. (Constable, London), 670 pp.

ASHBY, K. R. (1967) 'Studies on the ecology of field mice and voles (*Apodemus sylvaticus, Clethrionomys glareolus* and *Microtus agrestis*) in Houghall wood, Durham' *J. Zool.* **152**: 389–513.

ATWELL, G. (1964) 'Wolf predation on calf moose' *J. Mammal.* **45**: 313–14.

AULERICH, R. J., HOLCOMB, L. C., RINGER, R. K., and SCHAIBLE, P. J. (1963) 'Influence of photoperiod on reproduction in mink' *Q. Bull. Mich. St. Univ. agric. Exp. Stn.* **46**: 132–8.

AVERILL, R. L. W. (1955) 'Fertility of the ewe' *Stud. Fert.* **7**: 139–48.

—— (1959) 'Ovulatory activity in mature Romney ewes in Otago' *N.Z. Jl agric. Res.* **2**: 575–83.

—— (1964a) 'Variability in ovulatory activity in 5–7 year Romney ewes in New Zealand' *Proc. Vth. Int. Congr. Anim. Reprod. (Trento)* **3**: 206–11.

—— (1964b) 'Ovulatory activity in mature Romney ewes in New Zealand' *N.Z. Jl Agric. Res.* **7**: 514–24.

—— (1965) 'Variability in ovulatory activity in mature Romney ewes in New Zealand' *Wld. Rev. Anim. Prod.* **3**: 51–7.

BAEVSKY, U. B. (1963) 'The effect of embryonic diapause on the nuclei and mitotic activity of mink and rat blastocysts', chapter in *Delayed Implantation*, A. C. Enders, Ed. (Univ. of Chicago Press, Chicago), 318 pp.

BAILEY, E. D. (1966) 'Social interaction as a population-regulating mechanism in mice' *Can. J. Zool.* **44**: 1007–12.

BAKER, J. R. (1930) 'The breeding season in British wild mice' *Proc. zool. Soc. Lond.* **1930**, pt. 1: 113–26.

—— (1938a) 'Evolution of breeding seasons', chapter in *Evolution Essays on Aspects of Evolutionary Biology Presented to Professor E. S. Goodrich* G. R. deBeer, Ed. (Clarendon Press, Oxford), 350 pp.

—— (1938b) 'The relationship between latitude and breeding seasons in birds' *Proc. zool. Soc. Lond.* **108**: 557–82.

—— and BAKER, Z. (1936) 'The seasons in a tropical rainforest (New Hebrides) Part 3, Fruit bats (Pteropidae)' *J. Linn. Soc.* **40**: 123–41.

—— and BIRD, T. F. (1936) 'The seasons in a tropical rainforest (New Hebrides) Part 4, Insectivorous bats (Vespertilionidae and Rhinolophidae) *J. Linn. Soc.* **40**: 143–61.

—— and RANSON, R. M. (1932) 'Factors affecting the breeding of the field mouse (*Microtus agrestis*) Part I – Light' *Proc. R. Soc. (B)* **110**: 313–22.

—— —— (1933a) 'Factors affecting the breeding of the field mouse (*Microtus agrestis*) Part II, Temperature and food' *Proc. R. Soc. (B)* **112**: 39–46.

—— —— (1933b) 'Factors affecting the breeding of the field mouse (*Microtus agrestis*) Part III – Locality' *Proc. R. Soc. (B)* **113**: 486–95.

BAKKO, E. B. and BROWN, L. N. (1967) 'Breeding biology of the white-tailed prairie dog, *Cynomys leucurus*, in Wyoming' *J. Mammal.* **48**: 100–12.

BALSER, D. S. (1964) 'Management of predator populations with antifertility agents' *J. Wildl. Mgmt.* **28**: 352–8.

BARKALOW, F.S. (1962) 'Latitude related to reproduction in the cottontail rabbit' *J. Wildl. Mgmt.* **26**: 32–7.

BARNETT, S. A. (1962) 'Total breeding capacity of mice at two temperatures' *J. Reprod. Fert.* **4**: 327–35.

—— and COLEMAN, E. M. (1959) 'The effect of low environmental temperature on the reproductive cycle of female mice' *J. Endocr.* **19**: 232–40.

BASIROV, E. B. (1960) '(Basic causes of seasonal variation in quantity and quality of bull and buffalo bull semen on Azerbaijan)' *Zivotnovodstvo* **22**: 24–7 (from *Anim. Breed. Abstr.* **28**: 1294).

BATCHELOR, R. (1963) 'Evidence of yearling pregnancies in the Roosevelt elk' *J. Mammal.* **44**: 111–12.

BATTEN, C. A. and BERRY, R. J. (1967) 'Prenatal mortality in wild-caught house mice' *J. Anim. Ecol.* **36**: 453–63.

BEDFORD, duke of, and MARSHALL, F. H. A. (1942) 'On the incidence of the breeding season in mammals after transference to a new latitude' *Proc. R. Soc.* (*B*) **130**: 396–9.

BEER, J. R. and MACLEOD, C. F. (1961) 'Seasonal reproduction in the meadow vole' *J. Mammal.* **42**: 483–9.

—— —— and FRENZEL, L. D. (1957) 'Prenatal survival and loss in some cricetid rodents' *J. Mammal.* **38**: 392–402.

BEILHARZ, R. G. (1968) 'Effect of stimuli associated with the male on litter size in mice' *Aust. J. Biol. Sci.* **21**: 583–5.

BELJAEV, D. K. and KLOCKOV, D. V. (1965) '(Increasing fertility in mink by additional light)' *Krolikov. Zverov.* **8**: 2–5 (from *Anim. Breed. Abstr.* **34**: 1449).

—— —— and ZELEZOVA, A. I. (1963) '(The effect of light on the reproduction and fertility of mink (*Mustela vison* Schr.))' *Byull. mosk. Obshch. Ispyt. Prir. Otd. biol.* **68**: 107–25 (from *Anim. Breed. Abstr.* **34**: 523).

BERGER, P. J. (1966) 'Eleven-month embryonic diapause in a marsupial' *Nature, Lond.* **211**: 435–6.

BERGSTEDT, B. (1965) 'Distribution, reproduction, growth and dynamics of the rodent species, *Clethrionomys glareolus*, *Apodemus flavicollis* and *Apodemus sylvaticus* in southern Sweden' *Oikos* **16**: 132–60.

BERNARD, R. F. and GAERTNER, R. A. (1964) 'Some effects of DDT on reproduction in mice' *J. Mammal.* **45**: 272–6.

BIDER, J. R. (1961) 'An ecological study of the hare *Lepus americanus*' *Can. J. Zool.* **39**: 81–103.

BIRKENHOLZ, D. E. (1963) 'A study of the life history and ecology of the round tailed muskrat (*Neofiber alleni* True) in north-central Florida' *Ecol. Monogr.* **33**: 255–80.

BISHOP, M. W. H. and WALTON, A. (1960) 'Spermatogenesis and the structure of mammalian spermatozoa', chapter in *Marshall's Physiology of Reproduction* vol. 1, part 2, A. S. Parkes, Ed. (Longmans, London), 877 pp.

BISSONNETTE, T. H. (1932) 'Modification of mammalian sexual cycles I: Reactions of ferrets (*Putorius vulgaris*) of both sexes to electric light added after dark in November and December' *Proc. R. Soc.* (*B*) **110**: 322–36.

—— (1935a) 'Modifications of mammalian sexual cycles II: Effects upon young male ferrets (*Putorius vulgaris*) of constant eight one-half hour days and of six hours illumination after dark, between November and June' *Biol. Bull. Mar. Biol. Lab., Woods Hole* **68**: 300–13.

—— (1935b) 'Modifications of mammalian sexual cycles III: Reversal of the cycle in male ferrets (*Putorius vulgaris*) by increasing periods of exposure to light between October 2 and March 13' *J. exp. Zool.* **71**: 341–73.

—— (1941) 'Experimental modification of breeding cycles in goats' *Physiol. Zoöl.* **14**: 379–83.

—— and BAILEY, E. E. (1936) 'Litters from ferrets in January induced by increasing exposures to light after nightfall' *Am. Nat.* **70**: 454–8.

—— and CSECH, A. G. (1937) 'Modifications of mammalian sexual cycles VII: Fertile matings of raccoons in December instead of February induced by increasing daily periods of light' *Proc. R. Soc.* (*B*) **122**: 246–54.

—— —— (1939) 'A third year of modified breeding behaviour with raccoons' *Ecology* **20**: 156–62.

BLOOD, P. A. (1963) 'Some aspects of behaviour of a bighorn herd' *Can. Fld Nat.* **77**, 77–94.

BLUS, L. J. (1966) 'Relationship between litter size and latitude in the golden mouse' *J. Mammal.* **47**: 546–7.

BODENHEIMER, F. S. (1938) *Problems of Animal Ecology* (Oxford Univ. Press, Oxford), 183 pp.

—— (1949) *Problems of Vole Populations in the Middle East: Report on the Population Dynamics of the Levant Vole* (Microtus guentheri D. et A.) (Research Council of Israel, Jerusalem), 77 pp.

—— (1954) 'Problems of physiology and ecology of desert animals', chapter in *Biology of Deserts*, J. L. Cloudsley-Thompson, Ed. (Inst. of Biol., London), 224 pp.

—— (1957) 'The ecology of mammals in arid zones', chapter in *Arid Zone Research VIII. Human and Animal Ecology* (UNESCO – Paris), 244 pp.

—— and SULMAN, F. (1946) 'The estrous cycle of *Microtus guentheri* D. and A., and its ecological implications' *Ecology* **27**: 253–62.

BOOKHOUT, T. A. (1965) 'Breeding biology of snowshoe hares in Michigan's upper peninsula' *J. Wildl. Mgmt.* **29**: 296–303.

BOULIÉRE, F. (1954) *The Natural History of Mammals* (A. Knopf, New York), 363 pp.

BRADSHAW, G. V. R. (1962) 'Reproductive cycle of the California leaf-nosed bat *Macrotus californicus*' *Science, N.Y.* **136**: 645–6.

BRAMBELL, F. W. R. (1944) 'The reproduction of the wild rabbit *Oryctolagus cuniculus* (L)' *Proc. zool. Soc. Lond.* **114**: 1–45.

—— (1948) 'Prenatal mortality in mammals' *Biol. Rev.* **23**: 370–407.

—— and DAVIS, D. H. S. (1941) 'Reproduction of the multimammate mouse (*Mastomys erythroleucus* Temn.) of Sierra Leone' *Proc. zool. Soc. Lond.* **111**: 1–11.

—— and HALL, K. (1939) 'Reproduction of the field vole, *Microtus agrestis hirtus* Bellamy' *Proc. zool. Soc. Lond.* **109**: 133–8.

—— and MILLS, I. H. (1948) 'Studies on sterility and prenatal mortality in wild rabbits IV: The loss of embryos after implantation' *J. exp. Biol.* **25**: 241–69.

—— and ROWLANDS, I. W. (1936) 'Reproduction of the bank vole (*Evotomys glareolus* Schreber). 1. The oestrus cycle of the female' *Phil. Trans. R. Soc.* (B) **226**: 71–97.

BREAKEY, D. R. (1963) 'The breeding season and age structure of feral house mouse populations near San Francisco Bay, California' *J. Mammal.* **44**: 153–68.

BREED, W. G. (1967) 'Ovulation in the genus *Microtus*' *Nature, Lond.* **214**: 826.

BRODY, S. (1955) *Bioenergetics and Growth* (Hafner Publishing Co. Inc., New York), 1,023 pp.

BRONSON, F. H. (1963) 'Some correlates of interaction rate in natural populations of woodchucks' *Ecology* **44**: 637–43.

—— and ELEFTHERIOU, B. E. (1965) 'Behavioural, pituitary and adrenal correlates of controlled fighting (defeat) in mice' *Physiol. Zoöl.* **38**: 406–11.

——, ——, and GARICK, E. I. (1964) 'Effects of intra- and inter-specific social stimulation on implantation in deermice' *J. Reprod. Fert.* **8**: 23–7.

—— and MARSDEN, H. M. (1964) 'Male-induced sychrony of estrus in deermice' *Gen. comp. Endocrin.* **4**: 634–7.

BROWN, H. L. (1945) 'Evidence of winter breeding of *Peromyscus*' *Ecology* **26**: 308–9.

BROWN, L. N. (1964) 'Reproduction of the brush mouse and white-footed mouse in the central United States' *Am. Midl. Nat.* **72**: 226–40.

—— (1966) 'Reproduction of *Peromyscus maniculatus* in the Laramie basin of Wyoming' *Am. Midl. Nat.* **76**: 183–9.

—— and CONAWAY, C. H. (1964) 'Persistance of corpora lutea at the end of the breeding season in *Peromyscus*' *J. Mammal.* **45**: 260–5.

BROWN, R. Z. (1953) 'Social behaviour, reproduction and population changes in the house mouse (*Mus musculus* L)' *Ecol. Monogr.* **23**: 217–40.

BRUCE, H. M. (1960) 'A block to pregnancy in the mouse caused by the proximity of strange males' *J. Reprod. Fert.* **1**: 96–103.

—— (1961) 'Time relations in the pregnancy block induced in mice by strange males' *J. Reprod. Fert.* **2**: 138–42.

—— (1963) 'Olfactory block to pregnancy among grouped mice' *J. Reprod. Fert.* **6**: 451–60.

—— (1967) 'Effects of olfactory stimuli on reproduction in mammals', chapter in *Effect of External Stimuli on Reproduction*. G. E. W. Wolstenholme and M. O'Connor, Eds. (Churchill, London), 107 pp.

BUCKNER, C. H. (1966) 'Populations and ecological relationships of shrews in tamarack bogs in south-eastern Manitoba' *J. Mammal.* **47**: 181–94.

BUECHNER, H. K. (1961a) 'Territorial behaviour in Uganda kob' *Science, N.Y.* **133**: 698–9.

—— (1961b) 'Regulation of numbers of pronghorn antelope in relation to land use' *Terre Vie* **2**: 266–85.

—— (1963) 'Territoriality as a behavioural adaptation to environment in Uganda kob' *Proc. XVIth. Int. Congr. Zool.* **3**: 59–63.

—— and SWANSON, C. V. (1955) 'Increased natality resulting from lowered population density among elk in southwestern Washington' *Trans. N. Am. Wildl. Conf.* **20**: 560–7.

BULL, P. C. (1961) 'Population dynamics of the wild rabbit, *Oryctolagus cuniculus* (L)' *Sci. Rev.* **20**: 23–6.

BULLOUGH, W. S. (1951) *Vertebrate Sexual Cycles* (Methuen, London), 123 pp.

BURKHARDT, J. (1947) 'Transition from anoestrus in the mare and the effects of artificial lighting' *J. agric. Sci., Camb.* **37**: 64–8.

BUSS, I. O. and JOHNSON, O. W. (1967) 'Relationships of Leydig cell characteristics and intra-testicular testosterone levels to sexual activity in the African elephant' *Anat. Rec.* **157**: 191–6.

—— and SAVIDGE, J. M. (1966) 'Change in population number and reproductive rate of elephants in Uganda' *J. Wildl. Mgmt.* **30**: 791–809.

BUXTON, P. A. (1936) 'Breeding rates of domestic rats trapped in Lagos, Nigeria and certain other countries' *J. Anim. Ecol.* **5**: 53–66.

CALABY, J. H. (1959) 'Notes on the little eagle, with particular reference to rabbit predation' *Emu* **51**: 33–56.

CALDWELL, L. D. and GENTRY, J. B. (1965) 'Natality in *Peromyscus polionotus* populations' *Am. Midl. Nat.* **74**: 168–75.

CAMERON, A. W. (1967) 'Breeding behaviour in a colony of western Atlantic gray seals' *Can. J. Zool.* **45**: 161–73.

CANIVENC, R. (1966) 'A study of progestation in the European badger *Meles meles* L.' *Symp. zool. Soc. Lond.* **15**: 15–26.

CARPENTER, C. R. (1942) 'Sexual behaviour of free ranging rhesus monkeys (*Macaca mulatta*)' *J. comp. Psychol.* **33**: 113–67.

CARRICK, R., CSORDAS, S. E., and INGHAM, S. E. (1962) 'Studies on the southern elephant seal *Mirounga leonina* (L.) IV: Breeding and development' *C.S.I.R.O. Wildl. Res.* **7**: 161–97.

—— and INGHAM, S. E. (1960) 'Ecological studies of the southern elephant seal, *Mirounga leonina* (L.) at Macquarie Island and Heard Island' *Mammalia* **24**: 325–42.

CHANCE, M. R. A. (1956) 'Environmental factors influencing gonadotrophic assay in the rat' *Nature, Lond.* **177**: 228–9.

CHAPMAN, B. M., CHAPMAN, R. F., and ROBERTSON, I. A. D. (1959) 'The growth and breeding of the multimammate rat *Rattus* (*Mastomys*) *natalensis* (Smith) in Tanganyika territory' *Proc. zool. Soc. Lond.* **133**: 1–9.

CHEATUM, E. L. and MORTON, G. H. (1946) 'Breeding season of white-tailed deer in New York' *J. Wildl. Mgmt.* **10**: 249–63.

—— and SEVERINGHAUS, C. W. (1950) 'Variation in fertility in white-tailed deer related to range conditions' *Trans. N. Am. Wildl. Conf.* **15**: 170–90.

CHERNIAVSKI, F. B. (1962) '(On the reproduction and growth of the snow sheep (*Ovis nivicola* Esch.))' *Zool. Zh.* **41**: 1556–66.

CHEW, R. M. and BUTTERWORTH, B. B. (1964) 'Ecology of rodents in Indian cave (Mojave Desert) Joshua tree national monument, California' *J. Mammal.* **45**: 203–25.

CHIPMAN, R. K. and FOX, K. A. (1966) 'Factors in pregnancy blocking: Age and reproductive background of females; numbers of strange males' *J. Reprod. Fert.* **12**: 399–403.

——, HOLT, J. A. and FOX, K. A. (1966) 'Pregnancy failure in laboratory mice after multiple short-term exposure to strange males' *Nature, Lond.* **210**: 653.

CHITTLEBOROUGH, R. G. (1958) 'The breeding cycle of the female humpback whale, *Megaptera nodesa* (Bonnaterre)' *Aust. J. Mar. Freshwat. Res.* **9**: 1–18.

CHITTY, D. (1952) 'Mortality among voles (*Microtus agrestis*) at Lake Vyrnwy, Montgomeryshire in 1936–1939' *Phil. Trans. R. Soc.* (B) **236**: 505–52.

—— (1955) 'Adverse effects of population density upon the viability of later generations', chapter in *The Numbers of Man and Animals*, J. B. Cragg and N. W. Pirie, Eds. (Oliver and Boyd, Edinburgh), 152 pp.

—— (1962) 'Population processes in the vole and their relevance to general theory' *Can. J. Zool.* **38**: 99–113.

CHITTY, H. and AUSTIN, C. R. (1957) 'Environmental modification of oestrus in the vole' *Nature, Lond.* **179**: 592–3.

—— and CHITTY, D. (1962) 'Body weight in relation to population phase in *Microtus agrestis*' *Symp. Theriol.* Brno, 1962 (Proc. Internat. Symp. Methods of Mammalogical Investigation): 77–86.

CHRISTIAN, J. J. (1955) 'Effect of population size on the adrenal glands and reproductive organs of male mice in populations of a fixed size' *Am. J. Physiol.* **182**: 292–300.

—— (1956) 'Adrenal and reproductive responses to population size in mice from freely growing populations' *Ecology* **37**: 258–73.

—— (1963) 'The pathology of overpopulation' *Milit. Med.* **128**: 571–603.

—— and DAVIS, D. E. (1964) 'Endocrines, behaviour and population' *Science, N.Y.* **146**: 1550–60.

——, LLOYD, J. A. and DAVIS, D. E. (1965) 'The role of endocrines in the self-regulation of mammalian populations' *Recent Prog. Horm. Res.* **21**: 501–78.

CHU, E. W. (1965) 'Effect of environmental illumination on the estrous cycles of rodents' *Acta cytol.* (*Philad.*) **9**: 221–7.

CLARKE, J. R. (1955) 'Influence of numbers on reproduction and survival in two experimental vole populations' *Proc. R. Soc.* (B) **144**: 68–85.

CLEGG, M. T., COLE, H. H., and GANONG, W. F. (1965) 'The role of light in the regulation of cyclical estrous activity in sheep' in *Proceedings of a Conference on Estrous Cycle Control in Domestic Animals*, Misc. Publ. U.S. dept. Agric. **1005**: 96–103.

—— and GANONG, W. F. (1959) 'Environmental factors, other than nutrition, affecting reproduction', chapter in *Reproduction in Domestic Animals*, H. H. Cole and P. T. Cupps, eds. (Academic Press, New York), 450 pp.

COETZEE, C. G. (1965) 'The breeding season of the multimammate mouse, *Praomys* (*Mastomys*) *natalensis* (A. Smith) in the Transvaal high veld' *Zool. Afric.* **1**: 29–39.

CONAWAY, C. H. (1959) 'The reproductive cycle of the eastern mole' *J. Mammal.* **40**: 180–94.

——, BASKETT, T. S., and TOLL, J. E. (1960) 'Embryo resorption in the swamp rabbit' *J. Wildl. Mgmt.* **24**: 197–202.

CONAWAY, C. H. and KOFORD, C. B. (1964) 'Estrous cycles and mating behaviour in a free-ranging band of rhesus monkeys' *J. Mammal.* **45**: 577–98.

—— and WIGHT, H. M. (1962) 'Onset of reproductive season and first pregnancy of the season in cottontails' *J. Wildl. Mgmt.* **26**: 278–90.

——, ——, and SADLER, K. C. (1963) 'Annual production by a cottontail population' *J. Wildl. Mgmt.* **27**: 171–5.

COOP, I. E. (1966) 'Effect of flushing on reproductive performance of ewes' *J. agric. Sci., Camb.* **67**: 305–23.

COOPER, K. J. and HAYNES, N. B. (1967) 'Modification of the oestrous cycle of the under-fed rat associated with the presence of the male' *J. Reprod. Fert.* **14**: 317–20.

CORBET, G. B. (1966) *The Terrestrial Mammals of Western Europe* (G. T. Foulis, London), 264 pp.

COULSON, J. C. and HICKLING, G. (1964) 'The breeding biology of the gray seal *Halichoerus grypus* (Fab.), on the Farne island, Northumberland' *J. Anim. Ecol.* **33**: 485–512.

COWAN, I. MCT. (1956) 'Life and times of the coast black-tailed deer', chapter in *The Deer of North America* W. P. Taylor, Ed. (Stackpole Co., Harrisburg, Penn. and Wildlife Management Inst., Washington), 668 pp.

—— and ARSENAULT, M. G. (1955) 'Reproduction and growth in the creeping vole (*Microtus oregoni serpens* Merriam)' *Can. J. Zool.* **32**: 198–208.

COWGILL, U. M. (1966a) 'The season of birth in man' *Man* **1**: 232–40.

—— (1966b) 'The season of birth in man. Contemporary situation with special reference to Europe and the southern hemisphere' *Ecology* **47**: 614–23.

—— (1966c) 'Historical study of the season of birth in the city of York, England' *Nature, Lond.* **209**: 1067–70.

—— (1966d) 'The season of birth in man; the northern new world' *Kroeber Anthropol. Soc. Pap.* **35**: 1–21.

——, BISHOP, A., ANDREW, R. J., and HUTCHINSON, G. E. (1962) 'An apparent lunar periodicity in the sexual cycle of certain prosimians' *Proc. natn. Acad. Sci. U.S.A.* **48**: 238–41.

COX, T. J. (1965) 'Seasonal change in the behaviour of the western pipistrelle because of lactation' *J. Mammal.* **46**: 703.

CROOKE, J. H. and GARTLAN, J. S. (1966) 'Evolution of primate societies' *Nature, Lond.* **210**: 1200–3.

CROWCROFT, P. (1964) 'Note on the sexual maturation of shrews (*Sorex araneus*) in captivity' *Acta theriol.* **8**: 89–93.

—— and ROWE, F. P. (1957) 'The growth of confined colonies of the wild house mouse (*Mus musculus* L.)' *Proc. zool. Soc. Lond.* **129**: 359–70.

—— —— (1958) 'The growth of confined colonies of the wild house mouse (*Mus musculus* L.): The effect of dispersal on female fecundity' *Proc. zool. Soc. Lond.* **131**: 357–65.

CUNNINGHAM, D. J. (1905) 'Cape hunting dogs (*Lycaeon pictus*) in the gardens of the Royal Zoological Society of Ireland' *Proc. R. Soc. Edinb.* **25**: 843–8.

DAISH, C. B. (1954) *Light—to Advanced and Scholarship Level* (English Univ. Press, London), 292 pp.

DALE, H. E., RAGSDALE, A. C., and CHENG, C. S. (1959) 'Effect of constant environmental temperatures, 50° and 80° F., on appearance of puberty in beef calves' *J. Anim. Sci.* **18**: 1363–6.

DANIEL, M. J. (1963) 'Early fertility of red deer hinds in New Zealand' *Nature, Lond.* **200**: 380.

DASMANN, R. F. and MOSSMAN, A. S. (1962a) 'Population studies of impala in Southern Rhodesia' *J. Mammal.* **43**: 375–95.

—— —— (1962b) 'Reproduction in some ungulates in southern Rhodesia' *J. Mammal.* **43**: 533–7.

DASMANN, R. F. and TABER, R. D. (1956) 'Behaviour of Columbian white-tailed deer with reference to population ecology' *J. Mammal.* **37**: 143–64.

DAVIES, J. L. (1953) 'Colony size and reproduction in the grey seal' *Proc. zool. Soc. Lond.* **123**: 327–32.

—— (1957) 'The geography of the gray seal' *J. Mammal.* **38**: 297–310.

DAVIS, D. E. (1951a) 'The relation between level of population and pregnancy of Norway rats' *Ecology* **32**: 459–61.

—— (1951b) 'A comparison of reproductive potential of two rat populations' *Ecology* **32**: 469–75.

—— (1953) 'The characteristics of rat populations' *Q. Rev. Biol.* **28**: 373–401.

——, CHRISTIAN, J. J., and BRONSON, F. (1964) 'Effects of exploitation on birth, mortality and movement rates in a woodchuck population' *J. Wildl. Mgmt.* **28**: 1–9.

DAWBIN, W. H. (1966) 'The seasonal migratory cycle of humpback whales', chapter in *Whales, Dolphins and Porpoises*, K. S. Norris, Ed. (Univ. of California Press, Berkeley), 789 pp.

DAWSON, A. B. (1941) 'Early estrus in the cat following increased illumination' *Endocrinology* **28**: 907–10.

de COURSEY, G. E. (1957) 'Identification, ecology and reproduction of *Microtus* in Ohio' *J. Mammal.* **38**: 44–52.

DELANY, M. J. and BISHOP, I. R. (1960) 'The systematics, life history and evolution of the bank vole *Clethrionomys* Tilesus in north-west Scotland' *Proc. zool. Soc. Lond.* **135**: 409–22.

de LONG, K. T. (1966) 'Population ecology of feral house mice: Interference by *Microtus*' *Ecology* **47**: 481–4.

DEMPSEY, E. W., MEYERS, H. I., YOUNG, W. C., and JENNISON, D. B. (1934) 'Absence of light and the reproductive cycle in the guinea pig' *Am. J. Physiol.* **109**: 307–11.

de VORE, L. (1965) 'Male dominance and mating behaviour in baboons' chapter in *Sex and Behaviour* F. A. Beach, Ed. (John Wiley, New York), 592 pp.

de VOS, A. (1960) 'Behaviour of barrenground caribou on their calving grounds' *J. Wildl. Mgmt.* **24**: 250–8.

——, BROKX, P., and GEIST, V. (1967) 'A review of social behaviour of the North American cervids during the reproductive period' *Am. Midl. Nat.* **77**: 390–417.

—— and DOWSETT, R. J. (1966) 'The behaviour and population structure of three species of the genus *Kobus*' *Mammalia* **30**: 30–55.

DIXON, J. (1929) 'The breeding season of the pocket gopher in California' *J. Mammal.* **10**: 327–8.

DODDS, D. G. (1958) 'Observations of pre-rutting behaviour in Newfoundland moose' *J. Mammal.* **39**: 412–16.

—— (1965) 'Reproduction and productivity of snowshoe hares in Newfoundland' *J. Wildl. Mgmt.* **29**: 303–15.

DOMINIC, C. J. (1967) 'Effect of ectopic pituitary grafts on the olfactory block to pregnancy in mice' *Nature, Lond.* **213**: 1242.

DONOHOE, R. W. (1966) 'Muskrat reproduction in areas of controlled and uncontrolled water level units' *J. Wildl. Mgmt.* **30**: 320–6.

DONOVAN, B. T. and HARRIS, G. W. (1956) 'The influence of pituitary stalk section on light induced oestrus in the ferret' *J. Physiol.* **131**: 102–14.

—— —— (1966) 'Neurohormonal mechanisms in reproduction' chapter in *Marshall's Physiology of Reproduction* vol. 3, A. S. Parkes, Ed. (Longmans, London), 1,168 pp.

—— and VAN DER WERFF TEN BOSCH, J. J. (1959a) 'The relationship of the hypothalamus to oestrus in the ferret' *J. Physiol.* **147**: 93–108.

—— —— (1959b) 'The hypothalamus and sexual maturation in the rat' *J. Physiol.* **147**: 78–92.

DRORI, D. and FOLMAN, Y. (1964) 'Effects of cohabitation on the reproductive system, kidneys and body composition of male rats' *J. Reprod. Fert.* **8**: 351–9.

DUDZINSKI, M. L. and MYKYTOWYCZ, R. (1960) 'Analysis of weights and growth of an experimental colony of wild rabbits *Oryctolagus cuniculus*' *C.S.I.R.O. Wildl. Res.* **5**: 102–15.

DUKELOW, W. R. (1966) 'Variations in gestation length of mink (*Mustela vison*)' *Nature, Lond.* **211**: 211.

DUNCAN, D. L. and LODGE, G. A. (1960) 'Diet in relation to reproduction and the viability of the young Part III: pigs' *Comm. Bur. Anim. Nut. Tech. Comm.* **21**: 106 pp.

DUNMIRE, W. W. (1960) 'An altitudinal survey of reproduction in *Peromyscus maniculatus*' *Ecology* **41**: 174–82.

—— (1961) 'Breeding season of three rodents on White Mountain, California' *J. Mammal.* **42**: 489–93.

DUNNETT, G. M. (1956) 'A live-trapping study of the brush-tailed possum (*Trichosurus vulpecula* Kerr (Marsupialia)' *C.S.I.R.O. Wildl. Res.* **1**: 1–18.

DUTT, R. H. (1963) 'Critical period for early embryo mortality in ewes exposed to high ambient temperatures' *J. Anim. Sci.* **22**: 713–19.

—— (1964) 'Detrimental effect of high ambient temperatures on fertility and early embryo survival in sheep' *Int. J. Bioclim. Biomet.* **8**: 47–56.

—— and BUSH, L. F. (1955) 'The effect of low environmental temperature on initiation of the breeding season and fertility in sheep' *J. Anim. Sci.* **14**: 885–96.

——, ELLINGTON, E. F., and CARLTON, W. W. (1959) 'Fertilization rate and early embryo survival in sheared and unsheared ewes following exposure to elevated air temperature' *J. Anim. Sci.* **18**: 1308–18.

EADIE, W. R. and HAMILTON, W. J. (1958) 'Reproduction in the fisher in New York' *N.Y. Fish Game J.* **5**: 77–83.

EALEY, E. H. M. (1963) 'The ecological significance of delayed implantation in a population of the hill kangaroo (*Macropus robustus*)', chapter in *Delayed Implantation* A. C. Enders, Ed. (Univ. of Chicago Press, Chicago), 318 pp.

EDEY, T. N. (1966) 'Nutritional stress and pre-implantation embryonic mortality in Merino sheep' *J. agric. Sci., Camb.* **67**: 287–93.

EDWARDS, R. Y. and RITCEY, R. W. (1958) 'Reproduction in a moose population' *J. Wildl. Mgmt.* **22**: 261–8.

EHRLICH, S. (1966) 'Ecological effects of reproduction in nutria *Myocastor coypus* Mol.' *Mammalia* **30**: 144–52.

EINARSON, A. E. (1956) 'Life of the mule deer', chapter in *The Deer of North America*, W. P. Taylor, Ed. (Stackpole Co., Harrisburg, Penn. and Wildlife Management Inst., Washington), 668 pp.

EISENBERG, J. F. (1966) 'The social organizations of mammals' *Handbuch der Zoologie* **8**: (10, 7), 92 pp.

—— (1967) 'A comparative study of rodent ethology with emphasis on evolution of social behaviour' *Proc. U.S. Natn. Mus.* **122**: (3597), 51 pp.

—— and ISAAC, D. E. (1963) 'The reproduction of heteromyid rodents in captivity' *J. Mammal.* **44**: 61–7.

ELEFTHERIOU, B. E., BRONSON, F. H., and ZARROW, M. X. (1962) 'Interaction of olfactory and other environmental stimuli on implantation in the deermouse' *Science, N.Y.* **137**: 764.

ENDERS, A. C. (1966) 'The reproductive cycle of the nine-banded armadillo' *Symp. zool. Soc. Lond.* **15**: 295–310.

ENDERS, R. K. (1952) 'Reproduction in the mink (*Mustela vison*)' *Proc. Am. phil. Soc.* **96**: 691–755.

—— and PEARSON, O. P. (1943) 'Shortening gestation by inducing early implantation with increased light in the marten' *Am. Fur Breed.* **15**: 18.

ERICKSON, A. W., NELLOR, J. E., and PETRIDES, G. A. (1964) 'The black bear in Michigan'
  Mich. St. Univ. Agr. Expt. Sta. Res. Bull. **4**: 102 pp.
ERRINGTON, P. L. (1951) 'Concerning fluctuations in populations of the prolific and
  widely distributed muskrat' *Am. Nat.* **85**: 273–92.
—— (1963) *Muskrat Populations* (Iowa State Univ. Press, Ames), 665 pp.
——, SIGLIN, R. J., and CLARK, R. C. (1963) 'The decline of a muskrat population' *J. Wildl.
  Mgmt.* **27**: 1–8.
ESTES, R. D. (1966) 'Behaviour and life history of the wildebeest (*Connochaetes taurinus*
  Burchell)' *Nature, Lond.* **212**: 999–1000.
—— (1967) 'The comparative behaviour of Grant's and Thomson's gazelles' *J. Mammal.*
  48: 189–209.
—— and GODDARD, J. (1967) 'Prey selection and hunting behaviour of the African wild
  dog' *J. Wildl. Mgmt.* **31**: 52–70.
EVANS, R. D., SADLER, K. C., CONAWAY, C. H., and BASKETT, T. S. (1965) 'Regional
  comparisons of cottontail reproduction in Missouri' *Am. Midl. Nat.* **74**: 176–84.
EVELETH, P. B. (1966) 'Eruption of permanent teeth and age at menarche of American
  children living in the tropics' *Hum. Biol.* **38**: 60–70.
EWING, L. L., GREEN, P. M., and STEBLER, A. M. (1965) 'Metabolic and biochemical
  changes in testis of cotton rats (*Sigmodon hispidus*) during breeding cycle' *Proc. Soc. exp.
  Biol. Med.* **118**: 911–13.
FALLON, G. R. (1961) 'Reproductive patterns in Australian purebred Hereford cattle'
  Proc. IVth. Int. Congr. Anim. Reprod. (The Hague) **2**: 180–5.
FARNER, D. S. (1961) 'Comparative physiology: Photoperiodicity' *A. Rev. Physiol.* **23**:
  71–96.
—— (1964) 'The photoperiodic control of reproductive cycles in birds' *Am. Scient.* **52**:
  137–56.
FEDOSEEV, G. A. (1965) '(A note on the ecology of seal reproduction in the northern sea of
  Okhotsk)' *Izv. Tikhoojeanskogo Nauchissled Inst. Rybn. Khoz. Okeanogr.* **59**: 212–16
  (from *Biol. Abstr.* **48**: 63246).
FINDLEY, J. S. and JONES, C. (1964) 'Seasonal distribution of the hoary bat' *J. Mammal.*
  45: 461–70.
FISKE, V. M. (1941) 'Effect of light on sexual maturation, estrous cycles and anterior
  pituitary of the rat' *Endocrinology* **29**: 189–96.
FITCH, H. S. (1947) 'Ecology of a cottontail rabbit (*Sylvilagus auduboni*) population in
  central California' *Calif. Fish Game* **33**: 159–84.
FLOWER, S. S. (1932) 'Notes on the recent mammals of Egypt, with a list of the species
  recorded from that kingdom' *Proc. zool. Soc. Lond.* **1932** (1): 369–450.
FLUX, J. E. C. (1964) 'Hare reproduction in New Zealand' *N.Z. Jl Agric.* **109**: 483–6.
—— (1965a) 'Incidence of ovarian tumours in hares in New Zealand' *J. Wildl. Mgmt.* **29**:
  622–4.
—— (1965b) 'Timing of the breeding season in the hare *Lepus europaeus* (Pallas) and
  rabbit *Oryctolagus cuniculus* (L.)' *Mammalia* **29**: 557–62.
—— (1967) 'Reproduction and body weights of the hare *Lepus europaeus* in New Zealand'
  N.Z. Jl Sci. **10**: 357–401.
FOA, C. (1900) 'La greffe des ovaires, en relation avec quelques questions de biologie
  generale' *Arch. ital. Biol.* **34**: 43–73.
FOREMAN, D. (1962) 'The normal reproductive cycle of the female prairie dog and the
  effects of light' *Anat. Rec.* **142**: 391–406.
FRASER, A. F. (1968) *Reproductive behaviour in ungulates* (Academic: London, New York),
  202 pp.
—— and LAING, A. H. (1966) 'Preliminary report on the induction of oestrus in sheep
  with fixed periods of dark-housing' *Vet. Rec.* **78**: 430–1.

FRENCH, C. E., MCEWEN, L. C., MAGRUDER, N. D., RADAR, T., LONG, T. A., and SWIFT, R. W. (1960) 'Responses of white-tailed bucks to added artificial light' *J. Mammal.* **41**: 23–9.

FRENCH, N. R. (1964) 'Analysis of reproduction in a black-tailed jackrabbit population' *Proc. 16th. Int. Congr. Zool.* **1**: 258.

——, MCBRIDE, R. and DITMER, J. (1965) 'Fertility and population density of the black-tailed jackrabbit' *J. Wildl. Mgmt.* **29**: 14–26.

FRIEDMAN, M. H. and FRIEDMAN, G. S. (1939) 'Gonadotrophic extracts from green leaves' *Am. J. Physiol.* **125**: 486–90.

—— —— (1940) 'The relation of diet to the restitution of the gonadotrophic hormone content of the discharged rabbit pituitary' *Am. J. Physiol.* **128**: 493–9.

FRITH, H. J. and SHARMAN, G. B. (1964) 'Breeding in wild populations of the red kangaroo, *Megaleia rufa*' *C.S.I.R.O. Wildl. Res.* **9**: 86–114.

FULLAGAR, P. J., JEWELL, P. A., LOCKLEY, R. M., and ROWLANDS, I. W. (1963) 'The Skomer vole *Clethrionomys glareolus skomerensis* and long-tailed field mouse *Apodemus sylvaticus* on Skomer island, Pembrokeshire in 1960' *Proc. zool. Soc. Lond.* **140**: 295–314.

GEIST, V. (1964) 'On the rutting behaviour of the mountain goat' *J. Mammal.* **45**: 551–68.

—— (1968) 'On delayed social and physical maturation in mountain sheep' *Can. J. Zool.* **46**: 899–904.

GILBERT, F. F. and BAILEY, E. D. (1967) 'The effect of visual isolation on reproduction in the female ranch mink' *J. Mammal.* **48**: 113–18.

GODDARD, J. (1966) 'Mating and courtship of the black rhinoceros (*Diceros bicornis* L.)' *East Afric. Wildl. J.* **4**: 69–75.

GODLEY, W. C., WILSON, R. L., and HURST, V. (1966) 'Effect of controlled environment on the reproductive performance of ewes' *J. Anim. Sci.* **25**: 212–16.

GOERTZ, J. W. (1965) 'Reproductive variation in cotton rats' *Am. Midl. Nat.* **74**: 329–40.

GOLLEY, F. B. (1961) 'Interaction of natality, mortality and movement during one annual cycle in a *Microtus* population' *Am. Midl. Nat.* **66**: 152–9.

GOODING, C. D. and LONG, J. L. (1957) 'Some fluctuations within rabbit populations in Western Australia' *J. Aust. Inst. agric. Sci.* **23**: 334–6.

GRAHN, D. and KRATCHMAN, J. (1963) 'Variation in neonatal death rate and birth weight in the United States and possible relations to environmental radiation, geology and altitude' *Am. J. hum. Genet.* **15**: 329–52.

GREENWALD, G. S. (1957) 'Reproduction in a coastal California population of the field mouse *Microtus californicus*' *Univ. Calif. Publs. Zool.* **54**: 421–41.

GREER, K. R., HAWKINS, W. W., and CATLIN, J. E. (1968) 'Experimental studies of controlled reproduction in elk (wapiti)' *J. Wildl. Mgmt.* **32**: 368–76.

GRINDELAND, R. E. and FOLK, G. E. (1962) 'Effects of cold exposure on the oestrous cycle of the golden hamster *Mesocricetus auratus*' *J. Reprod. Fert.* **4**: 1–6.

GUNN, R. M. C., SANDERS, R. N., and GRANGER, W. (1942) 'Studies in fertility in sheep. II. Seminal changes affecting fertility in rams' *Comm. Aust. Counc. sci. ind. Res. Bull.* **148**: 140 pp.

HADJIPIERIS, G. and HOLMES, W. (1966) 'Studies on feed intake and feed utilization by sheep 1. Voluntary feed intake of dry, pregnant and lactating ewes' *J. agric. Sci., Camb.* **66**: 217–23.

HAFEZ, E. S. E. (1952a) 'Studies on the breeding season and reproduction of the ewe, Part I. The breeding season in different environments' *J. agric. Sci., Camb.* **42**: 189–99.

—— (1952b) 'Studies on the breeding season and reproduction of the ewe, Part II. The breeding season in one locality' *J. agric. Sci., Camb.* **42**: 199–231.

—— (1952c) 'Studies on the breeding season and reproduction of the ewe, Part III. The breeding season and artificial light, Part IV. Studies on the reproduction of the ewe, Part V, Mating behaviour and pregnancy diagnosis' *J. agric. Sci., Camb.* **42**: 232–65.

HAFEZ, E. S. E. (1964) 'Environment and reproduction in domesticated species' *Int. Rev. gen. exp. Zool.* **1**: 113–64.

HALL, A. B. (1964) 'Musk-oxen in Jameson land and Scoresby land, Greenland' *J. Mammal.* **45**: 1–11.

HAMILTON, W. J. (1937) 'The biology of microtine cycles' *J. agric. Res.* **54**: 779–90.

—— (1941) 'Reproduction of the field mouse (*Microtus pennsylvanicus* Ord.)' *Mem. Cornell Univ. agric. Exp. Stn.* **237**: 1–23.

—— (1958) 'Early sexual maturity in the female long-tailed weasel' *Science, N.Y.* **127**: 1057.

—— (1962) 'Reproductive adaptations in the red tree mouse' *J. Mammal.* **43**: 486–504.

—— and EADIE, W. R. (1964) 'Reproduction in the otter *Lutra canadensis*' *J. Mammal.* **45**: 242–52.

HAMLETT, G. W. D. (1935) 'Delayed implantation and discontinuous development in mammals' *Q. Rev. Biol.* **10**: 432–7.

HAMMOND, J. (1949) 'Physiology of reproduction in relation to nutrition' *Br. J. Nutr.* **3**: 79–83.

—— and MARSHALL, F. H. A. (1952) 'The life cycle', chapter in *Marshall's Physiology of Reproduction* vol. 2, A. S. Parkes, Ed. (Longmans, London), 880 pp.

HAMMOND, J. JNR. (1944) 'On the breeding season in the sheep' *J. agric. Sci., Camb.* **34**: 97–105.

—— (1951) 'Control by light of reproduction in ferrets and mink' *Nature, Lond.* **167**: 150–1.

—— (1953) 'Photoperiodicity in animals: the role of darkness' *Science, N.Y.* **117**: 389–90.

HANNEY, P. (1964) 'The harsh-furred rat in Nyasaland' *J. Mammal.* **45**: 345–8.

—— (1965) 'The Muridae of Malawi' *J. Zool.* **146**: 577–633.

—— and MORRIS, B. (1962) 'Some observations on the pouched rat in Nyasaland' *J. Mammal.* **43**: 238–48.

HANSEN, R. M. (1960) 'Age and reproductive characteristics of mountain pocket gophers in Colorado' *J. Mammal.* **41**: 323–35.

HANSSON, A. (1947) 'The physiology of reproduction in mink (*Mustela vison* Schreb.) with special reference to delayed implantation' *Acta zool. Stockh.* **28**: 1–136.

—— (1956) 'Influence of rearing intensity on body development and milk production' *Proc. Br. Soc. Anim. Prod.* **1956**: 51–6.

HARRISON, J. L. (1952) 'Breeding rhythms of Selangor rodents' *Bull. Raffles Mus.* **24**: 109–31.

—— (1954) 'The moonlight effect on rat breeding' *Bull. Raffles Mus.* **25**: 166–70.

HARRISON, R. J. (1963) 'A comparison of factors involved in delayed implantation in badgers and seals in Great Britain', chapter in *Delayed Implantation*, A. C. Enders, Ed. (Univ. of Chicago Press, Chicago), 318 pp.

HARRISON-MATTHEWS, L. (1956) 'The hibernation of mammals' *Smithsonian Rep.* **1955**: 407–17.

HART, D. S. (1950) 'Photoperiodicity in Suffolk sheep' *J. agric. Sci., Camb.* **40**: 143–9.

—— (1951) 'Photoperiodicity in the female ferret' *J. exp. Biol.* **28**: 1–12.

HART, J. S., HEROUX, O., COTTLE, W. H., and MILLS, C. A. (1961) 'The influence of climate on metabolic and thermal responses of infant caribou' *Can. J. Zool.* **39**: 845–57.

HEALEY, M. C. (1967) 'Aggression and self-regulation of population size in deermice' *Ecology* **48**: 377–92.

HEAP, F. C., LODGE, G. A., and LAMMING, G. E. (1967) 'The influence of plane of nutrition in early pregnancy on the survival and development of embryos in the sow' *J. Reprod. Fert.* **13**: 269–79.

HEDIGER, J. (1952) 'Observations on reproductive behaviour in zoo animals', chapter in *Hormones, Psychology and Behaviour and Hormone Administration*, G. E. W. Wohlstenholme, Ed. (CIBA – London).

HEINSHOHN, G. E. (1966) 'Ecology and reproduction of the Tasmanian bandicoots (*Perameles gunni* and *Isoodon obesulus*)' *Univ. Calif. Publ. Zool.* **80**, 96 pp.

HELMREICH, R. L. (1960) 'Regulation of reproductive rate by intrauterine mortality in the deermouse' *Science, N.Y.* **132**: 417–18.

HEWER, H. R. (1960) 'Behaviour of the gray seal (*Halichoerus grypus* Fab.) in the breeding season' *Mammalia* **24**: 400–21.

—— and BACKHOUSE, K. M. (1960) 'A preliminary account of a colony of gray seals (*Halichoerus grypus* Fab.) in the southern inner Hebrides' *Proc. zool. Soc. Lond.* **134**: 157–95.

HEWSON, R. (1965) 'Population changes in the mountain hare, *Lepus timidus* L.' *J. Anim. Ecol.* **34**: 587–600.

HILL, M. and PARKES, A. S. (1933) 'Studies on the hypophysectomised ferret V. Effect of hypophysectomy on the response of the female ferret to additional illumination during anoestrus' *Proc. R. Soc.* (B) **113**: 537–40.

—— —— (1934) 'Effect of absence of light on the breeding season of the ferret' *Proc. R. Soc.* (B) **115**: 14–17.

HINKLEY, R. (1966) 'Effects of plant extracts in the diet of male *Microtus montanus* on cell types of the anterior pituitary' *J. Mammal.* **47**: 396–400.

HOCK, R. J. (1960) 'Seasonal variation in physiologic functions of Arctic ground squirrels and black bears' *Bull. Mus. comp. Zool. Harv.* **124**: 155–69.

HOFFMAN, R. A., HESTER, R. J., and TOWNS, C. (1965) 'Effect of light and temperature on the endocrine system of the golden hamster (*Mesocricetus auratus*)' *Comp. Biochem. Physiol.* **15**: 525–34.

—— and REITER, R. J. (1965a) 'Pineal gland: Influence on gonads of male hamsters' *Science, N.Y.* **148**: 1609–11.

—— —— (1965b) 'Influence of compensatory mechanisms and the pineal gland on dark-induced gonadal atrophy in male hamsters' *Nature, Lond.* **207**: 658–9.

HOFFMAN, R. A. (1958) 'The role of reproduction and mortality in population fluctuations of voles (*Microtus*)' *Ecol. Monogr.* **28**: 79–109.

HOFFMEISTER, D. F. (1963) 'The yellow-nosed cotton rat *Sigmodon ochrognathus* in Arizona' *Am. Midl. Nat.* **70**: 429–41.

HOLCOMB, L., SCHAIBLE, J., and RINGER, R. (1962) 'The effects of varied lighting on reproduction in mink' *Q. Bull. Mich. St. Univ. agric. Exp. Stn.* **44**: 666–78.

HOYTE, H. M. D. (1955) 'Observations on reproduction in some small mammals of Arctic Norway' *J. Anim. Ecol.* **24**: 412–25.

HUGHES, R. L. (1962) 'Reproduction in the macropod marsupial *Potorous tridactylus* (Kerr)' *Aust. J. Zool.* **10**: 193–224.

—— (1964) 'Sexual development and spermatozoon morphology in the male macropod marsupial *Potorous tridactylus* (Kerr)' *Aust. J. Zool.* **12**: 42–51.

—— and ROWLEY, I. (1966) 'Breeding season of female wild rabbits in natural populations in the Riverina and Southern Tableland districts of New South Wales' *C.S.I.R.O. Wildl. Res.* **11**: 1–10.

HUNT, T. P. (1959) 'Breeding habits of the swamp rabbit with notes on its life history' *J. Mammal.* **40**: 82–91.

HUNTER, G. L. (1961) 'Some effects of plane of nutrition on the occurrence of oestrus in Merino ewes' *Proc. IVth. Int. Congr. Anim. Reprod.* (*The Hague*) **2**: 197–201.

—— (1964) 'Effects of season and mating on oestrus and fertility in the ewe' *Proc. S. Afr. Soc. Anim. Prod.* **3**: 196–206.

IMANSHI, K. (1960) 'Social organization of subhuman primates in their natural habitat' *Curr. Anthrop.* **1**: 393–407.

INGLES, L. G. (1961) 'Natural history observations on the Audubon cottontail' *J. Mammal.* **22**: 227–50.

U

IURGENSON, P. B. '(Ecology of the lynx in forests of the central zone of the U.S.S.R.)'
    Zool. Zh. **34**: 609–20.
JACKSON, W. B. (1965) 'Litter size in relation to latitude in two murid rodents' Am. Midl.
    Nat. **73**: 245–7.
JACZEWSKI, Z. (1958) 'Reproduction of the European bison Bison bonasus (L.) in reserves'
    Acta theriol. **1**: 333–76.
JAMESON, E. W. (1955) 'Some factors affecting fluctuations of Microtus and Peromyscus'
    J. Mammal. **36**: 206–9.
JENKINS, D., WATSON, A., and MILLER, G. R. (1963) 'Population studies on red grouse,
    Lagopus lagopus scoticus (Lath.) in north-east Scotland' J. Anim. Ecol. **32**: 317–76.
JEWELL, P. A. (1966) 'Breeding season and recruitments in some British mammals confined
    on small islands' Symp. zool. Soc. Lond. **15**: 89–116.
JÖCHLE, W. (1956) '(On the effect of light on sexual activity and periodicity of rats)'
    Endokrinologie **33**: 129–38.
—— (1964) 'Trends in photophysiologic concepts' Ann. N.Y. Acad. Sci. **117**: 88–104.
JOHNSON, D. E. (1951) 'Biology of the elk calf, Cervus canadensis nelsoni' J. Wildl. Mgmt.
    15: 396–410.
JOHNSON, E. L. (1924) 'The relation of sheep to climate' J. agric. Res. **29**: 491–500.
JOHNSTON, R. F. and RUDD, R. L. (1957) 'Breeding of the salt marsh shrew' J. Mammal.
    38: 157–63.
JONES, E. C. and KROHN, P. L. (1961) 'The relationships between age, numbers of oocytes
    and fertility in virgin and multiparous mice' J. Endocrin. **21**: 469–95.
JONKEL, C. J. and WECKWIRTH, R. P. (1963) 'Sexual maturity and implantation of
    blastocysts in the wild pine marten' J. Wildl. Mgmt. **27**: 93–8.
JORDAN, P. A., SHELTON, P. C., and ALLEN, D. L. (1967) 'Numbers, turnover, and social
    structure of the Isle Royale wolf population' Am. Zool. **7**: 233–52.
JOUBERT, D. M. (1954) 'The influence of winter nutritional depressions on the growth,
    reproduction and production of cattle' J. agric. Sci., Camb. **44**: 5–66.
—— (1963) 'Puberty in farm animals' Anim. Breed. Abstr. **31**: 295–306.
JULANDER, O., ROBINETTE, W. L., and JONES, D. A. (1961) 'Relation of summer range
    condition to mule deer productivity' J. Wildl. Mgmt. **25**: 54–60.
KACZMARSKI, F. (1966) 'Bioenergetics of pregnancy and lactation in the bank vole'
    Acta theriol. **11**: 409–17.
KALABUKHOV, N. I. (1960) 'Comparative ecology of hibernating mammals' Bull. Mus.
    comp. Zool. Harv. **124**: 45–74.
KALELA, O. (1957) 'Regulation of reproductive rate in subarctic populations of the vole
    Clethrionomys rufocanus (Sund.)' Suomal. Tiedeakat. Toim. (Ser. IV) **34**: 1–60.
—— (1961) 'Seasonal change of habitat in the Norwegian lemming, Lemmus lemmus (L.)'
    Suomal. Tiedeakat. Toim. (Ser. IV) **55**: 1–72.
KALLEN, F. C. and WIMSATT, W. A. (1962) 'Circulating red cell volume in little brown
    bats in summer' Am. J. Physiol. **203**: 1182–4.
KEELEY, K. (1962) 'Prenatal influence on behaviour of offspring of crowded mice' Science,
    N.Y. **135**: 44–5.
KEITH, L. B., RONSTAD, O. J., and MESLOW, E. C. (1966) 'Regional differences in repro-
    ductive traits of the snowshoe hare' Can. J. Zool. **44**: 953–61.
KELLAS, L. M. (1955) 'Observations on the reproductive activities, measurements and
    growth rates of the dikdik (Rhynchotragus kirkii thomasi Newmann)' Proc. zool. Soc.
    Lond. **124**: 751–84.
KELSALL, J. P. (1958) 'Continued barren ground caribou studies' Wildl. Mgmt. Bull.,
    Ottawa **12**: 1–148.
—— (1968) The Migratory Barren-ground Caribou of Canada (Queen's Printer, Ottawa),
    340 pp.

KELSALL, J. P. and LOUGHREY, A. G. (1957–8) 'The barren-ground caribou, 1957–1958' *Canad. Wildl. Serv. Coop. Invest. Rep. No.* 2.

KERR, M. A. (1965) 'The age at sexual maturity in male impala' *Arnoldia* 1: 1–6.

KEYS, A., BROZEK, J., HENSCHEL, A., MICKELSON, O., and TAYLOR, H. L. (1960) *The Biology of Human Starvation* (Univ. of Minnesota Press, Minneapolis, Minn.), 763 pp.

KIHLSTROM, J. E. (1958) 'On some effects of light upon the activity of the male accessory glands in rabbits' *Arkiv. fur Zool.* 11: 379–82.

KIRKPATRICK, T. H. and MCEVOY, J. S. (1966) 'Studies of Macropodidae in Queensland 5. Effect of drought on reproduction in the grey kangaroo (*Macropus giganteus*)' *Qld. J. Agric. Anim. Sci.* 23: 439–42.

KLINE, P. D. (1962) 'Vernal breeding of cottontails in Iowa' *Proc. Iowa Acad. Sci.* 69: 244–52.

KNOWLTON, F. F. and MICHAEL, E. D. (1965) 'Placental ingestion and possible parturant mortality in wild white-tailed deer' *J. Mammal.* 46: 107.

KNUDSEN, B. (1962) 'Growth and reproduction of housemice at three different temperatures' *Oikos* 13: 1–14.

KOFORD, C. B. (1958) 'Prairie dogs, whitefaces and blue grama' *Wildl. Monogr.* No. 3: 78 pp.

—— (1965) 'Population dynamics of rhesus monkeys on Cayo Santiago', chapter in *Primate Behaviour; Field Studies of Monkeys and Apes*, I. de Vore, Ed. (Holt, Rinehart, and Winston, New York), 654 pp.

—— (1966) 'Population changes in rhesus monkeys: Cayo Santiago, 1960–1964' *Tulane Stud. Zool.* 13: 1–7.

KOSHKINA, T. V. (1965) '(Population density and its importance in regulating the abundance of the red vole)' *Bull. Moscow Soc. Nat. Biol. Sec.* 70: 5–19.

KOTT, E. and ROBINSON, W. L. (1963) 'Seasonal variation in litter size of the meadow vole in southern Ontario' *J. Mammal.* 44: 467–70.

KREBS, C. J. (1963) 'Lemming cycle at Baker Lake, Canada, during 1959–62' *Science, N.Y.* 140: 674–6.

—— (1964) 'The lemming cycle at Baker Lake, Northwest Territories, during 1959–1962' *Tech. Pap. Arct. Inst. N. Am.* 15: 104 pp.

—— (1966) 'Demographic changes in fluctuating populations of *Microtus californicus*' *Ecol. Monogr.* 36: 239–73.

KUHN, L. W., WICK, W. Q., and PEDERSON, R. J. (1966) 'Breeding nests of Townsends mole in Oregon' *J. Mammal.* 47: 239–49.

LAMMING, G. E., SALISBURY, G. W., HAYS, R. L., and KENDALL, K. A. (1954) 'The effect of incipient vitamin A deficiency on reproduction in the rabbit, Part I Decidua, ova and fertilization, Part II Embryonic and fetal development' *J. Nutr.* 52: 217–39.

LAMOND, D. R. (1959) 'Effect of stimulation derived from other animals of the same species on oestrous cycle in mice' *J. Endocr.* 18: 343–9.

—— (1962) 'Effect of season on hormonally induced ovulation in Merino ewes' *J. Reprod. Fert.* 4: 111–20.

—— (1964) 'Seasonal changes in the occurrence of oestrus following progesterone suppression of ovarian function in the Merino ewe' *J. Reprod. Fert.* 8: 101–14.

LANCASTER, J. B. and LEE, R. B. (1965) 'The annual reproductive cycle in monkeys and apes', chapter in *Primate Behaviour: Field Studies of Monkeys and Apes* I. de Vore, Ed. (Holt, Reinhart, and Winston, New York), 634 pp.

LAURIE, E. M. O. (1946) 'The reproduction of the house mouse (*Mus musculus*) living in different environments' *Proc. R. Soc.* (B) 133: 248–81.

LAVAUDEN, L. (1927) 'Quelques effects de la secheresse sur les vertebres superieurs de l'Afrique du nord' *C.R. Acad. Sci., Paris* 185: 1210–12.

LAWS, R. M. (1956) 'The elephant seal, III The physiology of reproduction' *Falkland Is. Depend. Surv. Sci. Rep. No.* **15**. 66 pp.

—— (1959) 'On the breeding season of the southern hemisphere fin whale, *Balaenoptera physalus* (Linn.)' *Norsk Hvalfangsttid.* **48**: 329–51.

—— (1961) 'Reproduction, growth and age of southern fin whales' *'Discovery' Rep.* **31**: 327–486.

—— (1966) 'Age criteria for the African elephant, *Loxodonta a. africana*' *East Afric. Wildl. J.* **4**: 1–37.

—— and CLOUGH, G. (1966) 'Observations on reproduction in the hippopotamus, *Hippopotamus amphibius* Linn.' *Symp. zool. Soc. Lond.* **15**: 117–40.

LECHLEITNER, R. R. (1959) 'Sex ratio and age classes and reproduction of the black-tailed jackrabbit' *J. Mammal.* **40**: 63–81.

LECYK, M. (1962a) 'The effect of the length of daylight on reproduction in the field vole *Microtus arvalis* (Pall.)' *Zoologica Pol.* **12**: 189–221.

—— (1962b) 'The dependence of breeding in the field vole (*Microtus arvalis* Pall.) on light intensity and wave length' *Zoologica Pol.* **12**: 255–68.

—— (1963) 'The effect of short daylength on sexual maturation in young individuals of the vole *Microtus arvalis* Pall.' *Zoologica Pol.* **13**: 77–86.

LEEGE, T. A. and WILLIAMS, R. M. (1967) 'Beaver productivity in Idaho' *J. Wildl. Mgmt.* **31**: 326–32.

LEES, J. L. (1966) 'Variations in the time of onset of the breeding season in clun ewes' *J. agric. Sci., Camb.* **67**: 173–9.

LESLIE, P. H., PERRY, J. S., and WATSON, J. S. (1946) 'The determination of the median body weight at which female rats reach maturity' *Proc. zool. Soc. Lond.* **115**: 473–88.

——, VENABLES, U. M., and VENABLES, L. S. V. (1952) 'The fertility and population structure of the brown rat (*Rattus norvegicus*) in cornricks and some other habitats' *Proc. zool. Soc. Lond.* **122**: 187–238.

LEUTHOLD, W. (1966) 'Variations in territorial behaviour in Uganda kob (*Adenota kob thomansi* Neumann 1896)' *Behaviour* **27**: 214–58.

LIDICKER, W. Z. (1965) 'Comparative study of density regulation in confined populations of four species of rodents' *Researches Popul. Ecol. Kyoto Univ.* **7**: 57–72.

—— (1966) 'Ecological observations on a feral house mouse population declining to extinction' *Ecol. Monogr.* **36**: 27–50.

LINTVAREVA, N. I. (1955) '(The effect of light on reproduction)' *Trudy. Vsesoyuz. Nauch. Issledovatel. Inst. Konev. Moscow Seljhozgig.*: 44–65 (from *Anim. Breed. Abstr.* **24**: 1491).

LLOYD, C. W. and WEISZ, J. (1966) 'Some aspects of reproductive physiology' *A. Rev. Physiol.* **28**: 267–310.

LLOYD, H. G. (1963) 'Intrauterine mortality in the wild rabbit (*Oryctolagus cuniculus* L.) in populations of low density' *J. Anim. Ecol.* **32**: 549–63.

—— (1964) 'Influences of environmental factors on some aspects of breeding in the wild rabbit *Oryctolagus cuniculus* (L.)' M.Sc. Thesis, University College of North Wales, Bangor.

—— (1967) 'Variations in fecundity in wild rabbit populations', chapter in *Effects of External Stimuli on Reproduction*, G. E. W. Wolstenholme and M. O'Connor, Eds. (Churchill, London), 107 pp.

LLOYD, M. (1967) 'Mean crowding' *J. Anim. Ecol.* **36**: 1–30.

LODGE, G. A., ELSLEY, F. W. H., and MACPHERSON, R. M. (1966) 'The effects of level of feeding of sows during pregnancy. 1. Reproductive performance' *Anim. Prod.* **8**: 29–38.

—— and HARDY, B. (1968) 'The influence of nutrition during oestrus on ovulation rate in the sow' *J. Reprod. Fert.* **15**: 329–32.

LONG, C. A. (1964) 'Comments on reproduction in the deer mouse of Wyoming' *Trans Kans. Acad. Sci.* **67**: 149–53.

LORD, R. D. (1960) 'Litter size and latitude in North American mammals' *Am. Midl. Nat.* **64**: 488–99.

—— (1961) 'Magnitudes of reproduction in cottontail rabbits' *J. Wildl. Mgmt.* **25**: 28–33.

LOWE, V. P. W. (1966) 'Observations on the dispersal of red deer on Rhum' *Symp. zool. Soc. Lond.* **18**: 211–28.

LUDBROOK, J. V. (1963) 'Desertion of a buffalo calf' *Puku* **1**: 216.

LUGG, D. J. (1966) 'Annual cycle of the Weddell seal in the Vestfold hills, Antarctica' *J. Mammal.* **47**: 317–22.

MACFARLANE, W. V., PENNYCUIK, P. R., YEATES, N. T. M., and THRIFT, E. (1959) 'Reproduction in hot environments', chapter in *Recent Progress in the Endocrinology of Reproduction*, C. W. Lloyd, Ed. (Academic Press, New York).

—— and SPALDING, D. (1960) 'Seasonal conception rates in Australia' *Med. J. Aust.* **23**: 121–4.

MAIN, A. R., SHIELD, J. W., and WARING, H. (1959) 'Recent studies on marsupial ecology' chapter in *Biogeography and Ecology in Australia*, A. Keast, Ed. (Junk, The Hague), 640 pp.

MANN, T., ROWSON, L. E. A., SHORT, R. V., and SKINNER, J. D. (1967) 'The relationship between nutrition and androgenic activity in pubescent twin calves, and the effect of orchitis' *J. Endocrin.* **38**: 455–68.

MANSFIELD, A. W. (1958) 'The breeding behaviour and reproductive cycle of the Weddell seal (*Leptonychotes weddelli* Leison)' *Falkland Is. Depend. Surv. Sci. Rep.* No **18**: 41 pp.

MAQSOOD, M. and PARSONS, U. (1954) 'Influence of continuous light or darkness on sexual development in the male rabbit' *Experientia* **10**: 188–9.

MARBURGER, R. G. and THOMAS, J. W. (1965) 'A die-off in white-tailed deer of the central mineral region of Texas' *J. Wildl. Mgmt.* **29**: 706–16.

MARSDEN, H. M. and BRONSON, F. H. (1965a) 'Strange male block to pregnancy: its absence in inbred mouse strains' *Nature, Lond.* **207**: 878.

—— —— (1965b) 'The synchrony of oestrus in mice: relative roles of the male and female environments' *J. Endocr.* **32**: 313–19.

—— and CONAWAY, C. H. (1963) 'Behaviour and the reproductive cycle in the cottontail' *J. Wildl. Mgmt.* **27**: 161–70.

MARSHALL, A. J. (1947) 'The breeding cycles of an equatorial bat (*Pteropus giganteus*) of Ceylon' *Proc. Linn. Soc.* **159**: 103–11.

—— (1967) 'Origin of delayed implantation in marsupials' *Nature, Lond.* **216**: 192–3.

—— and CORBET, P. S. (1959) 'The breeding biology of equatorial vertebrates: Reproduction of the bat *Chaerophon hindei* Thomas at 0° 26′ N' *Proc. zool. Soc. Lond.* **132**: 607–16.

—— and WILKINSON, O. (1956) 'Reproduction in the Orkney vole (*Microtus orcadensis*) under a six hour daylength and other conditions' *Proc. zool. Soc. Lond.* **126**: 391–5.

MARSHALL, F. H. A. (1937) 'On the change over in oestrous cycles in animals after transference across the equator, with further observations on the incidence of the breeding seasons and the factors controlling sexual periodicity' *Proc. R. Soc.* (B) **122**: 413–28.

—— (1940) 'The experimental modification of the oestrous cycle in the ferret by different intensities of light irradiation and other methods' *J. exp. Biol.* **17**: 139–46.

—— (1942) 'Exteroceptive factors in sexual periodicity' *Biol. Rev.* **17**: 68–89.

—— and BOWDEN, F. P. (1934) 'The effect of irradiation with different wavelengths on the oestrous cycle of the ferret, with remarks on the factors controlling sexual periodicity' *J. exp. Biol.* **11**: 409–22.

—— —— (1936) 'The further effects of irradiation on the oestrous cycle of the ferret' *J. exp. Biol.* **13**: 383–6.

MARTINET, L. (1963) 'Etablissement de la spermatogenese ches le campagnol de champs (*Microtus arvalis*) en fonction de la duree quotidienne d'eclairement' *Ann. Biol. anim. Biochim. Biophys.* **3**: 343–52.

MATHISEN, O., BAADE, R. T., and LOPP, R. J. (1962) 'Breeding habits, growth and stomach contents of the Steller sea lion in Alaska' *J. Mammal.* **43**: 469–77.

MAULEON, P. and ROUGEOT, J. (1962) Régulation des saisons sexuelles chez des brebis de races différentes au moyen de divers rythmes lumineux. *Ann. Biol. anim. Biochim. Biophys.* **2**: 209–22.

MAZESS, R. B. (1965) 'Neonatal mortality and altitude in Peru' *Am. J. phys. Anthrop.* **23**: 209–14.

MCCANCE, I. and ALEXANDER, G. (1959) 'The onset of lactation in the Merino ewe and its modification by nutritional factors' *Aust. J. agric. Res.* **10**: 699–719.

MCCARLEY, H. (1966) 'Annual cycle, population dynamics and adaptive behaviour of *Citellus tridecemlineatus*' *J. Mammal.* **47**: 294–316.

MCCLURE, T. J. (1961a) 'An apparent nutritional lacational stress in fertility in dairy herds' *N.Z. vet. J.* **9**: 107–12.

—— (1961b) 'Uterine pathology of temporary fasted pregnant mice' *J. comp. Path. Ther.* **71**: 16–19.

—— (1961c) 'Pathogenesis of early embryonic mortality caused by fasting pregnant rats and mice for short periods' *J. Reprod. Fert.* **2**: 381–6.

—— (1962) 'Infertility in female rodents caused by temporary inanition at or about the time of implantation' *J. Reprod. Fert.* **4**: 241.

—— (1963) 'Infertility in mice caused by nutritional stress before mating' *Nature, Lond.* **199**: 504–5.

—— (1965) 'Experimental evidence for the occurrence of nutritional infertility in otherwise clinically healthy pasture-fed lactating dairy cows' *Res. vet. Sci.* **6**: 202–8.

—— (1966) 'Infertility in mice caused by fasting at about the time of mating I. Mating behaviour and littering rates' *J. Reprod. Fert.* **12**: 243–8.

—— (1967) 'Infertility in mice caused by fasting at about the time of mating' II. Pathological changes *J. Reprod. Fert.* **13**: 387–91.

MCCULLOCH, C. Y. and INGLIS, J. M. (1961) 'Breeding periods of the Ord kangaroo rat' *J. Mammal.* **42**: 337–44.

MCDONALD, R. L. (1966) 'Lunar and seasonal variations in obstetric factors' *J. genet. Psychol.* **108**: 81–8.

MCDOUGALL, W. A. (1946) 'An investigation of the rat problem in the Queensland cane fields 4. Breeding and life histories' *Qld. J. agric. Sci.* **3**: 1–43.

MCKEEVER, S. (1963) 'Seasonal changes in body weight, reproductive organs, pituitary, adrenal glands, thyroid gland and spleen of the Belding ground squirrel (*Citellus beldingi*)' *Am. J. Anat.* **113**: 153–67.

—— (1966) 'Reproduction in *Citellus beldingi* and *Citellus lateralis* in northeastern California' *Symp. zool. Soc. Lond.* **15**: 364–85.

MCLEOD, J. A. and BONDAR, J. F. (1952) 'Studies on the biology of the muskrat in Manitoba I. Oestrous cycle and breeding season' *Can. J. Zool.* **30**: 243–53.

MEANS, T. M., ANDREWS, F. M., and FONTAINE, W. E. (1959) 'Environmental factors in the induction of estrus in sheep' *J. Anim. Sci.* **18**: 1388–96.

MEASROCH, V. (1954) 'Growth and reproduction in the females of two species of gerbil. *Tatera brantsii* (A. Smith) and *Tatera afra* (Gray)' *Proc. zool. Soc. Lond.* **124**: 631–58.

MENAKER, M. and ESKIN, A. (1967) 'Circadian clock in photoperiodic time measurement: A test of the Bunning hypothesis' *Science, N.Y.* **157**: 1182–5.

MERCIER, E. and SALISBURY, G. W. (1947a) 'Seasonal variations in hours of daylight associated with fertility level of cattle under natural breeding conditions' *J. Dairy Sci.* **30**: 747–56.

MERCIER, E. and SALISBURY, G. W. (1947b) 'Fertility level in artificial breeding associated with season, hours of daylight and the age of cattle' *J. Dairy Sci.* **30**: 817–26.

MESCHAKS, P. and NORDKVIST, M. (1967) 'Sexual cycle in the reindeer male' *Acta vet. scand.* **3**: 151–62.

MEYER, B. J. and MEYER, R. K. (1944) 'The effect of light on maturation and the estrous cycle of the cotton rat, *Sigmodon hispidus hispidus*.' *Endocrinology* **34**: 276–81.

MILLER, M. A. (1946) 'Reproductive rates and cycles in the pocket gopher' *J. Mammal.* **27**: 335–58.

MILLER, R. S. (1963) 'Weights and color phases of black bear cubs' *J. Mammal.* **44**: 129.

MILLS, C. A. (1939) *Medical Climatology* (C. C. Thomas, Springfield, Ill.), 296 pp.

MITCHELL, G. J. (1967) 'Minimum breeding age of female pronghorn antelope' *J. Mammal.* **48**: 489–90.

MITCHELL, O. G. (1959) 'The reproductive cycle of the male Arctic ground squirrel' *J. Mammal.* **40**: 45–53.

MOORE, C. R., SIMMONS, G. F., WELLS, L. J., ZALESKY, M., and NELSON, W. O. (1934) 'On the control of reproductive activity in an annually breeding mammal *Citellus tridecemlineatus*' *Anat. Rec.* **60**: 279–89.

MORLEY, F. H. W. (1954) 'Prenatal loss in Merino sheep' *Aust. vet. J.* **30**: 125–8.

MORRIS, D. (1965) *The Mammals* (Hodder & Stoughton, London), 448 pp.

MORRIS, R. D. (1968) 'Effects of endrin feeding on survival and reproduction in the deer-mouse *Peromyscus maniculatus*' *Can. J. Zool.* **46**: 951–8.

MORTON, G. H. and CHEATUM, E. L. (1946) 'Regional differences in breeding potential of white-tailed deer in New York' *J. Wildl. Mgmt.* **10**: 242–8.

MORTON-BOYD, J., LOCKIE, J. D., and HEWER, H. R. (1962) 'The breeding colony of gray seals on North Rona, 1959' *Proc. zool. Soc. Lond.* **138**: 257–77.

MOULE, G. R. (1950a) 'Some problems of sheep breeding in semi-arid tropical Queensland' *Aust. vet. J.* **26**: 29–37.

—— (1950b) 'The influence of a rapid decrease in the hours of daylight on the sexual desire of Merino rams' *Aust. vet. J.* **26**: 84–7.

—— (1954) 'Observations on mortality amongst lambs in Queensland' *Aust. vet. J.* **30**: 153–71.

—— (1963) 'Postpubertal nutrition and reproduction in the male' *Aust. vet. J.* **39**: 299–304.

——, BRADEN, A. W. H., and LAMOND, D. R. (1963) 'The significance of oestrogens in pasture plants in relation to animal production' *Anim. Breed. Abstr.* **31**: 139–57.

MOUNTFORD, M. D. (1968) 'The significance of litter size' *J. Anim. Ecol.* **37**: 363–7.

MOUSTGAARD, J. (1959) 'Nutrition and reproduction in domestic animals', chapter in *Reproduction in Domestic Animals*, H. H. Cole and P. T. Cupps, Eds. (Academic Press, New York), 451 pp.

MULLER-VELTEN, H. (1966) 'Uber den angstgeruch bei der Housmaus' *Z. vergl. Physiol.* **52**: 401–29.

MURPHY, D. A. and COATES, J. A. (1966) 'Effects of dietary protein on deer' *Trans. N. Am. Wildl. Conf.* **31**: 129–39.

MURRAY, K. F. (1965) 'Population changes during the 1957–58 vole (*Microtus*) outbreak in California' *Ecology* **46**: 163–71.

MUTERE, F. A. (1967) 'The breeding biology of equatorial vertebrates: reproduction in the fruit bat *Eidolon helvum* at latitude 0° 20' N.' *J. Zool.* **153**: 153–61.

MYERS, K. (1958) 'Further observations on the use of field enclosures for the study of the wild rabbit, *Oryctolagus cuniculus* (L.)' *C.S.I.R.O. Wildl. Res.* **3**: 40–9.

—— (1964) 'Influence of density on fecundity, growth rates and mortality in the wild rabbit' *C.S.I.R.O. Wildl. Res.* **9**: 134–7.

—— (1966) 'The effects of density on sociality and health in mammals' *Proc. ecol. Soc. Aust.* **1**: 40–64.

MYERS, K. and PARKER, B. S. (1965) 'A study of the biology of the wild rabbit in climati-
cally different regions in Eastern Australia 1. Patterns of distribution.' C.S.I.R.O.
Wildl. Res. 10: 1–32.

—— and POOLE, W. E. (1959) 'A study of the biology of the wild rabbit, Oryctolagus cuniculus (L.) in confined populations 1. The effects of density on home range and the
formation of breeding groups' C.S.I.R.O. Wildl. Res. 4: 14–26.

—— —— (1962) 'A study of the biology of the wild rabbit, Oryctolagus cuniculus (L.) in
confined populations. III. Reproduction' Aust. J. Zool. 10: 225–67.

—— and SCHNEIDNER, E. C. (1964) 'Observations on reproduction, mortality and behaviour in a small, free-living population of wild rabbits' C.S.I.R.O. Wildl. Res. 9: 138–43.

MYKYTOWYCZ, R. (1959) 'Social behaviour of an experimental colony of wild rabbits,
Oryctolagus cuniculus (L.) II. First breeding season' C.S.I.R.O. Wildl. Res. 4: 1–13.

—— (1960) 'Social behaviour of an experimental colony of wild rabbits, Oryctolagus
cuniculus (L.) III. Second breeding season' C.S.I.R.O. Wildl. Res. 5: 1–20.

—— (1961) 'Social behaviour of an experimental colony of wild rabbits, Ocryctolagus cuniculus (L.) IV. Conclusion: Outbreak of myxomatosis, third breeding season and starvation
C.S.I.R.O. Wildl. Res. 6: 142–55.

—— (1964) 'Territoriality in rabbit populations' Aust. Natur. Hist. 14: 326–9.

—— (1965) 'Further observations on the territorial function and histology of the submandibular (chin) glands in the rabbit, Oryctolagus cuniculus (L.)' Anim. Behav. 13:
400–12.

—— and GAMBALE, S. (1965) 'A study of the interwarren activities and dispersal of wild
rabbits, Oryctolagus cuniculus (L.) living in a 45-acre paddock' C.S.I.R.O. Wildl. Res. 10:
111–23.

NEAL, B. J. (1965) 'Reproductive habits of round-tailed and Harris antelope ground
squirrels' J. Mammal. 46: 200–6.

NEGUS, N. C. and PINTER, A. J. (1966) 'Reproductive responses of Microtus montanus to
plants and plant extracts in the diet' J. Mammal. 47: 596–601.

NEWSOME, A. E. (1965) 'Reproduction in natural populations of the red kangaroo Megaleia
rufa in central Australia' Aust. J. Zool. 13: 735–59.

—— (1966) 'The influence of food on breeding in the red kangaroo in central Australia'
C.S.I.R.O. Wildl. Res. 11: 187–96.

NEWSON, J. (1964) 'Reproduction and prenatal mortality of snowshoe hares in Manitoulin
Island, Ontario' Can. J. Zool. 42: 987–1005.

NEWSON, R. (1963) 'Differences in numbers, reproduction and survival between two
neighbouring populations of bank voles (Clethrionomys glareolus)' Ecology 44: 110–20.

NEVO, E. and AMIR, E. (1964) 'Geographic variation in reproduction and hibernation
patterns of the forest dormouse' J. Mammal. 45: 69–87.

NIEVERGELT, B. (1966) 'Unterschiede in der Setzzeit bein Alpensteinbock (Capra ibex
L.)' Rev. Suiss. Zool. 73: 446–54.

NISHIKAWA, Y. and HORIE, T. (1952) '(Studies on the effect of daylength on the reproductive function in horses II. Effect of daylength on the function of the testes)' Bull. nat.
Inst. agric. Sci. (Japan) Ser. G. 3: 45 (from Anim. Breed. Abstr. 22: 439).

——, SUGIE, T., and HARADA, N. (1952) '(Studies on the effect of daylength on reproductive function in horses I. Effect of daylength on the function of the ovary)' Bull. nat.
Inst. agric. Sci. (Japan) Ser. G. 3: 35–44 (from Anim. Breed. Abstr. 22: 436).

OLSEN, P. F. (1959) 'Muskrat breeding biology at Delta, Manitoba' J. Wildl. Mgmt. 23:
40–52.

ORLOVA. A. F. (1955) '(The influence of a dry summer on the times of propagation of the
Siberian suslik)' C.R. Acad. Sci. U.S.S.R. 105: 1368–70.

ORTAVANT, R. (1961) 'Responses spermatogenetiques du belier a differentes durees
d'eclairment' Proc. IVth. Int. Congr. Anim. Reprod. (the Hague) 2: 236–42.

ORTAVANT, R., MAULEON, P., and THIBAULT, C. (1964) 'Photoperiodic control of gonadal and hypophyseal activity in domestic animals' *Ann. N.Y. Acad. Sci.* **117**: 157–93.

ORTIZ, E. (1947) 'Postnatal development of the reproductive system of the golden hamster (*Cricetus auratus*) and its reactivity to hormones' *Physiol. Zoöl.* **20**: 45–66.

OSBORNE, V. E. (1966) 'Analysis of the pattern of ovulation as it occurs in the annual reproductive cycle of the mare in Australia' *Aust. vet. J.* **42**: 149–54.

PARKES, A. S. (1924) 'Fertility in mice' *Br. J. exp. Biol.* **1**: 21–31.

—— (1963) 'The exocrinology of reproduction', chapter in *Perspectives in Biology*, C. F. Cori, V. C. Foglia, L. F. Leloir, and S. Ochoa, Eds. (Elsevier, Amsterdam).

—— (1966) 'Activation of the gonads', chapter in *Marshall's Physiology of Reproduction* vol. 3, A. S. Parkes, Ed. (Longman's, London), 1,168 pp.

—— and BRUCE, H. M. (1961) 'Olfactory stimuli in mammalian reproduction' *Science, N.Y.* **134**: 1049–54.

—— —— (1962) 'Pregnancy-block in female mice placed in boxes soiled by males' *J. Reprod. Fert.* **4**: 303–8.

PARSONS, S. D. and HUNTER, G. L. (1967) 'Effect of the ram on duration of oestrus in the ewe' *J. Reprod. Fert.* **14**: 61–70.

PATRIC, E. F. (1962) 'Reproductive characteristics of red-backed mouse during years of differing population densities' *J. Mammal.* **43**: 200–5.

PEARSON, O. P. (1963) 'History of two local outbreaks of feral house mice' *Ecology* **44**: 540–9.

—— and ENDERS, R. K. (1944) 'Duration of pregnancy in certain mustelids' *J. exp. Zool.* **95**: 21–35.

PERRY, J. S. (1945) 'Reproduction of the wild brown rat (*Rattus norvegicus* Erzleben)' *Proc. zool. Soc. Lond.* **115**: 19–46.

—— and ROWLANDS, I. W. (1962) 'The ovarian cycle in vertebrates', chapter in *The Ovary*, vol. 1, S. Zuckerman, Ed. (Academic Press, New York), 619 pp.

PETRUSEWICZ, K. (1958) 'Investigation of experimentally induced population growth' *Ekol. pol.* (Ser. A) **5**: 281–309.

PILTON, P. E. (1961) 'Reproduction in the great grey kangaroo' *Nature, Lond.* **189**: 984–5.

PIMLOTT, D. H. (1959) 'Reproduction and productivity of Newfoundland Moose' *J. Wildl. Mgmt.* **23**: 381–401.

—— (1967) 'Wolf predation and ungulate populations' *Am. Zool.* **7**: 267–78.

PINTER, A. J. and NEGUS, N. C. (1965) 'Effects of nutrition and photoperiod on reproductive physiology of *Microtus montanus*' *Am. J. Physiol.* **208**: 633–38.

POLIKARPOVA, E. F. (1945) '(The influence of external factors on the maturation of ova in sows)' *Ref. Rab. Ucrezden. Otd. biol. Nauk. Akad. Nauk. U.S.S.R.* (**1941–43**): 214–16 (from *Anim. Breed. Abstr.* **14**: 159).

POOLE, W. E. (1960) 'Breeding of the wild rabbit, *Oryctolagus cuniculus* (L.) in relation to the environment' *C.S.I.R.O. Wildl. Res.* **5**: 21–43.

PRAKASH, I. (1958) 'The breeding season of the monkey *Macaca mulatta* (Zimmerman) in Rajasthan' *J. Bombay nat. Hist. Soc.* **55**: 154.

—— (1960) 'Breeding of mammals in Rajasthan desert, India' *J. Mammal.* **41**: 386–9.

PRASAD, M. R. N. (1956) 'Reproductive cycle of the male Indian gerbille *Tatera indica cuvierii* (Waterhouse)' *Acta Zool. Stockh.* **37**: 87–122.

PRICE, A. B. and ORTIZ, E. (1944) 'Relation of age to reactivity in the reproductive system of the rat' *Endocrinology* **34**: 215–39.

PRICE, E. O. (1966) 'Influence of light on reproduction in *Peromyscus maniculatus gracilis*' *J. Mammal.* **47**: 343–4.

PRUITT, W. O. (1961) 'On post-natal mortality in Barren-ground caribou' *J. Mammal.* **42**: 550–1.

QUAY, W. B. (1960) 'The reproductive organs of the collared lemming under diverse temperature and light conditions' *J. Mammal.* **41**: 74–89.

RACZYNSKI, J. (1964) 'Studies on the European hare V. Reproduction' *Acta theriol.* **9**: 305–52.

RADFORD, H. M. (1961) 'Photoperiodism and sexual activity in Merino ewes I. The effect of continuous light on the development of sexual activity II. The effect of equinotal light on sexual activity' *Aust. J. agric. Res.* **12**: 139–53.

—— and WATSON, R. H. (1957) 'Influence of rams on ovarian activity and oestrus in Merino ewes in the spring and early summer' *Aust. J. agric. Res.* **8**: 460–70.

RAJAKOSKI, E. (1961) 'Seasonal variations of the ovarian follicular system in cattle' *Proc. IVth. Int. Congr. Anim. Reprod. (the Hague)* **2**: 186–9.

RAND, R. W. (1955) 'Reproduction in the female fur seal, *Arctocephalus pusillus* (Schreber)' *Proc. zool. Soc. Lond.* **124**: 717–40.

RANSOM, A. B. (1966) 'Breeding seasons of white-tailed deer in Manitoba' *Can. J. Zool.* **44**: 59–62.

—— (1967) 'Reproductive biology of white-tailed deer in Manitoba' *J. Wildl. Mgmt.* **31**: 114–23.

RASEK, A. (1964) '(Reproduction of the kulan, *Equus hemionus onager* Boddaert, on Barsa-Keljmes island (Aral sea))' *Vestn. csl. Spolec. zool.* **28**: 89–95 (from *Anim. Breed. Abstr.* **33**: 2013).

RASHEK, V. L. (1963) '(Notes on the reproduction of the saiga on Barsa-Kelmes island in the Aral sea)' *T. Inst. Zool. Akad. Nauk. Kazakhsk. S.S.R.* **20**: 164–93 (from *Biol. Abstr.* **46**: 18994).

RAUSCH, R. (1950) 'Observations on a cyclic decline of lemmings (*Lemmus*) on the Arctic coast of Alaska in the spring of 1949' *Arctic* **3**: 166–77.

RAY, D. E. and MCCARTY, J. W. (1965) 'Effect of temporary fasting on reproduction in gilts' *J. Anim. Aci.* **24**: 660–3.

REYNOLDS, H. G. (1960) 'Life history notes on Merriams kangaroo rat in southern Arizona' *J. Mammal.* **41**: 48–58.

REYNOLDS, J. K. and STINSON, R. H. (1959) 'Reproduction of the European hare in southern Ontario' *Can. J. Zool.* **37**: 627–32.

RICE, D. W. (1957) 'Life history and ecology of *Myotis austroriparius* in Florida' *J. Mammal.* **38**: 15–32.

ROBERTS, J. C., HOCK, R. J., and SMITH, R. E. (1966) 'Seasonal metabolic responses of deermice (*Peromyscus*) to temperature and altitude' *Fed. Proc.* **25**: 1275–83.

ROBINETTE, W. L. (1956) 'Productivity – the annual crop of mule deer', chapter in *The Deer of North America*, W. P. Taylor, Ed. (Stackpole Co., Harrisburg, Penn. and Wildlife Management Inst., Washington), 668 pp.

—— and CHILD, G. F. T. (1964) 'Notes on the biology of the lechwe, *Kobus leche*' Puku **2**: 84–117.

—— and GASHWILER, J. S. (1950) 'Breeding season, productivity and fawning period of the mule deer in Utah' *J. Wildl. Mgmt.* **14**: 457–69.

——, ——, JONES, D. A., and CRANE, H. S. (1955) 'Fertility of mule deer in Utah' *J. Wildl. Mgmt.* **19**: 115–36.

——, ——, and MORRIS, O. W. (1961) 'Notes on cougar productivity and life history' *J. Mammal.* **42**: 204–17.

ROBINSON, T. J. (1950) 'The control of fertility in sheep Part I. Hormonal therapy in the induction of pregnancy in the anoestrous ewe' *J. agric. Sci., Camb.* **40**: 275–307.

—— (1951) 'Reproduction in the ewe' *Biol. Rev.* **26**: 121–57.

RONGSTAD, O. J. (1965) 'A life history study of the thirteen-lined ground squirrel in southern Wisconsin' *J. Mammal.* **46**: 76–87.

ROSSDALE, P. D. and SHORT, R. V. (1967) 'The time of foaling of thoroughbred mares' *J. Reprod. Fert.* **13**: 341–3.

ROWAN, W. and KEITH, L. B. (1956) 'Reproductive potential and sex ratios of snowshoe hares in northern Alberta' *Can. J. Zool.* **34**: 273–81.

ROWE, F. P., TAYLOR, E. J., and CHUDLEY, A. H. J. (1964) 'The effect of crowding on the reproduction of the house mouse (*Mus musculus* L.) living in corn-ricks' *J. Anim. Ecol.* **33**: 477–83.

ROWLANDS, I. W. and HEAP, R. B. (1966) 'Histological observations on the ovary and progesterone levels in the coypu, *Myocastor coypus*' *Symp. zool. Soc. Lond.* **15**: 335–52.

—— and PARKES, A. S. (1935) 'The reproductive processes of certain mammals VIII Reproduction in foxes (*Vulpes spp.*)' *Proc. zool. Soc. Lond.* (**1935**): 823–41.

RUSSELL, F. C. (1948) 'Diet in relation to reproduction and the viability of the young, Part I Rats and other laboratory animals' *Common. Bur. Anim. Nut. Tech. Comm.* No. **16**: 99 pp.

RUSSELL, K. R. (1966) 'Effects of a common environment on cottontail ovulation rates' *J. Wildl. Mgmt.* **30**: 819–27.

SADLEIR, R. M. F. S. (1961) 'Age estimation by joeys of the euro (*Macropus robustus*) in Western Australia' *Aust. J. Zool.* **11**: 241–9.

—— (1965a) 'Reproduction in two species of kangaroo (*Macropus robustus* and *Megaliea rufa*) in the arid Pilbara region of Western Australia' *Proc. zool. Soc. Lond.* **145**: 239–61.

—— (1965b) 'The relationship between agonistic behaviour and population changes in the deermouse, *Peromyscus maniculatus* (Wagner)' *J. Anim. Ecol.* **34**: 331–52.

—— and SHIELD, J. W. (1960) 'Delayed birth in marsupial macropods – The euro, the tammar and the marloo' *Nature, Lond.* **185**: 335.

SAKAI, T. (1963) '(Reproductive organs in rats housed in continuous darkness)' *J. Physiol. Soc., Japan* **25**: 132–9 (from *Anim. Breed. Abstr.* **32**: 2335).

SAMKOV, J. A. (1964) '(Group breeding of the Arctic fox)' *Krolikovodstro i. Zverovodstvo* **7**: 5. (from *Anim. Breed. Abstr.* **33**: 1600).

SCHENKEL, R. (1966a) 'On sociology and behaviour in impala (*Aepyceros melampus* Lichtenstein)' *E. Afric. Wildl. J.* **4**: 99–114.

—— (1966b) 'Play, exploration and territoriality in the wild lion' *Symp. zool. Soc. Lond.* **18**: 18–22.

SCHILLER, E. L. (1956) 'Ecology and health of *Rattus* at Nome, Alaska' *J. Mammal.* **37**: 181–8.

SCHINDLER, J. (1954) 'Seasonal fluctuations in fertility and other characteristics of bull semen in Israel' *Bull. Res. Coun. Israel (Jerusalem)* **4**: 184–7.

SCHINKEL, P. G. (1954) 'The effect of the ram on the incidence and occurrence of oestrus in ewes' *Aust. vet. J.* **30**: 189–95.

—— (1963) 'Nutrition and sheep production—a review' *Proc. Wld. Conf. Prod.* **1**: 199–248.

SCHULTZ, J. R., SPEER, V. C., HAYS, V. W., and MELAMPY, R. M. (1966) 'Influence of feed intake and progestogen on reproductive performance in swine' *J. Anim. Sci.* **25**: 157–60.

SCOTT, P. P. and SCOTT, M. G. (1964) 'Vitamin A and reproduction in the cat' *J. Reprod. Fert.* **8**: 270–1.

—— and LLOYD-JACOB, A. (1959) 'Reduction in the anoestrous period of laboratory cats by increased illumination' *Nature, Lond.* **184**: 2022.

SEALANDER, J. A. (1964) 'The influence of body size, season, sex, age, and other factors upon some blood parameters in small mammals' *J. Mammal.* **45**: 598–616.

SEVERINGHAUS, C. W. and CHEATUM, E. L. (1956) 'Life and times of the white-tailed deer', chapter in *The Deer of North America*, W. P. Taylor, Ed. (Stackpole Co., Harrisburg, Penn. and Wildlife Management Inst., Washington), 668 pp.

SHALASH, M. R. and SALAMA, A. (1961) 'Seasonal variation in the ovarian activity of the buffalo-cow' *Proc. IVth. Int. Congr. Anim. Reprod. (the Hague)* **2**: 190–1.

SHARMAN, G. B., CALABY, J. H., and POOLE, W. E. (1966) 'Patterns of reproduction in female diprotodont marsupials' *Symp. zool. Soc. Lond.* **15**: 205–32.

—— and CLARK, M. J. (1967) 'Inhibition of ovulation by the corpus luteum in the red kangaroo, *Megaleia rufa*' *J. Reprod. Fert.* **14**: 129–37.

——, FRITH, H. J., and CALABY, J. (1964) 'Growth of the pouch young, tooth eruption and age determination in the red kangaroo, *Megaleia rufa*' *C.S.I.R.O. Wildl. Res.* **9**: 20–49.

SHELTON, M. (1960) 'Influence of the presence of a male goat on the initiation of estrous cycling and ovulation of Angora does' *J. Anim. Sci.* **19**: 368–75.

SHEVCHENKO, V. L. (1963) '(Reproduction and change in population of *Lagurus lagurus* (Myomorpha) in the Urals)' *Zool. Zh.* **42**: 114–24.

SHEPPE, W. (1963) 'Population structure of the deermouse, *Peromyscus*, in the Pacific northwest' *J. Mammal.* **44**: 180–5.

SHIELD, J. W. (1964) 'A breeding season difference in two populations of the Australian macropod marsupial *Setonix brachyurus*' *J. Mammal.* **45**: 616–25.

—— and WOOLLEY, P. (1961) 'Age estimation by measurement of pouch young of the quokka, *Setonix brachyurus*' *Aust. J. Zool.* **9**: 14–23.

—— —— (1963) 'Population aspects of delayed birth in the quokka' *Proc. zool. Soc. Lond.* **141**: 783–9.

SHIPP, E., KEITH, K., HUGHES, R. L., and MYERS, K. (1963) 'Reproduction in a free-living population of domestic rabbits *Oryctolagus cuniculus* (L.) on a sub-Antarctic island' *Nature, Lond.* **200**: 858–60.

SHORT, R. V. and HAY, M. F. (1966) 'Delayed implantation in the roe deer, *Capreolus capreolus*' *Symp. zool. Soc. Lond.* **15**: 173–94.

—— and MANN, T. (1966) 'The sexual cycle of a seasonally breeding mammal, the roebuck, *Capreolus capreolus*' *J. Reprod. Fert.* **12**: 337–51.

SHUBIN, N. G. (1964) '(Reproduction of the chipmunk in the Tom river basin)' *Zool. Zh.* **43**: 910–17.

SIMKIN, D. W. (1965) 'Reproduction and productivity of moose in northwestern Ontario' *J. Wildl. Mgmt.* **29**: 740–50.

SINCLAIR, A. N. (1950) 'A note on the effect of the presence of rams on the incidence of oestrus in maiden Merino ewes during spring mating' *Aust. vet. J.* **26**: 37–9.

SINHA, A. A., CONAWAY, C. H., and KENYON, K. W. (1966) 'Reproduction in the female sea otter' *J. Wildl. Mgmt.* **30**: 121–30.

SITTMAN, D. B., ROLLINS, W. C., SITTMAN, K., and CASADY, R. B. (1964) 'Seasonal variation in reproductive traits of New Zealand white rabbits' *J. Reprod. Fert.* **8**: 29–37.

SKINNER, J. D. (1967) 'An appraisal of the eland as a farm animal in Africa' *Anim. Breed. Abst.* **35**: 177–86.

—— and BONSMA, J. C. (1964) 'The effect of early introduction of vasectomised bulls upon the sexual activity of the breeding herd' *Proc. S. Afr. Soc. Anim. Prod.* **3**: 60–2.

—— and BOWEN, J. (1968) 'Puberty in the Welsh stallion' *J. Reprod. Fert.* **16**: 133–5.

SLONAKER, J. R. (1931) 'The effect of different per cents of protein in the diet V. Reproduction' *Am. J. Physiol.* **97**: 322–8.

SMELSER, G. K., WALTON, A., and WHETHAM, E. O. (1934) 'The effect of light on ovarian activity in the rabbit' *J. exp. Biol.* **11**: 352–63.

SMITH, I. D. (1962) 'Reproductive wastage in a Merino flock in central western Queensland' *Aust. vet. J.* **38**: 500–7.

—— (1964a) 'The effect of seasonal variations in body weight upon oestrus activity in Merino ewes' *Proc. Vth. Int. Congr. Anim. Reprod. (Trento)* **3**: 481–4.

—— (1964b) 'Ovine neonatal mortality in western Queensland' *Proc. Aust. Soc. Anim. Prod.* **5**: 100–6.

SMITH, I. D. (1964c) 'Postparturient anoestrus in the Peppin Merino in western Queensland' *Aust. vet. J.* **40**: 199–201.

—— (1964d) 'Reproduction in Merino sheep in tropical Australia' *Aust. vet. J.* **40**: 156–60.

—— (1965a) 'Reproductive wastage in Merino sheep in semi-arid tropical Queensland. Observations on flocks mated during the spring and summer' *Aust. J. exp. Agric. Anim. Husb.* **5**: 110–14.

—— (1965b) 'The influence of level of nutrition during winter and spring upon oestrous activity in the ewe' *Wld. Rev. Anim. Prod.* **4**: 95–102.

—— (1966) 'The onset of the breeding season in Southdown ewes in subtropical Australia' *J. agric. Sci., Camb.* **66**: 295–6.

SMITH, N. B. and BARKALOW, F. S. (1967) 'Precocious breeding in the gray squirrel' *J. Mammal.* **48**: 328–30.

SMYTH, M. (1966) 'Winter breeding in woodland mice, *Apodemus sylvaticus*, and voles, *Clethrionomys glareolus* and *Microtus agrestis*, near Oxford' *J. Anim. Ecol.* **35**: 471–85.

SNYDER, B. D. (1963) 'The effects of endrin on vole (*Microtus pennsylvanicus*) reproduction in blue grass meadows' Ph.D. thesis, Ohio State Univ. (not seen. as quoted in *Wildl. Rev.* **116**: 6).

SNYDER, R. L. (1962) 'Reproductive performance of a population of woodchucks after a change in sex ratio' *Ecology* **43**: 506–15.

SOLIEN DE GONZALEZ, N. L. (1964) 'Lactation and pregnancy; a hypothesis' *Am. Anthrop.* **66**: 873–8.

SOPER, J. D. (1941) 'History, range and home life of the northern bison' *Ecol. Monogr.* **11**: 347–412.

SORENSEN, A. M., THOMAS, W. B., and GOSSETT, J. W. (1961) 'A further study on the influence of level of energy intake and season on reproductive performance of gilts' *J. Anim. Sci.* **20**: 347–51.

SOUTHERN, H. N. and HOOK, O. (1963) 'Notes on breeding of small mammals in Uganda and Kenya' *Proc. zool. Soc. Lond.* **140**: 503–15.

SOUTHWICK, C. H. (1955a) 'The population dynamics of confined house mice supplied with unlimited food' *Ecology* **36**: 212–25.

—— (1955b) 'Regulatory mechanisms of house mouse populations: social behaviour affecting litter survival' *Ecology* **36**: 627–34.

—— (1958) 'Population characteristics of house mice living in English corn ricks: Density relationships' *Proc. zool. Soc. Lond.* **131**: 163–75.

——, BEQ, M. A. and SIDDIQI, M. R. (1965) 'Rhesus monkeys in northern India', chapter in *Primate Behaviour; Field Studies of Monkeys and Apes*, I. de Vore, Ed. (Holt, Reinhart, and Winston, New York), 654 pp.

SOWLS, L. K. (1957) 'Reproduction in the Audubon cottontail in Arizona' *J. Mammal.* **38**: 234–43.

SPENCER, A. W. and STEINHOFF, H. W. (1968) 'An explanation of geographic variation in litter size' *J. Mammal.* **49**: 281–6.

SPENCER, B. (1896) *Report on the work of the Horn scientific expedition to central Australia II. Zoology, Mammalia* (London), 52 pp.

SPIES, H. G., MENZIES, C. S., SCOTT, S. P., COON, L. L., and KIRACOFE, G. H. (1965) 'Effects of forced exercise and cooling on reproductive performance of finewool ewes bred during the summer' *J. Anim. Sci.* **24**: 9–12.

STAPLES, R. E. (1967) 'Behavioural induction of ovulation in the oestrous rabbit' *J. Reprod. Fert.* **13**: 429–35.

STEVENS, M. N. (1952) 'Seasonal observations on the wild rabbit *Oryctolagus cuniculus* (L.) in west Wales' *Proc. zool. Soc. Lond.* **122**: 417–34.

STEVENS, M. R. (1957) *The Natural History of the Otter* (Univ. Fed. Anim. Welf., London), 88 pp.

STODART, E. and MYERS, K. (1964) 'A comparison of behaviour, reproduction, and mortality of wild and domestic rabbits in confined populations' *C.S.I.R.O. Wildl. Res.* **9**: 144–59.

—— —— (1966) 'The effects of different foods on confined populations of wild rabbits. *Oryctolagus cuniculus* (L.)' *C.S.I.R.O. Wildl. Res.* **11**: 111–24.

STONES, R. C. and WEIBERS, J. E. (1965) 'Seasonal changes in food consumption of little brown bats held in captivity at a "neutral" temperature of 92° F.' *J. Mammal.* **46**: 18–22.

STOTT, G. H. and WILLIAMS, R. J. (1962) 'Causes of low breeding efficiency in dairy cattle associated with seasonal high temperatures.' *J. Dairy Sci.* **45**: 1369–75.

STRECKER, R. L. (1954) 'Regulatory mechanisms in house mouse populations: The effect of limited food supply on an unconfined population' *Ecology* **35**: 249–57.

—— and EMLEN, J. T. (1953) 'Regulatory mechanisms in house mouse populations: The effect of limited food supply on a confined population' *Ecology* **34**: 375–85.

SUGIYAMA, Y., YOSHIBA, K., and PATHASARATHY, M. D. (1965) 'Home range, mating season, male group and inter-troop relations in Hanuman langurs (*Presbytis entellus*)' *Primates* **6**: 73–106.

SYKES, J. F. and COLE, C. L. (1944) 'Modification of mating season in sheep by light treatment' *Q. Bull. Mich. St. Univ. agric. Exp. Stn.* **26**: 250–6.

SYMINGTON, R. B. and OLIVER, J. (1966) 'Observations on the reproductive activity of tropical sheep in relation to photoperiod' *J. agric. Sci., Camb.* **67**: 7–12.

TABER, R. D. (1953) 'Studies of black-tailed deer reproduction on three chaparral cover types' *Calif. Fish Game* **39**: 177–86.

—— and DASMANN, R. F. (1957) 'The dynamics of three natural populations of the deer (*Odocoileus hemionus columbianus*)' *Ecology* **38**: 233–46.

TAMSITT, J. R. and MEJIA, C. A. (1962) 'The reproductive status of a population of a neotropical bat *Artibeus jamaicanensis*, at Providencia' *Carrib. J. Sci.* **2**: 139–44.

—— and VALDIVIESO, D. (1963) 'Reproductive cycle of the big fruit-eating bat *Artibeus literatus*' *Nature, Lond.* **198**: 104.

—— —— (1965a) 'The male reproductive cycle of the bat, *Artibeus literatus*' *Am. Midl. Nat.* **73**: 150–60.

—— —— (1965b) 'Reproduction of the female big fruit-eating bat, *Artibeus literatus palmarum*, in Colombia' *Carrib. J. Sci.* **5**: 157–66.

TAYLOR, J. M. (1963) 'Reproductive mechanisms of male grasshopper mouse' *J. exp. Zool.* **154**: 109–23.

TENER, J. S. (1965) *Muskoxen in Canada* (Queen's Printer, Ottawa), 166 pp.

TERMAN, C. R. (1965) 'A study of population growth and control exhibited in the laboratory by prairie deermice' *Ecology* **46**: 890–5.

TERRY, W. A. and MEITES, J. (1951) 'The effects of continuous light or darkness on thyroid and reproductive function in ewes' *J. Anim. Sci.* **10**: 1081–2.

THIBAULT, C., COURIT, M., MARTINET, L., MAULEON, P., DU MESNIL DU BUISSON, F., ORTAVANT, P., PELLETIER, J., and SIGNORET, J. P. (1966) 'Regulation of breeding season and oestrous cycles by light and external stimuli in some mammals' *J. Anim. Sci.* **25**: (*Suppl.*): 119–42.

THOMPSON, A. P. D. (1951) 'Relation of retinal stimulation to oestrus in the ferret' *J. Physiol.* **113**: 425–33.

—— (1954) 'The onset of oestrus in normal and blinded ferrets' *Proc. R. Soc. (B).* **142**: 126–35.

—— and ZUCKERMAN, S. (1954) 'The effect of pituitary stalk section on light induced oestrus in ferrets.' *Proc. R. Soc. (B).* **142**: 437–51.

THOMPSON, J. N., HOWELL, J. McC., and PITT, G. A. J. (1964) 'Vitamin A and reproduction in rats' *Proc. R. Soc. (B).* **159**: 510–35.

THOMSON, A. M. and THOMSON, W. (1949) 'Lambing in relation to the diet of the pregnant ewe' *Br. J. Nutr.* **2**: 290–305.

THOMSON, W. and AITKEN, F. C. (1959) 'Diet in relation to reproduction and the viability of the young. Part II, Sheep: World survey of reproduction and review of feeding experiments' *Comm. Bur. Anim. Nut. Tech. Comm. No.* **20**: 93 pp.

THORPE, D. H. (1967) 'Basic parameters in the reaction of ferrets to light' chapter in *Effects of External Stimuli on Reproduction* G. E. W. Wolstenholme and M. O'Connor, Eds. (Churchill, London), 107 pp.

THWAITES, C. (1965) 'Photoperiodic control of breeding activity in the Southdown ewe with particular reference to the effects of an equatorial light regime' *J. agric. Sci., Camb.* **65**: 57–64.

TRUSCOTT, B. L. (1944) 'Physiological factors in hypophyseal-gonadal interaction 1. Light and the follicular mechanism of the rat' *J. exp. Zool.* **95**: 291–305.

TWIGG, C. I. (1965) 'Studies of *Holochilus sciureus berbicensis*, a cricetine rodent from the coastal region of British Guiana' *Proc. zool. Soc. Lond.* **145**: 263–83.

ULBERG, L. C. (1958) 'Influence of high temperature on reproduction' *J. Hered.* **49**: 62–4.

UNITED STATES NAVAL OBSERVATORY (1945) 'Tables of sunrise, sunset, and twilight' Supplement to the *American Ephemeris*, 1946 (U.S. Govt. Printing Off., Washington), 196 pp.

VAN BRUNT, E. E., SHEPERD, M. D., WALL J. R., GANONG, W. F., and CLEGG, M. T. (1964) 'Penetration of light into the brain of mammals' *Ann. N.Y. Acad. Sci.* **117**: 217–24.

VANDENBERGH, J. G. and VESSEY, S. (1968) 'Seasonal breeding of free-ranging Rhesus monkeys and related ecological factors' *J. Reprod. Fert.* **15**: 71–9.

VAN WIJNGAARDEN, A. (1960) 'The population dynamics of four confined populations of the continental vole, *Microtus arvalis* (Pallas)' *Verslag. Landbouwk. Onderzock.* **66**: (22): 28 pp.

VAUGHAN, T. A. (1962) 'Reproduction in the plains pocket gopher in Colorado' *J. Mammal.* **43**: 1–13.

VERME, L. J. (1962) 'Mortality of white-tailed deer fawns in relation to nutrition' *Proc. 1st. Natn. White-tailed deer Disease Symp.* (University of Athens, Georgia) **1**: 15–28.

—— (1965) 'Reproduction studies on penned white-tailed deer' *J. Wildl. Mgmt.* **29**: 74–9.

VESSEY, S. (1967) 'Effects of chlorpromazine on aggression in laboratory populations of wild house mice' *Ecology* **48**: 367–76.

VOLCANI, R. (1953) 'Seasonal variations in spermatogenesis of some farm animals under the climatic conditions of Israel' *Bull. Res. Coun. Israel (Jerusalem)* **3**: 123–6.

VORANGER, J. (1953) 'Influence de la meteorologie et la mortalite sur les naussances' *Population, Paris* **8**: 93–102.

WADE, P. (1958) 'Breeding season amongst mammals in the lowland rain forest of north Borneo' *J. Mammal.* **39**: 429–33.

WATSON, J. S. (1950) 'Some observations on the reproduction of *Rattus rattus* (L.)' *Proc. zool. Soc. Lond.* **120**: 1–12.

—— (1957) 'Reproduction of the wild rabbit, *Oryctolagus cuniculus* (L.) in Hawke's bay, New Zealand' *N.Z. Jl Sci. Technol.* **38**: 451–82.

WATSON, R. H. and GAMBLE, L. C. (1961) 'Puberty in the Merino ewe with special reference to the influence of season of birth upon its occurrence' *Aust. J. agric. Res.* **12**: 124–38.

—— and RADFORD, H. M. (1960) 'Influence of rams on the onset of oestrus in Merino ewes in the spring' *Aust. J. agric. Res.* **11**: 65–71.

WELLS, L. J. (1935) 'Seasonal sexual rhythm and its experimental modification in the male of the thirteen-lined ground squirrel (*Citellus tridecemlineatus*)' *Anat. Rec.* **62**: 409–47.

WELLS, L. J. and ZALESKY, M. (1940) 'Effects of low temperature environment on the reproductive organs of male mammals with annual aspermia' *Am. J. Anat.* **66**: 429–38.

WHITAKER, W. L. (1940) 'Some effects of artificial illumination on reproduction in the white-footed mouse, *Peromyscus leucopus noveboracensis' J. exp. Zool.* **83**: 33–60.

WHITTEN, W. K. (1956) 'Modification of the oestrous cycle of the mouse by external stimuli associated with the male' *J. Endocr.* **13**: 399–404.

—— (1958) 'Modification of the oestrous cycle of the mouse by external stimuli associated with the male; Changes in the oestrous cycle determined by vaginal smears' *J. Endocr.* **17**: 307–13.

—— (1959) 'Occurrence of anoestrus of mice caged in groups' *J. Endocr.* **18**: 102–7.

—— (1966) 'Pheromones and mammalian reproduction' *Adv. Reprod. Physiol.* **1**: 155–77.

WIGHT, H. M. and CONAWAY, C. H. (1961) 'Weather influences on the onset of breeding in Missouri cottontails' *J. Wildl. Mgmt.* **25**: 87–9.

WILLIAMS, C. E. and CASKEY, A. L. (1965) 'Soil fertility and cottontail fecundity in southeastern Missouri' *Am. Midl. Nat.* **74**: 211–24.

WILLIAMS, R. G., CARMON, J. L., and GOLLEY, F. B. (1965) 'Effect of sequence of pregnancy on litter size and growth in *Peromyscus polionotus' J. Reprod. Fert.* **9**: 257–60.

WIMSATT, W. A. (1960) 'Some problems of reproduction in relation to hibernation in bats' *Bull. Mus. comp. Zool. Harv.* **124**: 249–63.

—— (1963) 'Delayed implantation in the Ursidae, with particular reference to the black bear (*Ursus americanus* Pallas)' chapter in *Delayed Implantation* A. C. Enders, Ed. (Univ. of Chicago Press, Chicago), 318 pp.

WIRTZ, W. O. (1968) 'Reproduction, growth, and development, and juvenile mortality in the Hawaiian monk seal' *J. Mammal.* **49**: 229–38.

WODZICKA-TOMASZEWSKA, M., HUTCHINSON, J. C. D., and BENNETT, J. W. (1967) 'Control of the annual rhythm of breeding in ewes: effect of an equatorial day-length with reversed thermal seasons' *J. agric. Sci., Camb.* **68**: 61–7.

WODZICKI, K. (1950) 'Introduced mammals of New Zealand' *D.S.I.R. N.Z. Res. Bull.* **98**: 250 pp.

—— and DARWIN, J. H. (1962) 'Observations on the reproduction of the wild rabbit (*Oryctolagus cuniculus* L.) at varying latitudes and altitudes in New Zealand' *N.Z. Jl Sci.* **5**: 463–74.

WOOLLEY, P. (1966) 'Reproduction in *Antechinus spp.* and other dasyurid marsupials' *Symp. zool. Soc. Lond.* **15**: 281–94.

WRIGHT, P. L. (1963) 'Variations in reproductive cycles in North American mustelids' chapter in *Delayed Implantation,* A. C. Enders, Ed. (Univ. of Chicago Press, Chicago), 318 pp.

—— (1966) 'Observations on the reproductive cycle of the American badger *Taxidea taxus' Symp. zool. Soc. Lond.* **15**: 27–45.

—— and COULTER, M. W. (1967) 'Reproduction and growth in Maine fishers' *J. Wildl. Mgmt.* **31**: 70–87.

YEATES, N. T. M. (1949) 'The breeding season of the sheep with particular reference to its modification by artificial means using light' *J. agric. Sci., Camb.* **39**: 1–42.

—— (1954) 'Daylight changes', chapter in *Recent Progress in the Physiology of Farm Animals* J. Hammon, Ed. (Butterworths, London).

—— (1956a) 'The effect of high air temperature on pregnancy and birth weight in Merino sheep' *Aust. J. agric. Res.* **7**: 435–9.

—— (1956b) 'The effect of light on the breeding season, gestation and birth weight of Merino sheep' *Aust. J. agric. Res.* **7**: 440–6.

—— (1965) *Modern Aspects of Animal Production* (Butterworths, London), 371 pp.

YOSHIOKA, Z., AWASAWA, T., and SUZUKI, S. (1952) '(Effect of the short day treatment on the modification of the breeding season in goats)' *Bull. Nat. Inst. agric. Sci. (Japan) Ser C.* **1**: 105–11.

YOUATT, W. G., VERME, L. J., and ULLREY, D. E. (1965) 'Composition of milk and blood in nursing white-tailed does and blood composition of their fawns' *J. Wildl. Mgmt.* **29**: 79–84.

ZACHARIAS, L. and WURTMAN R. J. (1964) 'Blindness: its relation to age of menarche' *Science, N.Y.* **144**: 1154–5.

ZEJDA, J. (1961) 'Age structure in population of the bank vole, *Clethrionomys glareolus* Ischeber, 1780' *Zool. Listy* **10**: 249–64.

—— (1962) 'Winter breeding in the bank vole, *Clethrionomys glareolus* Schreb.' *Zool. Listy* **11**: 309–22.

—— (1964) 'Development of several populations of the bank vole, *Clethrionomys glareolus* Schreb., in a peak year' *Zool. Listy* **13**: 15–30.

—— (1966) 'Litter size in *Clethrionomys glareolus* Schreber 1780' *Zool. Listy* **15**: 193–206.

—— (1967) 'Mortality of a population of *Clethrionomys glareolus* Schreb. in a bottomland forest in 1964' *Zool. Listy* **16**: 221–38.

ZEUNER, F. E. (1963) *A History of Domesticated Animals* (Hutchinson, London), 560 pp.

ZIMMERMAN, D. R., SPIES, H. G., SELF, H. L., and CASIDA, L. E. (1960) 'Ovulation rate in swine as affected by increased energy intake just prior to ovulation' *J. Anim. Sci.* **19** (1): 295–301.

Indexes

Taxonomic Index

General Index